THE NEW WHY YOU DON'T NEED MEAT

PETER COX

BLOOMSBURY

First published 1992 by Bloomsbury Publishing Ltd, 2 Soho Square, London W1V 5DE

This paperback edition published 1994

Copyright © 1992 by Peter Cox

The moral right of the author has been asserted

A CIP record for this book is available from the British Library

ISBN 0 7475 1696 0

Typeset by Florencetype Ltd, Kewstoke, Avon
Printed by Cox & Wyman Ltd, Reading, Berks

CONTENTS

ACKNOWLEDGEMENTS

Many people have helped me, in various ways, to write this book. I would like to express my sincere appreciation to the following people for most generously offering their time and expertise:

Dr. Helen Grant; Colin and Lis Howlett; Barry and Sue Kew; Andrew Kimbrell of the Foundation on Economic Trends; Drs. G. and C. Langley; Dr. Alan Long; Philip L. Pick and the staff of the Jewish Vegetarian Society; Dr. David Ryde; Gregory Sams of the Realeat Company; Joyce D'Silva and Peter Stevenson of Compassion in World Farming; Sarah Starkey; Andrew Tyler; Michael Verney-Elliott and the staff of the Science Reference and Information Service at the British Library, London; the Sir Thomas Browne Library, Norwich; the Addenbrookes Hospital Library, Cambridge. I would also like to express my gratitude to everyone at Bloomsbury for the enthusiasm and talent they have brought to this project.

To Linda, my thanks for writing the foreword; to Paul and Linda, my gratitude for their courage in showing millions of people that there is a better way of living. And to the friends who have sustained me while writing this book, my heartfelt appreciation and love.

The information in this book is regularly updated in The Superliving! Letter, edited by Peter Cox. For a free copy, please send a first-class stamp to Superliving!, PO Box 1612, London NW3 1TD.

FOREWORD

For many years now some of us have been saying 'Stop eating animals' because we know that it is pointless and cruel. My friend Peter Cox is one of those who has been saying it loudest.

Unfortunately the majority of those who are steeped in the tradition of meat-eating either closed their ears or said they didn't care about the moral arguments.

But now, in this remarkable book, Peter has researched arguments that the majority cannot ignore – not if they care for their health. Or their lives.

His conclusions are – for the meat-eater – alarming. Those who believe that meat is somehow good for you should read the facts connecting it to heart disease, high blood pressure and cancer before swallowing another mouthful.

This is a book that will change lives and save lives. I hope that yours is one of them.

Linda McCartney

CHAPTER ONE
EVERYTHING YOU'RE NOT SUPPOSED TO KNOW

What you're about to read may change your life, maybe even save it. And you won't get this information anywhere else.

Why not?

Because it is forbidden knowledge. Not secret, but forbidden. Forbidden by commercial interests, by politics, and by a society which fundamentally misunderstands itself.

This book presents you with the information you need to start making your own connections. Once you make a few, your world will change permanently, and you'll wonder how it was that you didn't make these connections sooner. You'll also wonder why no-one ever told you these things before.

Well, I wonder that, too. I wonder why our doctors don't tell us, why our schools and colleges don't tell our young people, why governments don't tell their citizens.

Why, indeed. There are a good many reasons. Doctors aren't going to tell us, because most doctors don't know themselves. We'll discuss this dire state of affairs a little later.

Governments and their agencies aren't going to tell us, because politicians will always be politicians. You know what I mean by that.

And, of course, the meat industry certainly isn't going to tell us. After all, why should they cut their own throats? They have many other throats to cut.

So that leaves you with me.

YOU BE THE JUDGE

But why should you believe me?

That's a really tough question for me to answer. It's so tough that it left me absolutely dumb-struck when someone asked it recently at a workshop I gave. I had to think hard for many moments. Eventually, a few reasons started to surface.

The first thing I'd say is: Listen to the evidence. Use your intuition, see if it sounds right, then make up your mind. As you read this book, you'll see I give detailed references to all the research I mention. Now I have to be truthful and say that simply citing a reference to some research doesn't necessarily mean all that much. For example, I know that whenever I check other people's

1

references, I find that they're downright wrong every once in a while. However, all mine are bang on the nail – go ahead and check them for yourself and see.

The second thing is: Ask yourself, who's got something to gain? The meat moguls have plenty to gain by deceiving you (and, equally, plenty to lose if you find out). On the other hand, I have very little to gain. Your personal dietary habits are not going to make my fortune, one way or another. (So why am I writing this book? Simply because I believe in it. People still do things like that, sometimes.)

Thirdly: Ask yourself, what have you got to lose? Even if only 10 per cent of what I say is right, *what have you got to lose* by acting on it and changing your lifestyle for the better? Think about it.

I know from the response to the original *Why You Don't Need Meat* that some people found it extremely hard to accept that there could be a huge body of evidence, clearly implicating the meat-based lifestyle, that *no-one had ever told them about*. They figured that if they hadn't been told, it just couldn't be true. That's circular logic, of course. But I can understand people thinking that way. 'Surely someone should have told us?' is a common reaction to many scandals, after the event. The same reaction also manifests itself as 'If only someone had told us, we could have done something about it . . .' Well I'm telling you, now.

But by far the most depressing reaction comes from the person who says 'I hear what you're saying, but in another month or two the news will be full of experts saying exactly the opposite.' That's depressing, for two reasons. It reveals that the person concerned has almost entirely abdicated their own judgement in favour of the latest newspaper headlines. And it also proves that the meat industry's powerful and expensive lobbying machine has succeeded in confusing people so much that they don't know what to do, and end up doing nothing at all.

Now I need your undivided attention for a few minutes.

The News From America

The easiest way of comparing the health of meat-eaters to non-meat-eaters is just to watch them, and to see who dies of what. Basically, it's not too difficult to do, although it can be years before you get any results. From the scientist's point of view, the main danger is that you'll die before the experiment has finished. This branch of science is called 'epidemiology', meaning 'the study of the relationships of the various factors determining the frequency and distribution of diseases in a human community'.[1] To put it more simply, epidemiology is scientific detective work.

In some ways, epidemiology is a seriously overlooked discipline. It isn't as glamorous as the 'wet' sciences that make headlines with the latest high-tech brain transplant or AIDS cure, but because it concentrates on studying the way things actually are in the real world – rather than the way laboratory whitecoats might like them to be – it is capable of giving us extremely relevant insights into health and disease. You're going to see the results of some epidemiological studies now, and while you are considering them, please remember that the knowledge which these studies give us has been obtained at a high cost – hundreds of thousands of people have died to bring you these findings. If we don't learn anything from them, they will have died in vain.

Let's see how one such study was conducted. In 1978, a paper appeared in *The American Journal of Clinical Nutrition* written by Dr. Roland L. Phillips, one of America's most respected epidemiologists.[2] He and his team were very interested in a subgroup of the American population called Seventh-Day Adventists. This group was particularly fascinating because the Adventist church advocates a very different lifestyle to the typical meat-based American one. The first thing Dr. Phillips did was to locate a large number of Seventh-Day Adventists. We're not talking about a few dozen, or even a few hundred people here. Dr. Phillips' sample size was massive – 25,000 people. Obviously, the more people you study, the less likely it is that a few freak results are going to skew the analysis. In this case, the huge number of people involved makes the study very reliable indeed.

Then they waited. Every year, for six years, Dr. Phillips' team contacted each one of the people involved in the study, just to see if they were still alive. If a person had died, the death certificate was obtained, and the underlying cause of death determined. (Patience and tact are two key qualities for a good epidemiologist!) At the end of the six-year period, some highly significant results were emerging.

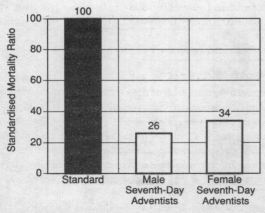

Figure 1: *Deaths from heart disease: Seventh-Day Adventists compared to the general population*

All the subjects lived in California. When compared to the average, meat-eating, Californian population, the results showed that the risk of dying from coronary heart disease among Adventists was far, far lower than normal. For every 100 non-Adventist Californians who died from heart disease, only 26 Adventist males had died – that's about one quarter the risk. Among females, the risk was one third. You can see this illustrated in Figure 1 (see page 3).

This is forceful evidence. It isn't theoretical, or hypothetical, or otherwise concocted. It is straightforward fact. Counting dead bodies is pretty convincing, even for the most hardened sceptics.

Now, the next question is: Why? Well one reason must be the fact that most Seventh-Day Adventists do not smoke. 'But,' say the sceptics, 'it's nothing to do with eating meat, it simply proves that smoking isn't healthy.' Unfortunately for the sceptics, that explanation doesn't hold up. You see, Dr. Phillips and his team had considered possibilities such as that, as indeed good epidemiologists should do. So next they compared deaths from heart disease amongst Seventh-Day Adventists to deaths from heart disease amongst a representative group of non-smokers, as studied by the American Cancer Society. Clearly, if Adventists were healthier purely because they didn't smoke, then death rates in these two groups should be the same. But they weren't – not by a long chalk. The cold figures showed that Adventists ran only half the risk of dying from heart disease, when compared to non-smokers (people identified by the American Cancer Society as *never* having smoked). Clearly, there was something else very special about the Adventist lifestyle.

What could it be? Perhaps people with religious faith die less from heart disease? Perhaps they have less stress in their lives? Perhaps they secretly take a

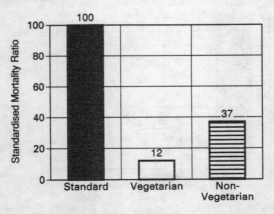

Figure 2: *Deaths from heart disease: Seventh-Day Adventist vegetarians compared to Seventh-Day Adventist non-vegetarians*

magic potion that protects them? A determined opponent could throw up any number of possibilities to explain away these findings. And that's where the sheer good science of Dr. Phillips' research really paid dividends. Because he realised that people might raise all kinds of possible explanations, such as those I've suggested, and he accounted for it. He realised that although the Adventist church advocated a vegetarian lifestyle, it wasn't compulsory. Some Adventists still ate meat. So he included that aspect in his research. He found that about 20 per cent of them ate meat four or more times a week, about 35 per cent ate it between one and three times a week, and the remaining 45 per cent never ate it at all. To a bright mind, that fact created a unique scientific opportunity. Why not simply compare the health of Adventists who never ate meat (i.e. vegetarians) to those Adventists who did eat it? In a flash, it would eliminate all other confounding factors. So that's what he did.

You can see the result in Figure 2 opposite. Among Adventist men who ate meat, the death rate from coronary heart disease was 37 per cent of the normal death rate – impressive in itself, and certainly proof that smoking is pretty lethal. But among those Adventists who were vegetarian, the death rate plummeted even further – right down to 12 per cent that of the normal population. *Twelve per cent!*

Now, one of the great things about large-scale studies such as this is the longer you are prepared to wait, the more interesting and more accurate the results become. So that's what happened next – they waited, and watched. For 20 long years. Eventually, Dr. Phillips' team published the final results of the study, a study which had literally observed people growing old and dying over two decades.[3] This landmark study provided the first ever scientific proof that the more meat you eat, the more at risk you are from heart disease.

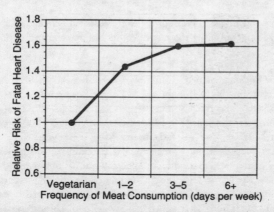

Figure 3: *Weekly meat consumption correlated to risk of fatal heart disease (males)*

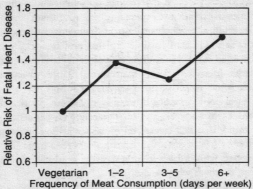

Figure 4: *Weekly meat consumption correlated to risk of fatal heart disease (females)*

Look at Figures 3 and 4 and you'll see a summary of the results. Here, you can see that the relative risk of fatal heart disease closely correlates with the frequency of eating meat. Those Adventist males who consumed meat on one or two days a week were 44 per cent more likely to die from heart disease. Those who consumed it between three and five times a week were 60 per cent more likely to die. And those who consumed it more than six times a week were 62 per cent more likely to die. For females, the increases are 38, 25 and 58 per cent respectively. The significant finding is that even a small amount of meat – once or twice a week – greatly elevates the risk. And for men in one particular age group – 45 to 54 – the risk is particularly high. For these people, prime candidates for heart disease, the risk when compared to vegetarians is 400 per cent greater!

I have covered this huge study in some detail because it is so extremely revealing. If your head's not spinning too much, could you manage another one?

The News From Japan

Inspired by the insights gained from the American Seventh-Day Adventist studies, scientists from the National Cancer Center Research Institute, Tokyo, embarked on similar research.[4] Similar, that is, in concept, but even broader in scope. In this case, the Japanese decided to follow not 25,000 people, but an astonishing 122,261, and to study them for over 16 years. The logistics alone must have been daunting – each man (they only studied males in this survey) had to be interviewed at home by specially trained public health nurses.

The Japanese study was so large that it was possible to divide the participants into various subgroups according to their dietary and lifestyle preferences. And this is the final conclusion, after much hard work and computing time was expended in analysis:

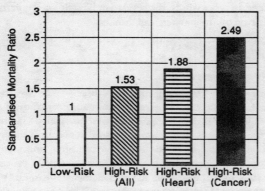

Figure 5: *Risky lifestyles – how two opposite lifetstyles compare*

Two lifestyles emerged as being very high-risk and very low-risk respectively. The high-risk lifestyle included smoking, drinking, meat consumption, and no green vegetables. The low-risk lifestyle was, not surprisingly, precisely the opposite. In Figure 5, you can see how the lifestyles compared. Deaths from all causes were elevated by 1.53 times amongst those who smoked, drank, ate meat and didn't eat green vegetables. The risk of heart disease was 1.88 times higher in this group, and the risk of any kind of cancer was 2.49 times higher.

So far so good – and probably just what you were expecting to see. But the statistical power of this huge study was able to reveal, for the first time, some extraordinary relationships between meat consumption and ill health. Let me summarise:

- The study showed that simply adding one factor – meat consumption – to an otherwise healthy lifestyle had a serious effect on mortality. The difference between the lowest-risk group (no smoking, no drinking, no meat eating, and lots of green vegetables) and those people who led a similar lifestyle, *except for eating meat*, was that the meat-eaters boosted their risk of dying from heart disease by 30 per cent. Just by adding meat to an otherwise healthy lifestyle!
- At the other end of the scale could be found the two most unhealthy groups. We generally think of smoking and drinking as being pretty unhealthy, and the study confirmed this – people who smoked and drank (but who consumed green vegetables and didn't eat meat) were 39 per cent more likely to die from any cause than the healthiest group. However, even more unhealthy were those people who smoked, drank, ate meat and didn't consume green vegetables. These people increased their risk of dying from any cause by a

further 14 per cent! In other words, the vegetarian lifestyle was conferring some protection even on the smokers and drinkers!

How about another?

The News From Germany

When the German Cancer Research Center advertised in *Der Vegetarier*, the German magazine for vegetarians, for participants in a similar study, they were following a rather different angle.[5] The scientists were particularly interested in the way a vegetarian diet seems able to protect against cancer. It is thought that nitrate consumption is linked to the development of cancer, and many vegetables contain nitrates. So why don't more vegetarians contract cancer? One possibility is that their overall diet contains other elements (vitamins A and C, for example) which protect them and lower their risk. This was one of the main areas which the researchers were keen to investigate. Eventually, a total of 1904 subjects were recruited.

After five years, the results began to emerge. Deaths from all causes were very low indeed – only 37 per cent of the number of deaths among the average meat-eating population. All forms of cancer were slashed to 56 per cent of the normal rate, and heart disease was down to 20 per cent.

Perhaps this was due to a lack of smoking? As in the two studies quoted above, the researchers had already taken this into account. Even when vegetarian smokers were compared to non-vegetarian smokers, it was found that the vegetarians' rate of heart disease was still only 40 per cent of the expected rate. Clearly, the vegetarian diet was playing a significant protective role.

Take a deep breath, now. Can you handle another?

The News From Britain

A study amongst 4671 British vegetarians, which tracked their health (and cause of death) for seven years also found very similar conclusions.[6] The death rate from all causes amongst the male vegetarians studied was 50 per cent compared to the death rate of the general population; 55 per cent amongst females. For heart disease, the male death rate was only 44 per cent of normal, and for female vegetarians 41 per cent.

The study also compared the vegetarians to a similar population group – people who were customers of health food shops – and found that the health food shoppers were also at less risk from heart disease – 60 per cent of the normal. Presumably, this reflected the greater interest of health food shoppers in their own health, and lack of smoking. However, when comparing the two groups, it was obvious that the vegetarians had reduced their risk of heart disease by a further third compared to the health food shoppers.

Still feeling courageous? Could you manage just one more?

The News From China

If the Japanese study was impressive in terms of the number of participants, the China Study is unprecedented in terms of the depth of information produced. So much so, in fact, that it made headline news in *The New York Times*.[7] Under the headline 'Huge Study of Diet Indicts Fat and Meat', the report began:

> 'Early findings from the most comprehensive large study ever undertaken of the relationship between diet and the risk of developing disease are challenging much of American dietary dogma. The study, being conducted in China, paints a bold portrait of a plant-based eating plan that is more likely to promote health than disease.'

A 'plant-based eating plan' . . . whatever could they mean? Surely not the 'V' word?

The Chinese study was truly massive, involving the collection of 367 detailed facts about the diet and lifestyle of 6500 participants across China, from 1983 onwards. Dr. Mark Hegsted, emeritus professor of nutrition at Harvard University and former administrator of human nutrition for the United States Department of Agriculture, commented:

> 'This is a very, very important study. It is unique and well done. Even if you could pay for it, you couldn't do this study in the United States because the population is too homogeneous. You get a lot more meaningful data when the differences in diet and disease are as great as they are in the various parts of China.'

The study consumed colossal amounts of manpower – 600 person-years of labour contributed by the Chinese government alone. Hundreds of trained workers were sent to take blood and urine samples from each participant, and to gather precise information on their diet. The study is still continuing, and it may be many years before a final analysis of the data has been accomplished (so far, the preliminary results fill a huge book of 920 pages). Here are the key findings to date:

- While 70 per cent of the protein in average Western diets comes from animals, only 7 per cent of Chinese protein does. While most Chinese suffer very little from the major killer diseases of the West, those affluent Chinese who consume similar amounts of animal protein to Westerners also have the highest rates of heart disease, cancer and diabetes.
- The Chinese consume 20 per cent more calories than Westerners do. This should mean that they are fatter than Westerners, but the reality is that Westerners are 25 per cent fatter than the Chinese! This is almost certainly due to a huge variation in diet – the Chinese eat only one third as much fat as Westerners, but eat twice the amount of complex carbohydrate.
- Current Western dietary guidelines suggest that we should reduce the fat in our diets to less than 30 per cent of our calorie consumption. The Chinese

study reveals that this is by no means enough to prevent heart disease and cancer effectively – it should be slashed to something closer to 10–15 per cent.

- You don't need to drink milk to prevent osteoporosis. Most Chinese consume no dairy products and instead get all their calcium from vegetables. While the Chinese consume only half the calcium Westerners do, osteoporosis is uncommon in China despite an average life expectancy of 70 years. 'Osteoporosis tends to occur in countries where calcium intake is highest and most of it comes from protein-rich dairy products,' says Dr. T. Colin Campbell, a nutritional biochemist from Cornell University and the American authority behind the study. 'The Chinese data indicate that people need less calcium than we think and can get adequate amounts from vegetables.'
- The study also reveals that meat-eating is not necessary to prevent iron-deficiency anaemia. The average Chinese adult, who shows no evidence of anaemia, consumes twice the iron Americans do, but the vast majority of it comes from plants.

And the conclusion? 'We're basically a vegetarian species,' says Dr. Campbell, 'and should be eating a wide variety of plant foods and minimising our intake of animal foods.' That's telling it like it is.

CONFUSED? YOU'RE MEANT TO BE!

Let's call a halt to all these studies, for the moment. The question that is probably uppermost in your mind at this point is: What happens to all this work? Surely the public ought to be given the life-saving information which studies such as these yield? Yes, they ought to be. But tragically, that doesn't happen. Instead, results of studies such as these are often trashed, ignored, or forgotten.

Trashed? Indeed. There is a whole industry available to rubbish the sort of evidence you've just seen.

I don't know whether you've ever met a really top-notch public relations person, but you should. They're extremely attractive people. If you're looking for a congenial person to have lunch with, or to go drinking with, find yourself a major public relations agency, and knock on their door. The chances are the place will be overflowing with people like that. They're bags of fun to be with. And they're excellent at their jobs, too. A top PR person can handle almost any impending media disaster – for a fee, of course. They will do precisely what you want them to do and say what you want them to say. If you want them to find a doctor who will stand up at a press conference and say: 'People who don't eat meat will die from moonbeam poisoning,' then they will find just such a doctor. They may have to send halfway round the world to get him, of course, but if your budget's big enough, it will be done – money can buy these things. And they will do it all with a grin on their bright little faces, and not one twinge of conscience in the place where their hearts used to be. That, incidentally, is pretty close to a definition of a psychopath.

The scientists who undertake epidemiological studies don't employ PR people, of course, so when their research is published in the professional journals, it rarely makes headline news. The headlines go to the blowhards.

DEVIL'S ADVOCATE INC.

If I was untroubled by matters of conscience, and was being paid a more-than-generous sum by the agribusiness industries to discredit research such as this, I would call my 'crisis management' team at DA Inc together, and something similar to the following would happen:

'Boys and girls,' I'd say, 'we've got a problem. I have advance information that a study is going to be published which contains bad news for one of our accounts in the meat business. What are we going to do about it?'

'Right,' says Carl, the young and thrusting account director. 'Our objective has to be to discredit the work and the people behind it. If we sow the seeds of doubt, that will be sufficient to baffle most consumers, so they won't change their buying patterns.'

'I've got a tame nutritionist,' says Gloria, the young and thrusting account manager, 'who's had media training, and she really comes over well. We'll put her up against the other side's scientists. I expect, like most scientists, they don't like publicity, and it'll be easy to make them seem bumbling and incompetent – the "mad scientist" caricature. I'll brief our nutritionist with some provocative questions whose answers might make the other side appear to be "extremists". If we can make them lose their temper, so much the better.'

'OK,' I'd say. 'Now what about the specifics?'

'Well,' says Tony, the young and thrusting account executive, 'we could release our own survey 24 hours before the other one is due to be published. So we'll grab the headlines, and no-one will want to write about the other one. And I'll brief the client to come out with a short statement saying there's "no conclusive proof" about the other one in any case.'

'But these are landmark studies,' I'd say. 'How are you going to discredit them?'

'Easy!' says Gloria. 'Everyone knows that statistics can be made to say anything. These studies don't mean a thing!'

'And specifically?' I ask.

'OK. The Chinese and Japanese are fundamentally different to Westerners. What's good for them isn't good for us. Everyone knows they eat all sorts of strange food over there. And as far as the Adventists are concerned, we'll exploit the religious angle, make them out to be a "weird religious sect", and imply that to get the same health benefits, you'd have to shave your head and renounce the world. That'll do the trick.'

And it usually does. All it takes is a little imagination, and a lot of money. And remember, they're not playing to win, they're just playing for a draw. Because all they have to do is to provide a sufficiently dense smoke screen so that people say 'Oh, what the heck, I don't know who to believe, I'll just carry on as I am.' *No change means they win.*

'QUASH THE HEALTH LOBBY!'

I've made some pretty strong assertions so far, so you'll want me to back them up. All right, here's a case history for you. Meet Derek Miller.

'Mr. Miller is a man totally committed to meat,' proudly announced the *Meat Trades Journal.*[8] 'Not because he knows it is good for the body. Not because the Meat Promotions Executive has taken him on as an advisor. But because he loves the stuff.' Mr. Miller was no ordinary hype-merchant. One of the world's top nutritionists, he occupied many senior positions as an advisor to governments, the United Nations, and other influential bodies. So when the meat industry succeeded in 'taking him on as an advisor,' they couldn't contain their glee:

'SHOWDOWN!' proclaimed the *Meat Trades Journal* in huge letters across its front page. 'Top nutritionist joins forces with the Meat Promotions Executive to quash the health lobby.' The story continued: 'One of the world's top nutritionists has joined forces with the Meat Promotions Executive in a bid to kick the health lobby's arguments into touch.'

So there you had it, in black and white. Miller's job was to 'quash the health lobby', and to 'kick their arguments into touch'. It couldn't be much plainer – this man, a world-respected scientist, was now going to suppress the truth about meat-eating and health.

Further into the story, an even more outrageous statement was made:

'He believes that meat is not only good for you, but that it is impossible to live without it.'

Impossible?

There's no risk of confusion here. No chance of differing interpretations, differences of opinion, differences of emphasis. A nutritionist of Miller's reputation and expertise would certainly be aware of studies similar to those you've just read about – he must have known that millions of vegetarians world-wide were living healthier lives than meat-eaters. So we're left with just one conclusion.

It was a lie.

As wicked a lie as you will ever come across. Coming from a man who should have known better. A man whose reputation as a nutritionist would guarantee him access to television, radio, the press – and whose expert status would rarely be questioned by ever-respectful journalists. What a great find for the meat industry, indeed. Said Mr. Miller:

'I personally am all in favour of having a go at the vegetarian lobby. Their moral arguments are not on and their nutritional arguments are rubbish.'

Moral arguments? Mr. Miller was singularly ill-qualified to talk about morals.

PRIME TIME

The subtle art of moulding the public's perception of your product can take many forms. Sometimes, it's as simple as changing the name of your product. For example, when the word 'fat' acquired a negative image amongst consumers, the meat trade simply decided to ban the word. 'Fat lambs are now being called prime lambs. Fatstock is known as primestock, and fattening cattle are known as finishing cattle,' reported the *Meat Trades Journal.*[9] Commented a livestock auctioneer: 'There's no doubt that fat had become a nasty word in many people's minds.'

And it's not just the 'F' word that arouses nasty associations, as the following news report makes clear:

'The editor in chief of the *Meat Trades Journal* today urged that the words "butcher" and "slaughterhouse" be eradicated and replaced by the American euphemisms "meat plant" or "meat factory". Alternatively, butchers could adopt the Irish word "victualler". This would distance consumers from awareness of the "bloodier side" of the meat trade. The editor argued that it was time for a review of meat trade vocabulary in recognition of "a growing away among younger meat buyers from the concept that meat ever comes from an animal". This was partly because these buyers did their shopping in the bloodless ambience of supermarkets. The meat trade's cause was not helped by the "blood-spattered whites" of Smithfield porters as they strolled "in front of the secretary birds". They and butchers should be put into velvet overalls. "It will reduce cleaning bills and any adverse reaction from the fainthearted". These days the word "butcher" was spread over newspaper headlines about the Ripper or the aftermath of bomb attacks. A change of nomenclature might only seem a verbal difference but it would "conjure up an image of meat divorced from the act of slaughter". "The public does not want to be made aware of the bloodier side of slaughter", he said. "Perhaps now is the time for changes to be made".'[10]

But the 'Newspeak' (should that be 'Meatspeak'?) doesn't stop there. The Meat and Livestock Commission now wants terms such as 'hormone-free', 'chemical-free' and 'additive-free' prohibited when used to describe organically produced meat, because they 'can be confusing and sometimes misleading and inaccurate', and lead to legal problems, bad publicity and lack of public confidence.[11]

It goes on and on. Pig farmers are now being encouraged to stop using the words 'growth promoters' to describe the drugs they give to their animals to (guess what?) promote growth. And the names given to the cells that these poor animals spend much of their lives in – 'flat-deck cages' and 'farrowing crates' – are now considered to be 'too emotive'. They're going to be replaced by 'nurseries' and 'maternity units'.[12]

Maternity units?
Come home, George Orwell, all is forgiven.

'DEATH OF A DIET MYTH!'

Confusion – sometimes deliberately created, sometimes accidental – abounds nowhere more than in this area of diet and health. Take a recent example involving animal fat. You must have seen the banner headlines – 'Death Of A Diet Myth' is how one newspaper described it, 'Fatty Food And Heart Attacks Not Linked', screamed another newspaper. What was it all about?

From reading the newspapers, you learnt that a scientific study had just been published which found that people who ate butter were less prone to heart attacks than people who ate polyunsaturated margarines. The study was performed by the government-funded Medical Research Council, which gave it good scientific credentials. And the media went to town. 'Butter Can Slice Heart Attack Risk', shrieked yet another newspaper.

What good news for the animal fat industry!

But behind the story, there was another story – the real one. First of all, a public relations company with, would you believe, the Butter Council as a client had seen the report three months before it was actually published. They then tried, unsuccessfully, to persuade the Medical Research Council to publicise the findings widely. The chief executive of the Butter Council was quoted as saying:

> 'We did not do anything of great importance, other than suggest that Dr. Elwood [scientific team leader] contact some medical journalists to offer them some information on his report.'[13]

Well, what's wrong with that? After all, the Butter Council is in business, just like everyone else, to sell their product. If there's some favourable publicity to be had, why not? The problem was that the results of the study simply were not adequate – they were raw data, which needed to be properly analysed before any meaningful conclusions could be drawn. But this didn't stop the media from going completely over the top. Hey, it was a good story, and the butter lobby knew it!

It took rather longer for the truth to emerge, and even then you wouldn't have seen it unless you happened to read the *British Medical Journal*.[14] What had happened was this: the raw data in the report showed that men who used polyunsaturated margarine had 1.9 times the heart attack rate seen in butter-eaters. From this, the media assumed that butter was better. Actually, if you look at other similar studies, you will see that this kind of result is not unique. Surveys do sometimes find that margarine-eaters have more heart attacks than butter-eaters. But what does this actually mean? Should we all start drowning our food in molten butter, in the hope of staving off a heart attack?

Nothing of the sort! A little applied common sense will suggest what's really happening here. For a decade or more, the margarine industry has been telling us to switch to polyunsaturated margarine, for the sake of our hearts. And that's precisely what many people have been doing – advertising works. So people with heart disease have been abandoning butter and switching to margarine. So although the survey *seems* to suggest that margarine-eaters get more heart disease, what it's *actually* telling us is that people with heart disease now prefer to eat margarine. And that is confirmed by the evidence. In the British Regional Heart Study, men who didn't eat butter were more than twice as likely as butter-eaters to have been previously diagnosed as having heart disease. As epidemiologist Professor Gerry Shaper wrote in the *British Medical Journal*:

> 'These findings strongly suggest that pre-existing illness (obesity and heart disease) is associated with choice of fat spread.'

A few days later, Dr. Elwood, leader of the team of scientists whose survey had sparked off all the publicity, wrote:

> 'We can certainly confirm what Professor Shaper and colleagues show in their own data. In retrospect, it was unwise for a brief description and preliminary analysis of [our] data to be included in a privately published report that attempted to summarise a wide range of work on cardiovascular disease. It was particularly unfortunate that the one small section on milk and fats was widely reported and interpreted uncritically in the popular press. I can only apologise for not having foreseen that this was likely to happen, and I now greatly regret not having withheld the data until a more adequate analysis could be submitted to the scientific press.'[15]

I wonder how many people, not having seen the conclusion of the story, now believe that 'butter slices your heart attack risk'?

MISLEADING IMPRESSIONS

Since the British meat industry has untold millions to spend on advertising and promotion, it is perhaps surprising that their track record isn't better than it is. Sometimes, their advertising slogans seem to be downright counter-productive – remember 'Where's The Meat?', which reminded millions of people that meat-eating was a declining habit, or 'Meat's Got The Lot', which emerged at the time that food poisoning, antibiotic and hormone contamination were also hitting the headlines. At other times, advertisements have seemed unconsciously humorous, such as the 'Slam In The Lamb' slogan which, as at least one wag pointed out, sounded suspiciously like a none-too-subtle Antipodean euphemism for sexual intercourse.

The American meat industry seems to be equally cursed. When they recently spent a fortune on a series of very high-profile advertisements featuring star names, they burnt their fingers not once, but twice. Cybill Shepherd was depicted as saying:

'Sometimes I wonder if people have a primal, instinctive craving for hamburgers. Something hot and juicy and so utterly simple you can eat it with your hands. I mean, I know some people who don't eat burgers. But I'm not sure I trust them.'

It is indeed gratifying when such fatuous copywriting gets its come-uppance, as it duly did when Shepherd was later reported as saying to *Family Circle* magazine that one of her own beauty tips was to try not to eat red meat.[16][17] Shepherd later denied making the statement, attributing the error to a misinformed publicist. Nevertheless, the beef barons who had paid for the $23 million campaign must have found the whole affair rather heartbreaking.

When James Garner agreed to appear promoting 'Real Food For Real People', his reward was even worse – prompt admission to hospital for heart surgery. Members of the Farm Animal Reform Movement thoughtfully sent him a vegetarian cookbook, a rather brilliant publicity coup which seemed to get more high-profile media coverage than Garner's original advertisements.[18] And to add insult to injury, the Beef Industry Council had a 'Hubbard award' (named after a nineteenth-century advertising shyster) bestowed on it by the Center for Science in the Public Interest (CSPI), for 'misleading, unfair, and irresponsible' advertising.[19] 'Popular beef products, such as hamburgers, are, by definition, not lean and contain large amounts of fat,' said Bonnie Liebman, director of nutrition for the CSPI, 'Real beef isn't so healthful when it's eaten by real people.'

One of the biggest Freudian slips of recent times was spotted when American college student Erik Pyontek saw a poster promoting meat products in his supermarket.[20] Entitled 'America's Meat Roundup', it depicted a tall, blond cowboy proudly holding the American flag, hand on hip, his firm-jawed gaze courageously meeting the horizon. Pyontek went away and dug up a picture in a high-school history textbook he'd been reminded of and, yes, there it was – a tall, blond Aryan proudly holding the Nazi flag, hand on hip, his firm-jawed gaze courageously meeting the horizon – the same all but for the swastika. 'We're not trying to send out any subliminal Nazi messages,' screeched a spokesperson for the ad agency that created it. Nevertheless, the common symbolism of the two images is very telling.

The art of writing advertising copy is a fine one. On the one hand, you have a responsibility to be accurate in what you say. On the other hand, you have to sell the product. Sometimes, the distinction between accuracy and salesmanship is blurred, as in the recent 'Meat To Live' advertising campaign. The advertisements feature a selection of male models doing typical he-man stunts, handstands, and so on, thus trying to create a masculine, athletic image for their product – all very predictable and bland. However, the accompanying text is more interesting. One of the advertisements claims:

'Without a regular supply [of iron], you could well suffer from listlessness or, in extreme cases, anaemia . . . This, on its own, is a powerful reason for eating meat.'

Is it? The Advertising Standards Authority considered that this turn of phrase might give the impression that meat was essential to a healthy diet, and warned the Meat and Livestock Commission not to create this impression in future advertisements.[21]

'Healthwise,' said another meat ad, 'it'll steel you against the elements too.' Again, the Advertising Standards Authority considered the wording to be ambiguous, and asked the Meat and Livestock Commission not to imply that eating meat could provide health benefits that couldn't be obtained by eating a balanced meat-free diet.

But if their public aspect has been less than irreproachable, at least the Meat and Livestock Commission appreciate the benefits of a vegetarian diet where it counts – at the very heart of their organisation. For when a journalist from *Marketing* magazine had lunch there recently, he was relieved to discover that 'the staff canteen offers a vegetarian option every day for those who prefer not to ingest what they sell.'[22]

THE BODY POLITIC: A DISSECTION

Not all the raised voices contributing to the general confused cacophony come from commercial interests – the following example comes from the political sphere. As more people have become aware that the intake of animal fat is closely connected to human disease, the opinions of a small but very vociferous opposition have been widely reported. One wonders sometimes how the long-suffering public ever manages to make up its mind what on earth it should be eating.

The 'Social Affairs Unit' is a self-described 'research and educational trust committed to the promotion of lively wide-ranging debate on social affairs'.[23] Behind this rather capacious definition there are various publications which give more of an insight into the Unit's real agenda: 'Breaking the Spell of the Welfare State', 'Set Fair: A Gradualist Proposal For Privatising Weather Forecasting', 'The Kindness That Kills: The Churches' Simplistic Response To Complex Social Issues' and other equally earnest fare. Quotations on its pamphlets credit the Unit with influencing ex-Prime Minister Margaret Thatcher. One of its main functions is to get publicity for the views of its authors. So far so good; a free democracy thrives on lively discussion and debate.

When the Social Affairs Unit produced a little pamphlet criticising healthy eating guidelines produced by the World Health Organization, it must have been very gratified by the media coverage the story received. The serious press gave it a good spread: 'Trivial Dietary Advice Attacked By Nutritionists', said one newspaper.[24] At the heart of the story was a familiar old theme, always guaranteed to get good media coverage:

'Dr. Petr Skrabanek [one of the authors], a community health specialist, says that the knowledge on which most dietary advice is formulated is weak. He sees no justification for cutting fat intake to 35 per cent of the total energy intake.'

Intrigued by the press coverage, I bought a copy of their booklet. It contains many surprising statements, such as:

'Blood cholesterol for practical purposes has no predictive value for the risk of a future heart attack.'

And:

'In fact, recommended cholesterol-lowering diets have been shown useless in a review of all controlled trials.'

I found these statements frankly amazing. These opinions – for that is all they are, even though they may subsequently be reported by the media as 'fact' – are so very, very different from the consensus amongst experts in these fields that I decided to examine just one of these assertions in detail. I decided to investigate Dr. Skrabanek's claim that 'cholesterol-lowering diets have been shown useless'.

Dr. Skrabanek cited a reference to a medical journal to support his claim – and that was my first problem, because the reference was incorrect, the page number didn't exist. With some trouble, I eventually tracked the study down. Published in the *British Medical Journal (BMJ)*, and entitled 'Dietary reduction of serum cholesterol: time to think again', its very first words were:

'Every 1 per cent reduction in serum cholesterol concentration reduces the risk of coronary events by about 1–2 per cent.'[25]

This, in itself, was rather ironic, because it appears to contradict Dr. Skrabanek's other claim that blood cholesterol levels can't predict the risk of a future heart attack.

I ploughed on. The authors of the study had reviewed the results of 16 scientific trials whose object was to reduce blood cholesterol by dietary means. Bearing in mind Dr. Skrabanek's claim, stated above, let me quote to you some extracts from the very report he cites allegedly to prove his point:

- 'Dietary change undoubtedly can lower serum cholesterol concentration, as shown by reductions averaging 12 per cent over one to five years with rigorous diets.'
- 'The correct conclusion from the Oslo study is that rigorous dietary intervention in male volunteers with very high serum cholesterol concentrations and very high dietary fat intake caused a substantial fall in serum cholesterol concentration.'
- 'These trials leave no doubt that modification of diet can lower serum cholesterol concentrations substantially . . .'

The picture is beginning to look rather different, isn't it? So is Dr. Skrabanek's claim untrue?

Well, not exactly. Here, we must examine Dr. Skrabanek's words very closely indeed. Look carefully, and you'll see that Skrabanek *isn't* saying that 'cholesterol-lowering diets are useless' – he is saying that '*recommended* cholesterol-lowering diets have been shown useless'. Everything hinges on what you understand by 'recommended', and on the evidence he refers to.

The *BMJ* report – which was very controversial, and sparked considerable correspondence, including a letter from Dr. Skrabanek himself – reviewed the success of certain cholesterol-lowering diets. The authors were certainly *not* saying that 'cholesterol can't be reduced by dietary means'. But they felt, after reviewing relevant studies, that the 'Step One' diet (the sort usually recommended as a first step to cholesterol reduction) didn't produce a large enough reduction in blood cholesterol to be useful in the treatment of people with raised cholesterol levels. They concluded:

> 'Guidelines should be reviewed to provide a more realistic estimate of the effect of a Step One diet and of the likely need for lipid lowering drugs.'

That's a long way from saying that 'cholesterol-lowering diets are useless'.

Subsequently, various doctors wrote to the *BMJ* to express their opinion on this paper. Two specialists wrote from Denmark to complain that the original paper had 'ignored a most important point'.[26] How well had the people in the studies complied with the diets prescribed for them? 'In trials with a high degree of dietary control,' wrote the Danish team, 'it has repeatedly been shown that serum cholesterol decreases substantially in subjects shifted from a typical Westernised diet to a Step One type diet.' And they gave evidence from their own work which showed that a 12 per cent reduction in cholesterol had been achieved in just one month on a Step One type diet.

Another doctor wrote to say that the authors of the original paper had ignored a study which he and colleagues had published in 1972, showing an average cholesterol reduction of 22 per cent, following the same type of Step One diet.[27] Yet another specialist wrote to point out that the original paper was wrong to imply that only extremely stringent diets could achieve worthwhile cholesterol reductions; the original paper had shown that six major studies had achieved an average of 14 per cent cholesterol reduction, and none of them was more intensive than the Step One diet.[28] 'Flawed analyses lead to faulty conclusions,' the writer noted. The authors of the original paper replied in the same issue, saying that while they agreed that the 14 per cent reduction in cholesterol was indeed worthwhile, they thought that the diets that produced this reduction were rather more rigorous than a Step One diet.

Dr. Skrabanek wrote too, commenting that 'lowering cholesterol is not synonymous with being effective in reducing mortality.'[29] This point was subsequently answered by the very people who wrote the original paper first cited by Dr. Skrabanek, who said:

'We are aware that reducing the cholesterol concentration has not lowered total mortality, and this important question needs to be resolved. Nevertheless, available evidence does indicate that reducing the cholesterol concentration prevents coronary events.'[30]

So here we have a real diversity of opinion. If you are feeling rather bemused by this stage, I can only say that I know just how you feel. Dr. Skrabanek's original statement is all too easy to misconstrue. And that, no doubt, is what many people may have done.

ESKIMOS' KNELL

In recent years, it has become accepted wisdom amongst a wide variety of people – doctors, health food shoppers, and even among some vegetarians – that fish – and particularly fish oil – is healthy. There's certainly no disputing that fish oil can make you very healthy indeed – if, that is, you happen to manufacture fish oil capsules! For several years, they've been the fastest-moving items in health food shops. According to the makers, fish oil can treat asthma, prevent cancer, lower your cholesterol level and banish arthritis. But what is the evidence? Here's a briefing which puts the other side of the picture:

- Fish oil is indeed a significant source of the omega-3 essential fatty acids. There are two important groups of essential fatty acids: omega-6 acids, found in abundance in corn, soy, safflower, and other vegetable oils, and omega-3 acids, found in fatty fish. Each group has distinct – and often antagonistic – physiological effects. But, contrary to popular opinion, fish is not the *only* source of omega-3 acids. Flaxseed (linseed) oil actually contains about twice as much omega-3 essential fatty acids as fish oil. According to nutritionist Ann Louise Gittleman, M.S., co-author of *Beyond Pritikin*, flaxseed oil's greatest attribute is its ranking as the vegetable source highest in omega-3 fatty acids. 'Fish is the best-known source of the omega-3s,' she says, 'but flaxseed oil contains 55 to 60 per cent omega-3 – about twice as much as is found in fish oil.'[31] Flaxseed is also rich in omega-6 fatty acids. It is a highly polyunsaturated oil, capable of providing the raw material necessary for the production of prostaglandins in the human body. Prostaglandins are vital, hormone-like compounds that regulate every function in the body at the molecular level. Without enough prostaglandins, our bodies cannot properly use the food we eat. (Note: Food that contains omega-3 oils goes bad easily because the unsaturated fatty acids attract oxygen and become oxidised or turn rancid.)
- Fish oil has been touted as the ultimate cure for heart disease. Some studies have indeed shown that large doses of fish oil can lower triglycerides (blood fats). But when continued over a longer period of time – six months or so – the initial triglyceride-lowering effect of fish oil in patients with high levels almost disappears.[32]

- Another study casts doubt on the benefits of fish oil for heart patients who have had angioplasty, a medical treatment for narrowed arteries. Because fish oil makes your blood thinner, it was thought that it could help keep clogged arteries open. And three small studies first hinted that it could. However, a larger study from Harvard Medical School and Beth Israel Hospital shows that people taking fish oil actually had a higher rate of recurrent narrowing of the arteries and more heart attacks than people taking olive oil![33]

- It has now passed into folklore that Eskimos have much less heart disease than other Westerners, and that this reduction in heart disease is due to the fish oil they consume. Actually, if you study almost any native population, you'll find they have much less heart disease than we do. In March 1990, the *American Journal of Public Health* published a review of previous scientific work on this subject. The author of the review wrote:

  'Several studies have reported that Arctic populations, which typically consume large amounts of fatty fish, have a low rate of atherosclerosis and cardiovascular disease. But a thorough examination of the methods used in these projects reveals that the evidence may not have been reliable. Two studies that reported causes of death used data from a modest number of autopsies that were performed without standard procedures by inadequately trained personnel.'[34]

  In fact, there have been persistent questions asked in scientific journals about the accuracy of the 'Eskimo' evidence. One medical critic has already pointed out the original study was seriously flawed, because far fewer deaths from cardiovascular disease were recorded than actually took place.[35]

- The same research that was supposed to demonstrate that fish oil could reduce deaths from heart disease also revealed that Eskimos were dying in greater numbers from cerebrovascular haemorrhages (strokes). Since fish oil is known to thin the blood, this is a perfectly possible consequence. But this finding has received very little publicity.

  A recent study conducted to assess the benefits of fish oil on young people with raised levels of fats in their blood ended up proving just how dramatic this blood-thinning effect can be. Of 11 patients, eight of them had nose-bleeds while taking the oil. 'It is concluded,' wrote the scientists, 'that the dose of fish oil necessary to reduce blood lipid levels may be associated with an extremely high risk of bleeding problems in adolescents.'[36]

- Can fish oil help arthritis? Again, the evidence is far less conclusive than the publicity indicates. A 1985 study found that people who took one specific omega-3 (known as EPA) reported less morning stiffness when compared to another group of people who didn't take the oil.[37] But there were no improvements in other areas, such as grip strength, exercise ability, fatigue or swelling. (Note: The people on fish oil didn't actually get any better, it was just that the people *not* taking fish oil got worse.)

- As far as cholesterol is concerned, the results are very mixed indeed. Some studies have shown that large doses of fish oil can lower cholesterol levels dramatically. But other studies have shown just the opposite – that it can, in fact, raise them, and in particular it can raise the level of 'bad' LDL cholesterol.

WHY DOCTORS DON'T TELL US

You've read enough now to appreciate that there is a whole ocean of confusion out there, sometimes deliberately created, other times the result of genuine differences of interpretation. But what about the healthcare professionals, the doctors? Surely they should know the truth?

You may well wonder why more doctors don't strongly advocate the vegetarian lifestyle. I wonder myself, constantly. After all, the research is published in medical and scientific journals, and you'd expect that most doctors would keep up to date with these things. The trouble is, an awful lot of other work gets published in scientific journals, too. Dr. Vernon Coleman, a well-known medical columnist, explains what happens to all this research:

'There are so many medical journals in existence that a new scientific paper is published somewhere in the world every 28 seconds . . . Because they know that they need to publish research papers if they are to have successful careers, doctors have become obsessed with research for its own sake. They have forgotten that the original purpose of research is to help patients . . . Believe it or not, much of the research work that has been done in the last 20 years has never been analysed. Somewhere, hidden deep in an obscure part of a medical library, there may be a new penicillin. Or a cure for cancer. You don't have to go far to find the evidence proving that many scientific papers go unread: approximately 20 per cent of all research is unintentionally duplicated because researchers haven't had the time to read all the published papers in their own specialised area.'[38]

So the first reason more doctors don't know the truth about the benefits of the vegetarian lifestyle is, simply, because they just don't come across the evidence. But even if they did, there are two further problems:

First, there's no-one to sell it to them. This may sound rather cynical, but the truth is that doctors respond to the information they are fed, and most of it comes from one direction – the drugs industry. Research has shown that by far the greatest influence over doctors' prescribing habits is the non-stop barrage of promotion which these companies produce.[39] By contrast, only 12 per cent of their prescribing decisions are influenced by articles in professional journals.

Secondly, doctors have traditionally focused on studying disease, rather than promoting health. As Dr. Joe Collier, a clinical pharmacologist who has studied and written about the drugs industry, puts it:

'Doctors fail patients because they are preoccupied with, even obsessed by, disease. Right from their earliest days at medical school, training concentrates on the recognition and treatment of disease, rather than its prevention . . . Disease is so much a part of a doctor's horizon that it may be difficult for a patient to escape the consulting room without an illness being diagnosed and at least one medicine being prescribed.'[40]

Then, we come up against 'the system' itself. The sad truth is, information from major studies such as those discussed earlier is rarely used to offer advice which will improve people's lives. When medical science comes across studies which show that vegetarians have less heart disease than meat-eaters, medical science doesn't respond by saying 'Great! Let's advise all our patients to go vegetarian!' Instead, it asks itself 'What is it about the meat-eaters that makes them so unhealthy?' This then generates a vast amount of research, as you will see.

Dr. T. Colin Campbell, the mastermind behind the China Study described earlier (see page 9), explains this mode of thinking like this:

'One line of investigation suggests that evidence is not sufficient for serious dietary recommendations until mechanisms are identified and understood. However, this logic is rather nihilistic. If this were necessary, then it should also be reasonable to require a full mechanistic accounting of the effect of the same food constituent upon other diseases as well. Such logic contradicts the true complexities of biology and discourages hope of public health progress ever being made.'[41]

In other words, it isn't necessary to understand *every last detail* of the cause-and-effect relationship between meat-eating and disease in order to start taking action now. Another expert, Dr. O. Turpeinen of Helsinki, who himself has produced some fascinating work which we will consider a little later, expressed it like this:

'It is not always judicious to wait for the final results and the irrefutable proof before taking action. Many lives could be saved and much good done by starting a little earlier. Although we do not yet have an absolute proof for dietary prevention of CHD, there is strong evidence for its effectiveness, and its safety.'[42]

So studies such as the five mentioned earlier usually go unpublicised, and serve to generate more theories, which are then explored and tested, often by use of animal experiments. You may be amazed to learn, as I was, that researchers have known *for decades* that feeding a naturally vegetarian species – such as rabbits – a meat diet will produce heart disease. And they've also known that in naturally carnivorous species – such as dogs – it is virtually impossible to produce clogging of the arteries, even when large amounts of cholesterol and saturated fat are fed to them.[43] Now for heaven's sake, doesn't this information tell us *something* about the sort of diet we humans should be eating? What have they been doing all this time? Why haven't they given us this vital information?

What they've been doing is yet more research. Looking in ever closer detail at the mechanisms of disease. And, oh yes, producing wonderfully profitable new ranges of drugs and medications to avoid heart disease, to treat heart disease, and to fight cholesterol.

'They Want People to Ignore Dieting'

'The ultimate wonder cure for a lousy lifestyle has arrived: the anti-cholesterol pill,' reports a British newspaper.[44] 'Take one a day and you can go back to junk food, throw away the running shoes, and even take up smoking again and still escape a heart attack.' Since Britain has one of the highest death rates from coronary heart disease in the world, the British market is certainly worth grabbing. Comments a stockbroker: 'The drug companies want people to ignore dieting, even though it is much more effective than drugs for 90 per cent of people. Ideally the industry would like to prescribe anti-cholesterol drugs to everyone with a family history of heart disease – the market is enormous.' And one doctor, who had just been whisked off to Rome for a lavish drugs company sales-pitch, adds: 'Anti-cholesterols are the hottest property in the drug world and people are being hounded into their massive use even before some of the long-term trials are completed. In theory they allow people to live on hamburgers and sausages and yet have the blood cholesterol of a Chinese peasant who eats rice and soybeans.'

A Doctor Speaks

This is all rather depressing. It suggests that although we already have a medicine which can prevent and treat heart disease and other major problems of our time – it's called the vegetarian diet – it will never become widely recognised or pre-scribed. When I went to interview a hospital dietician, whose job it is to help people with high cholesterol levels reduce them by dietary means before drug treatment is prescribed, I was amazed to find her including meat and other ani-mal products in the diet sheets she was giving out. 'Why aren't you encouraging people to go completely vegetarian?' I asked her. 'Surely you're aware of the weight of evidence in favour of the vegetarian lifestyle?' She replied dismissively. 'Oh, people would never do that,' she said. 'There's no point giving people diets which you know they just won't follow.' It seemed to me that she was denying her patients potentially life-saving information, based on little more than her own prejudice. As a result, many of them could be condemned to a lifetime of taking cholesterol-lowering drugs.

Luckily, some doctors don't share this dismal attitude. Dr. Bruce Kinosian, an assistant professor of medicine at the University of Maryland in Baltimore, is one. He says:

'If you can lower cholesterol with diet, why use drugs? There are clearly people who need drugs to lower their cholesterol, but there are other options out there that may be more cost-effective and are not being emphasised. There are a lot of people with high cholesterol levels in this country, and as a matter of social policy, you don't want to get in the habit of prescribing pills to everyone.'

So there are a few glimmers of hope out there. In a free society, it is difficult to suppress the truth for ever, particularly when it is something so eminently sensible as the vegetarian way of living.

I had heard about Dr. David Ryde, a family doctor, and I was curious to know if everything I'd heard was true, especially the revelation that he happened to be the lowest-prescribing general practitioner in the country. Dr. Ryde is in every respect a conventionally trained and qualified doctor, but he has gradually acquired a reputation for preferring to treat his patients through dietary means. So I visited him in his surgery. An athletic and vigorous man greeted me at the door, with a big grin. I later learnt that he is actually 20 years older than he looks. First, I asked him how he came to be vegetarian. He told me:

'The seeds were planted when I was walking home from school one day, and I saw some pigs being beaten. That set me thinking. Was it really necessary to inflict so much cruelty just to have bacon for breakfast? Anyway, at the age of 12, I stopped eating meat and fish, much to the horror of my parents. But they couldn't deny I was healthy enough – I was captain of athletics, rugby and swimming at school, and I could easily cycle 100 miles or more in a weekend.

'When I went to medical school, we were taught nothing about nutrition. They simply said there were two types of protein – "first class" and "second class". It was only years later that I began to understand that plant protein could be entirely satisfactory for human needs. I was still keenly interested in sport, playing rugby for the county, and for the United Hospitals.

'Eventually, I began to become interested in the science of nutritional medicine, and I started to offer my patients nutritional advice. Some patients simply didn't want to know – they'd take the attitude that they didn't want a lecture, they just wanted me to write a prescription for some pills – that's what they regarded as "proper" medicine. But other patients were more willing to try something new, and I started to get some extraordinary results.

'My first was a patient with severe angina. His condition had been deteriorating for about five years, and he'd been into hospital, was taking all the medication, and so on. But his condition was, frankly, almost terminal. It was a really pitiful sight to see him struggle to walk the few yards from the car to the surgery. Now a person in such a desperate state will listen, and they will try anything. So I suggested he try a strict vegetarian diet, actually a vegan one. Just one month later, he could walk one mile, from his home to my surgery. Three months later he could walk four miles, while carrying shopping. "It used to take him a quarter of an hour to climb three flights of steps," his daughter told me. "Now he's up in a few seconds!"

'That was my first success, and it encouraged me to try it with other patients. Another interesting case was a professor of medicine, actually the dean of a medical school. He had been taking anti-ulcer medication for four years, with little success. I suggested he try a vegan diet, and after three days, there was a remarkable improvement in pain reduction. A year later, he had lost about 10 pounds of weight, and he looked a new man, light-hearted and happy.

'Another interesting case was a woman with severe headaches, and a blood pressure of 185/120. I suggested she try a vegan diet, and the pressure soon came down to 115/75. Now you'd never have seen that kind of reduction using medication. And she felt fantastic! Which was another benefit, because anti-hypertensive medication often leaves patients feeling exhausted.

'I've seen results such as these in my patients too often to attribute them to coincidence. Really, this kind of treatment has no side-effects, and the benefits are so worthwhile that there's no reason not to try it.'

'What sort of reaction have you had from your colleagues?' I asked.

'In the early days, they used to warn me that I wasn't prescribing enough medication. When they charted the prescribing rates of GPs, I would always be right at the bottom, way off the graph. And I think that worried some people. But these days, I'm asked to give talks to colleagues, and to administrators. Obviously, my methods are far less costly to the health service than usual. I also feel strongly that we doctors need to examine more closely what actually goes on in the consulting room. You know, the truth is that patients don't usually come and see us because they're ill; they come because they're *worried*. They're anxious about some aspect of their health. Now, if all we do is simply send them away with a bottle of pills, we have actually reinforced their anxiety, which can make a cure harder.'

He paused, and smiled.

'Fundamentally, we must remember that we're not vending machines!'

Let's put a lid on the science for the moment, before we end up with a severe case of indigestion. All this evidence has basically been telling us one thing – that our 'vegetarian species', as Dr. Campbell calls it, is today eating the wrong sort of food. You've also seen that there are many reasons for this forbidden knowledge not to be widely understood or appreciated. Now, I want to address a question which has probably been formulating in your mind as you've been reading, and at the same time, we're going to examine a different kind of forbidden knowledge; an insight which goes right to the heart of our own individual and collective identities.

WHAT ARE WE?

Alarmed by the growth of vegetarianism among young people (the consumers of tomorrow), the meat industry is busy spending its vast resources launching its propaganda into schools and other places where young minds can be influenced. In its thinly disguised advertising material, you will find many astonishing statements, such as:

'Modern man does not need to hunt but he still needs a balanced diet – of which meat is an essential element.'[45]

One really wonders how it is possible for the meat lobby to get away with statements such as these. Meat is not 'an essential element' of a balanced diet, as millions of healthy vegetarians will testify. And, as a parent, I find it outrageous that the meat industry which claims to have 'established a good reputation among teachers for providing credible and well-balanced classroom resources'[46] should be allowed to go into schools with such misleading propaganda masquerading as fact.

Yet many people still mistakenly believe that humans are somehow 'genetically programmed' to eat flesh foods, and cannot thrive without them; that we are, in essence, carnivores. All right then – let's look at the evidence.

Scientific evidence suggests that our ancestors probably originated in the East African Rift Valley, which is a dry and desolate place today, but would have been very different 2–4 million years ago. The habitat was very lush then. There were large, shallow freshwater lakes, with rich, open grassland on the flood plains and dense woodland beside the rivers. Fossil evidence shows that foodstuffs such as *Leguminosae* (peas and beans) and *Anacardiaceae* (cashew nuts) were readily available, as were *Palmae* (sago, dates and coconuts). Evidence gained from the analysis of tooth markings indicates that our ancestors' diet was much the same as the Guinea Baboon's is today – hard seeds, stems, some roots, plant fibre – a typically tough diet requiring stripping, chopping and chewing actions.

Our ancestors also had very large molars, with small incisors, unsuited to meat consumption but ideal for consuming large quantities of vegetable matter. By 2.5 million years BC, however, evidence shows that the land began to dry out, forcing Australopithecus (the name of one of our early ancestors) to desert this idyllic Garden of Eden and to try to survive on the savannahs, where they were poorly prepared for the evolutionary struggle that was to come.

Before this crucial point, there is little doubt that our ancestors had largely followed a vegetarian diet, typical of primates. Recent studies of minute scratches on the dental enamel of Australopithecus suggest that their diet consisted largely of hard, chewy seeds and berries, although a few eggs and small animals may have been consumed too. Most scientists consider it unlikely that Australopithecus was a systematic hunter, or 'killer ape', as this species has sometimes been depicted.[47]

So we were forced by our rapidly changing environment to eat anything and everything we could get our hands on, which of course included some flesh. As our old habitat receded, we had to make some quick decisions. We had been used to eating a mainly fruit and nut diet. As this became increasingly scarce, we had to adapt to eating whatever we could find. There wasn't much. We found roots and grasses, and made do with them. We would have stumbled across some partly rotten carrion flesh, and gratefully ate what we could salvage. We would have chased easy-to-catch small game. We ate it all, no questions asked. Interestingly, we still preserve some ability to digest and utilise leaves and grasses, which recent scientific work has discovered and which probably dates from this period of our existence. We became not carnivores, but omnivores – actually, I would argue in favour of the word 'adaptivores', because it conveys a more accurate impression of what was going on at that point in our history. In his book *The Naked Ape*, zoologist Desmond Morris made an interesting observation about this period when he wrote:

> 'It could be argued that, since our primate ancestors had to make do without a major meat component in their diets, we should be able to do the same. We were driven to become flesh eaters only by environmental circumstances, and now that we have the environment under control, with elaborately cultivated crops at our disposal, we might be expected to return to our ancient primate feeding patterns.'[48]

If we as a species can be characterised by just one word, it would be 'adaptable' – we have learnt how to survive in almost any environment, no matter how seemingly hostile. It is our passport for success in any situation, no matter how desperate, and unquestionably the key to our survival. We were forced out of our original habitat, and miraculously we survived. We were forced to learn how to live on the plains in competition with other animals which were natural carnivores, and again we met the challenge.

So here we have a picture of a species which was originally vegetarian, which then due to force of circumstances adapted to become omnivorous. This reality is a long, long way from the 'meat is an essential element of the diet' myth propagated by the industry. It is clear from recent analyses of human remains that even during this period of our development, plant food was still by far the most important source of food. The level of strontium present in bones is an accurate guide to the amount of plant food consumed, and scientists at the University of Pisa, Italy, who have analysed the bones of early Europeans have found that they were eating an 'almost exclusively vegetarian diet' right up to the time agriculture was developed.[49]

To what extent should our omnivorous adaptation influence our modern food habits? The first point to understand here is that the word 'omnivore' does not mean 'carnivore', as some seem to think it does. The Meat and Livestock Commission says in the propaganda it gives out to our schoolchildren:

> 'We humans are biologically omnivores, and an omnivorous diet is one which includes a whole range of foods – meat, in various forms, prominent among them.'[50]

This is brilliantly misleading, for it implies that meat is an essential part of our diet. The fact is that meat is optional – we can choose to consume it, or not. Either way, we should know what the implications are.

The second point to understand is that our genetic constitution has changed very little for several tens of thousands of years. But, of course, our diet has changed – unfortunately, for the worse. Basically, our bodies are still in the Stone Age, and they expect the sort of nutrition they were getting then. They're just not used to getting the kind of junk food we give them today. No wonder so many diseases are related to our modern pattern of food consumption.

As you might imagine, modern Westernised humans consume vastly more animal flesh than we have ever done in the whole history of our species. And we don't even have to exercise to get it – the exertion of the chase has been replaced by the flick of the credit card as it slides from our wallet.

In 1912, the first ever medical observation was made of a heart attack. In less than a hundred years, heart disease has soared to become one of the leading killers of the Western world. But why? What has changed in such a comparatively short space of time? I put this question to a recognised authority in the field, Professor Michael Crawford, who told me:

> 'What has happened is that we all started from a common baseline of wild foods. This is the sort of primitive diet which humans have eaten throughout most of their evolution, over the past 5 million years. However, in the last few centuries, things have gone haywire. In Europe, our diets have gone in one direction, in Africa and India they've gone in a different direction. In Western Europe we've focused on consuming foods which are very rich in non-essential types of fat, but pretty miserable sources of essential fats. Our diets have also become rich in processed and refined carbohydrates. In fact, the problems are quite easy to identify – it's taking corrective action that seems to be difficult for some of us.'

All in all, it seems as if the human race has unwittingly been carrying out a huge experiment on itself over the past century. In the year 1860, about one quarter of our energy came from fat sources. By 1910, this had risen to one third, and by 1975 about 45 per cent of our total energy intake was coming from fat, much of it saturated animal fat. Thus, in no time at all, the amount of fat in our diet doubled. So it's hardly surprising if this new diet which we're eating today has some rather dreadful side-effects, in the form of diet-related diseases.

Modern food animals are bred to be fat: the carcass of a slaughtered animal can easily be 30 per cent fat or more. But the sort of animal that primitive people hunted was a wild animal – it had, on average, only 3.9 per cent fat on its carcass.[51] So today, even if we cut our meat consumption back to the greatly reduced amount that our ancestors consumed, we will still be taking in seven times more fat than they did!

But even this isn't the end of the story. The type of fat on the carcass of the animal that our ancestors ate was different, as well. Primitive meat had five times more polyunsaturated fat in it than today's meat does – which is high in saturated fat, but much lower in polyunsaturated. Also, our ancestral diet only had one sixth the amount of sodium (salt) that the modern diet contains. And because fresh food comprised such an important part of the diet, the primitive diet was much, much richer in natural vitamins. For example, there would have been nearly nine times as much Vitamin C in the primitive diet. And twice as much fibre. And three times as much total polyunsaturated fat. And so on . . .

So if you were worried that a meat-free diet might not be healthy, don't be. In

VEGETARIAN	FLESH-EATER	HUMAN
Hands / hoofs as appendages	Claws as appendages	Hands as appendages
Teeth flat	Teeth sharp	Teeth flat
Long intestines to digest nutrients in plant foods fully	Short intestines; rapidly excrete putrefying flesh	Long intestines to digest nutrients in plant foods fully; flesh foods cause constipation
Sweats to cool body	Pants to cool body	Sweats to cool body
Sips water	Laps water	Sips water
Vitamin C obtained solely from diet	Vitamin C manufactured internally	Vitamin C obtained solely from diet
Exists largely on a fruit and nut diet	Consumes flesh exclusively	Diet depends on environment; highly adaptable
Grasping hands capable of using tools or weapons	No manual dexterity	Grasping hands capable of using tools or weapons
Inoffensive excrement	Putrid excrement	Offensiveness of excrement depends on diet
Snack feeder	Large meals infrequently taken	Combines worst of both worlds
Predominantly sweet-toothed	Preference for salty / fatty food	Likes both sweet and salty / fatty food
Likes to savour food, experiment with variety, combine flavours	Bolts food down	Likes to savour food, experiment with variety, combine flavours
Large brains, able to rationalise	Small brains, less capable of adaptive behaviour	Large brains, able to rationalise (at least in laboratory studies)

point of fact, it's much closer to the kind of natural food that we've always eaten, and that our bodies have always been used to. In evolutionary terms, the meat we eat today is a *new* food for us, which means that we're actually conducting a huge experiment on our own bodies. And as you've started to see, the results don't look at all good.

Now, spend a moment looking at the table above. Here you can see typical characteristics of vegetarian animals compared to carnivorous animals. This straightforward evidence very clearly demonstrates the overwhelmingly vegetarian nature of our species.

WHO ARE WE?

Did you notice in the meat trade's propaganda quoted above that they spoke of 'Modern man', when they really meant to say 'Modern people'? Most people tend to dismiss unconscious sexism such as this as trivial, because it is so common. However, I now want to present you with yet more forbidden knowledge that goes straight to the heart of the modern myth of the red-blooded male meat-eater.

Many of us are conditioned by our upbringing to believe that 'man is a natural hunter and meat-eater'. Note that I, like the meat industry's propaganda quoted above, said 'man', not 'humans'. In the account of human evolution that most of us learn, women are mere appendages – accessories and mating-objects for the all-powerful hunting male. According to the conventional wisdom of anthropology, it is hunting that has made us what we are today – intelligent, because hunters must be wily; tool-makers, because hunters must have weapons; upright-walkers, because hunters must walk and run long distances; co-operative, because hunters must work with each other to ensure a kill; and masters of language, because hunters must communicate with each other. This is simplistic rubbish. But it is only in recent years that this ubiquitous stereotype has been challenged, by a few women anthropologists, who have become rather tired of the persistent denigration and omission of women from the accepted account of human history. Less than sexually secure males may wish to stop reading now. Adrienne Zihlman, professor of anthropology at the University of California, Los Angeles, explains:

> 'The most popular reconstruction of early human social behaviour is summarised in the phrase "man the hunter". In this hypothesis, meat-eating initiated man's separation from the apes. Males provided the meat, presumed to be the main item in early hominid diet, by inventing stone tools and weapons for hunting. Thus males played the major economic role, were protectors of females and young, and controlled the mating process. In this view of things, females fade into a strictly reproductive and passive role – a pattern of behaviour inconsistent with that of other primates or of modern gathering and hunting peoples. In fact, the obsession with hunting has long prevented anthropologists from taking a good look at the role of women in shaping the human adaptation.'[52]

The plain fact is that the sort of hunting that our ancestors practised was never a good enough way of providing food for everyone. Careful studies of societies who lead similar lifestyles to those of our ancestors – such as the Bush People of the Kalahari – reveal that the probability of obtaining meat on any one hunting day is about one in four.[53] Now, just how long do you think a society can exist, based on a 25 per cent success record? By contrast, the women always return from their gathering expeditions with food – a 100 per cent success rate. And the entire tribe could comfortably feed itself if each member put in a 15-hour week – rather better than our own society's achievement.

It is quite clear that in original societies such as these, hunting is only possible 'on the back' of an effective, dependable and reliable source of plant food. Once the tribe is certain of food, then those men who want to (about one third of the Kalahari males never hunted) can go off and gamble on a kill – nothing jeopardised if they come home empty-handed. And yet, many modern people, living entirely synthetic lives in wholly unnatural Western environments, still believe and behave as if meat-eating is the magic thread which keeps us in touch with the primitive, authentic humans we think we ought to be ('Real Food For Real People' is how the advertising slogan tries to exploit it). Modern people who have never been told of the absolutely crucial role of 'woman the gatherer' in human development are, to be blunt, profoundly ignorant. They are ignorant about the history of their own species, which makes them ignorant about their very own, personal identities. And ignorance leaves them open to exploitation.

THE SECRET CHRONICLE OF US

Let me present you with some snapshots from the secret family album of our species. 'Secret' because this is information not generally known – but one day, I hope, it will be. At that point, we will then have to ask ourselves some very difficult questions about our own identities; about the people who we were, and who we wish to become.

Women, being the principal gatherers, also became the first growers. There is a significant difference between horticulture (which came first, and involved the cultivation of wild plants) and agriculture (which came later, and involved ploughing the ground using domesticated animals). Whilst horticulture seemed to spring up almost simultaneously in many parts of the world, agriculture was never adopted in New World original societies (the Americas). And there are still some horticultural tribes in far-flung places, whose development never seems to have progressed to complete agriculture. In these tribes, such as the Australian aborigines, it is often the women who take responsibility for plant usage and cultivation, cutting the tops off wild yams, for example, and replanting them to produce a continually cropping plant in a perfectly balanced relationship with nature.

Why did horticulture first develop? Obviously, it represents a quantum leap in the amount of food that can be amassed for a given amount of effort. Instead of wandering and gathering, it was now possible to stay in a single spot and work continuously at harvesting grain. The transformation from gathering to cultivation seems to have taken place in locations where a high starch yielding plant was available. Grain being particularly easy to store when dry, it was now possible to work intensively at harvesting, and to accumulate an impressive store of food that would not spoil as it was kept. Modern experiments have shown that it is possible to harvest manually about five pounds of grain an hour. If four people worked continuously for the three weeks that wild wheat was ripe, they could produce about one ton of grain – enough to feed themselves for an entire year. Interestingly, this wild wheat was of a much higher protein content (about 24 per cent) than our modern, highly developed strains (about 14 per cent).

Now consider the crushing impact that ever-more-prolific female horticulture must have had on the male ego. 'Man the Provider' has always been a male-inspired, self-justifying myth (think of our phrase 'to bring home the bacon'). The reality of the traditional hunter/gatherer society was that it was held together primarily by the food-producing and child-rearing abilities of the females – not by the males, who contributed in total far less. With the advent of horticulture, women were further challenging the usefulness, indeed the whole *raison d'être* of the male role. They were steadily increasing the already large contribution they made to the group's food supplies. The male contribution, if anything, would have been diminishing at this point, for a fixed home base would have restricted the amount of wild animals within easy reach.

Perhaps the final insult to an already critically wounded hunter's ego may have come when women found that a variation on the male's symbol-laden axe (called a microlith) could be used to help her gather the harvest. For thousands of years, the roughly hewn stone axe had been a potent symbol of manliness. Tied on to the end of a stick, it was used as a spear with which to impale hunted animals. As a hand-held knife, it was used to sever the animal's arteries and cause death. Then it was used to dismember the creature and divide it up into the correct proportions for consumption. Once an object of respect and awe, this symbol of virility was now demoted in the male mind to become a mere female accessory – the sword had been turned into a ploughshare.

The hunter was redundant. What was the hunter to do? The response was predictably aggressive. There is only one thing more effective than destroying your enemy, and that is to dominate her. If the male hunter was threatened by the female farmer, then the hunter would take control of this new and highly effective method of food production. Perhaps he began to see a parallel in his own mind between the fertility of the female and the fertility of the earth – this was some kind of new magic that he didn't really understand. Since his earliest hunting days, man had always yearned for control – control of all the conditions that would make a successful hunt. Now it seemed only right that he should control this new process.

It is a strange thing, but the cultivation of plants is a rather difficult thing to control on old-fashioned, paternalistic principles. It just isn't naturally suited to it. For one thing, there are no 'best bits', no parts of the plant that are so much better than the rest. In the good old hunting days, certain parts of the dead animal were more highly prized than others, and tradition dictated that the best should go to the number-one hunter. The tail of a kangaroo, the trunk of an elephant, the tongue of a bison, the eyeball fat of a guanaco (a kind of llama) – all these things were considered to be prize delicacies in certain societies and, accordingly, should only be given to the very bravest hunter. But where were these perks in a plant? Search as you might, you just couldn't find them. It would seem that vegetable foods are innately egalitarian.

On the other hand, meat strongly reinforces the established pecking order. The smallest social divisions can be exaggerated and exploited, and great ego satisfaction can be obtained by comparing one's own position to someone further down the pecking order. Here is one fairly typical social hierarchy that anthropologists have identified in contemporary hunting societies[54] – those closer to the top receive the most highly prized cuts of meat:

- Active male hunters
- Net owners
- Helpers of net owners
- Spear owners
- Dog owners
- Fathers of dog owners
- The beaters
- Those who carried the meat
- Old people
- Sisters or sisters-in-law of the killer
- The children
- The women
- The dogs

All these people would receive meat in the quality and quantity that befits their station. In addition, tribal chiefs, 'house' chiefs and chiefs of confederacies would expect their dues as well. It is on this masculine set of values that our present society has largely modelled itself, rewarding as it does any successful display of aggression, competition, or social rivalry. It is only very recently that some women have started to realise that they have been tricked into supporting this pernicious ideology, and some of them, such as the writer Norma Benney, are starting to question it.[55] She considers that hierarchical structures such as these

> 'involve concepts of "higher" and "lower" in which the former inevitably exploits the latter. Feminist thinking challenges these hierarchies, and women are starting to realise that in the process of struggling for our own rights, we should not participate in the victimisation of those even worse off than ourselves in the patriarchal pecking order. We need to develop fresh ways of seeing the world if we are to get out of the habit of ignoring the realities of how other, non-human animals are living.'

It can be seen, then, that flesh consumption reinforces and indeed creates social divisions, and further celebrates the values upon which those divisions are based.

Plant cultivation, on the other hand, is stubbornly egalitarian. It is clear that if the system were changed, those with the most to lose would be those who occupied positions close to the top of the pile, those who received the tastiest treats, and those with the greatest social standing. With the advent of horticulture, there was less economic dependence on the hunter and his meat than ever before. So the hunter became a horticulturalist, then an agriculturalist, and brought with him the values and ideology of the hunt.

Horticulture is essentially a co-operative act with the earth. Seeds are given to the ground in an area that is likely to be well irrigated, and in return the earth will manifest her fertility. It is based on the great cycle of nature, what anthropologist Mircea Eliade calls 'the eternal return'.

Agriculture, however, has at its core an act of coercion – it is stamped with the symbolism of the hunter, even today. Female animals are made pregnant whenever the farmer so desires, sometimes by the use of an apparatus known as the 'rape rack' – whose function is precisely as it sounds. Even the crops in the fields are controlled by use of chemicals that 'wage war' on other plant species that have no commercial value to the farmer. And of course, in modern societies, agriculture is an operation almost exclusively controlled by males. Women have been relegated, once again, to the role of menials, labourers, child-rearers and food processors.

Pretty soon, animals were 'agriculturalised' too. It is likely that men had already formed something of a symbiotic relationship with a few types of wild species. Dogs may sometimes have been used to track and chase the hunters' quarry. Animals, both dead and alive, would have figured prominently in religious ceremonies designed to give men control over the species he intended to hunt. Young animals, orphaned when their parents were butchered by the hunt, would have been kept as pets. And it is likely that some animals served as substitute sex-objects for the male – even today in America, the Kinsey report estimated that one in 12 of all males had sexual relations with animals. Some animals, too, would have been kept as tame decoys, to allow the hunters to approach their quarry closely without alarming it. This practice still exists in some modern slaughterhouses, where so-called 'Judas sheep' are specially trained by the slaughtermen to lead the victims from the pens to the killing floor.

Man had therefore been involved in a symbiotic relationship with semi-wild animals for a considerable length of time prior to the development of agriculture. The status of animals and females may have been, in the collective male mind, remarkably similar. Superfluous female babies, like young animals, would be culled, sometimes by being buried alive. Women, like female animals, produced milk which men could drink. Women were (and still are in some societies) used as wet-nurses for young animals, particularly piglets.

Some anthropologists suggest that the presumed cult of the fertility goddess shows that men venerated and worshipped the female principle, but I doubt this. The whole point of evolving a religion was to better yourself, to gain control over some aspect of your existence. Early man did not worship the wild boar, the reindeer or the bear in the same way as modern people worship their God. He carved their likenesses, painted their outlines, performed magical ceremonies and made sacrifices for one main purpose – to gain control. In the same way, he sought (and achieved) control over the female.

So both women and animals became domesticated, enslaved to agriculture. And the new agriculture regularly and reliably produced food in more ready abundance than ever before. Nutritionally, there was less need now for flesh food than at any time previously. But culturally and symbolically, the ritual of meat production and consumption was now more essential than ever, serving as an embodiment and confirmation of the values of a society created around male dominion achieved through slaughter.

Meat-eaters perpetuate these repugnant values each time they consume animal flesh.

VICTORIAN VALUES – TOOTH AND CLAW

We have briefly touched upon the development of Western society from primate to hunter/gatherer, then to horticulturalist and finally to agriculturalist. Now we need to consider why, in a modern, post-industrialised society such as ours, the myth of the red-blooded masculine hunter-killer is still such a potent image for us.

There are two essential reasons. First, the historical record itself colours our judgement. The garbage that is generated as a result of eating meat is pretty permanent – bones last longer in the ground than husks or seeds. Scientists (usually males) have traditionally focused their attention on the tools and artefacts of hunting, rather than the easily overlooked remnants of horticulture. And this can produce some very misleading results indeed. For example, with only their rubbish-tips to go on, archaeologists studying the Bush People of the Kalahari would conclude that they were an almost exclusively meat-eating tribe, the very opposite of the truth.

But a further, and far more significant, reason is this: The science of anthropology began as a kind of natural history, a study of the peoples encountered along the frontiers of European expansion. Such peoples, invariably called primitives or savages, were often studied not so much for what they themselves were, but rather as a means of justifying Victorian culture's position at the apex of the evolutionary pyramid. The ideas of Darwin and Huxley, frequently misquoted and misunderstood, were similarly advanced as 'proof' of our culture's superiority over the savages; of Man's rightful dominion as lord and master of Nature; and of man's proper subjugation of woman. This was not what Darwin intended, but it was what happened. Darwin biographer Jonathan Howard writes:

'In late Victorian society, a peculiarly beastly form of social climbing, "Social Darwinism", was established under Herbert Spencer's slogan "The survival of the fittest". The evolutionary law was interpreted to mean victory to the strongest as the necessary condition for progress. As a prescription for social behaviour it justified the worst excesses of capitalist exploitation of labour, "reasoned savagery" as T.H. Huxley labelled it.'[56]

For many Victorians, 'Evolution' started to replace 'Religion' as the justification, the rationalisation, for the prevailing status quo. It was no longer necessary to believe that God had put Man at the top of the natural hierarchy; Man could now claim to have got there by his very own efforts. If Man was really only an animal, then he was the most successful animal – more aggressive, more dominant, more ruthless. In a fast-expanding industrial society, these values were prized beyond all others; 'female' values were never less visible. And it is precisely from this period in our recent history that many of our misconceptions about our origin date. Says Adrienne Zihlman:

'Our notion of women's and men's role in prehistory derives in part from currently perceived differences in status of the sexes. Popular pictures drawn of the past are too often little more than backward projections of cultural sex stereotypes onto humans who lived more than a million years ago. Themes of male aggression, dominance, and hunting have long pervaded reconstructions of early human social life; and this has led to a belief that present-day inequality of the sexes has its roots in an ancient lifestyle and in inherent biological differences between the sexes. . . . Beginning with Darwin's discussion of human evolution, the theme of male dominance and female passivity and the use of tools as weapons has run through thinking about evolution. The emphasis on hunting, as with male dominance, is an outcome of male bias, however unconscious it may be, and this bias pervades even studies of primate behaviour. In Darwin's case, given the values of Western society, especially Victorian England, and the nature of available evidence, his emphasis on males is not surprising.'[57]

It really is extraordinary that so many of our conceptions about the history of our own species, and our place relative to other animals and life forms, should still be so deeply biased by the values of Victorian England. Let us investigate some of them.

Tennyson's clichéd phrase, 'Nature red in tooth and claw', perfectly captures the prevailing ethos of the period, combining as it does Tennyson's own deep-rooted fear of the natural chaos and disorder he believed to exist in the natural world, together with the inference that it is the proper duty of Man to subdue and dominate this wild force. Today, it is still a powerful image in the minds of many people who, in other respects, would not wish to share the values and prejudices of their Victorian ancestors. Most of us do indeed take it for granted that nature in the raw is cruel and merciless, showing no compassion to those who are too weak to defend themselves. And it is certainly a convenient way for us to see the world, for it proves our claim to supremacy over all other creatures; and it excuses our actions towards them, no matter how barbaric.

Tennyson himself was a typical product of his era. Born in 1809, he had a secluded childhood in Lincolnshire, where his father was a minister of the church, a manic-depressive, an alcoholic, and frequently violent. Unable to form a close relationship with his father, the young boy became very shy, very insecure, and would often seek solace in the lonely churchyard, where he would fling himself down weeping among the graves, longing to die. He grew up with a sense of embitterment, and believed that life should have given him a better position than merely being a parson's son. He was a hypochondriac, and, according to those who knew him, constantly worried about his bowels.

His attitude towards women was equally characteristic of the period. 'Woman is the lesser man,' he believed, 'God made the woman for the man.' As far back as 1860, the feminist Emily Davies was poking fun at what she described as his 'bisexual theory of the human ideal'. Like many others, he was deeply worried by what he saw as the dangers of too much democracy. In 1865, there was a public outcry concerning the governor of Jamaica, E.J. Eyre. A small rebellion on the island led to Eyre taking savage retribution, hanging nearly 600 people, and flogging many more. There was an attempt to have Eyre prosecuted for murder, but Tennyson thought that Eyre's action was entirely justified, being 'the only method of saving English lives'. He even contributed to a fund set up to defend Eyre. 'Niggers are tigers,' growled Tennyson.

He wrote recruiting poems for the army, and held conventional views on the subject of Ireland, which has always been a problem for the English. 'Couldn't they blow up that horrible island with dynamite,' he asked, 'and carry it off in pieces – a long way off?'

All this begins to tell you something about the values of the man who invented that unpleasant phrase. It should come as no surprise to learn that Tennyson found 'Nature' quite horrifying. 'The lavish profusion in the natural world,' he wrote, 'appalls me, from the growths of the tropical forest, to the capacity of man to multiply, the torrent of babies.' The Victorians decided that they liked Tennyson, his poetry, and his values, enough to make him the Poet Laureate of his day.

Even so, some people may still feel that, bigot and racist though he was, Tennyson was essentially correct about nature, or at least, about other animals. They do kill and eat each other; and that justifies our own flesh-eating habit and the values it embraces. Certainly, the meat trade wishes to perpetuate this idea for its own commercial ends. In recent publicity material given to schoolchildren, it approvingly quoted the television naturalist David Attenborough:

'People have become divorced from the realities of nature in their urban environment. I hope to bring back in my programmes a clear understanding that we are part of that wider system, and that animals die and are eaten.'[58]

There are several points to make in response to this decidedly weird 'death is good' argument. First, natural carnivores – such as hyenas – certainly need to kill to stay alive; but as you have already seen, there is overwhelming evidence that humans are not carnivores.

Secondly, I should point out that, equally, many animals do *not* need to rip into other animals' flesh in order to survive. To argue that hyenas hunt their prey, therefore humans should symbolically do the same, is selective logic bordering on insanity. Why should humans behave like hyenas? Why not like the vegetarian elephant? Or the dik-dik? Or the lesser-spotted Patagonian nut cracker? If you're going to pretend to be another species, you may as well make it as exotic as possible, while you're waiting for the men in white coats to arrive.

If we are going to imitate other animals in our conduct, why not imitate good-natured ones? Television wildlife documentaries are often obsessed with the eating habits of carnivores, much to the satisfaction, no doubt, of the meat industry. But why don't they show us the highly developed altruistic behaviour which some species clearly demonstrate? Consider these remarkable examples:

• When dolphinaria were first becoming big business in the United States, the normal method of 'collecting' wild dolphins from the sea and bringing them in to captivity was to throw a charge of dynamite into the sea, amongst a school of dolphins, and to pick up those that had been stunned. Of course, this would kill many others, but that didn't matter to the people who owned the dolphinaria, there were plenty more of them in the sea. The men who were responsible for collecting the stunned dolphins in nets would frequently report other dolphins coming to the rescue of those that were unconscious. The normal practice would be for two dolphins to arrange themselves on either side of the unconscious one, and stay there until it recovered. This would enable it to continue to breathe, for dolphins are mammals and need air, otherwise they drown. 'That the action was deliberate,' said one report, 'is shown by the way the supporting dolphins, when they had to leave it to come up to breathe, swam in a wide arc to come back and continue to support it.' Unquestionably, this is altruistic behaviour of a very high order – the 'good Samaritan' dolphins could not have been reacting 'instinctively' to a distress call, of course, because the unconscious dolphin wouldn't be able to make one. The very latest research on dolphins again challenges the human conceit that only people are capable of showing love, enjoying sex and thinking creatively about abstractions such as the future and the past. 'I'm trying to tell people that these are cultural animals,' says naturalist Ken Norris, who researches spinner dolphins off Hawaii. 'We're dealing with an animal for whom co-operation with its fellows is life itself . . . they can carry on a discourse about things that don't exist, like the past and future and concepts. They also teach each other, which to me is the concourse of culture.'[59]

- In Tanzania, Africa, an elephant control officer is summoned with his gun to a village where elephants have been reported to be raiding the crops. He sees the bull elephant and fires, aiming at the brain. The bull falls wounded, but is not dead – the bullet has missed the brain, hitting the shoulder. Three other elephants move in on the prostrate bull, arranging themselves on each flank, one behind. Astonished, the officer does not fire again. 'They boosted him onto his feet,' he says. 'I was amazed by it.' He returns to the spot the next day, but there is no trace of the wounded male.

- In similar circumstances, another elephant control officer decides to shoot a bull elephant, raises his rifle and fires. He misses the brain, but breaks the bull's shoulder. The bull bellows in great pain, two cow elephants hear his calls and come running. They start to half carry, half drag the bull into the jungle, away from danger. The officer runs closer in to the bull, trying to get a final shot in to kill it. One of the cows angrily turns on him, and he shoots her point blank. She crumples up and dies. 'The remaining cow,' reports the officer, 'went sadly on her way, every few yards stopping to listen and look back.'

- Yet another officer is tracking three cow elephants and one bull. He finds them, and fires quickly at all four. The three cows drop dead, almost instantly. The bull does not, but is badly wounded and confused. To his horror, the officer now realises that the cows have baby elephant calves with them, which the long grass prevented his seeing. The calves rush to the bull, not for protection, but to arrange themselves on either side of him and to try to help him along.

All this is a very long way from the 'Nature red in tooth and claw' myth, demonstrating as it does compassion, altruism and courage on the part of non-human animals, and perhaps raising a gleam of hope for the future – a future based on shared values, shared experience and shared environment.

Yet another fragment of forbidden knowledge.

A PLATE FULL OF HATE

One of the saddest, most pernicious deceptions perpetrated on men today is the notion that if you are not able to kill (and what more potent symbol of killing is there than a slab of animal flesh on a plate?) then you are not really a man. This is how one modern man perfectly expresses this evil concept:

> 'The instinct of the hunter is one of the most deeply ingrained of our inheritances from the past. Could it be said that he who had no trace of such a feeling was somewhat lacking in virility?'[60]

And that man should know what he's talking about, having participated in the deaths of thousands of Earth's most magnificent mammals, not however without some stirring of conscience:

'A whale struggling in its death flurry is a really moving spectacle, even to the hardened eyes of a whaler. But no sound is heard from the whales. If they had vocal cords proportionate to their bulk, with which to express their suffering, there would undoubtedly be very few men who would have strong enough nerves to bear the last moments of a whale dying by the harpoon. A blue whale, mortally wounded by several harpoons, has been known to tow a modern "catcher" behind it for two hours before dying. Gunners themselves, who might be thought to be quite indifferent to the sufferings of their quarry, are generally affected by an obscure and uneasy feeling that we have all experienced when the "flurry" occurred.'

So are men forever destined by biology to be murderers of their fellow creatures? Of course not. As a man, I am outraged and enraged by those who tell me that the man who gazes back from the mirror is, at heart, an unrepentant and eternal killer. As the great writer and Nobel prize winner Isaac Bashevis Singer observed:

'People often say that humans have always eaten animals, as if this is a justification for continuing the practice. According to this logic, we should not try to prevent people from murdering other people, since this has also been done since the earliest of times.'

I also know that contact between our species and others does not have to be brutal and deadly.

In the 1970s, humans started to explore the alien world of whales and we first started to realise that we shared common bonds with these gentle sea creatures. Divers who have swum with them frequently report feeling as if the whales were protecting and taking care of them. In one amazing incident off Hawaii in March 1976, a female whale asked for human help. The 'White Bird' was carrying divers when a giant humpback whale knocked her head on the boat three or four times, diver Roy Nickerson reported. After each knock, she would withdraw, and raise herself to look up at those on deck. He donned his wetsuit and went down to investigate. He found she had aborted, and her baby calf was stillborn, but not free of her body. Other divers then went down, lassoed the dead calf, and pulled it clear. It was a sad incident, but illustrative of the co-operation that could exist between our species, if we wanted it.

But before that happens, we have first to understand, and then overcome, the doctrine of 'Meatismo' that corrupts the minds of many men. Here it is, perfectly expressed, with words so evil that they chill me each time I read them. Nazi philosopher Oswald Spengler spawned them:

'The beast of prey is the highest form of active life. It represents a mode of living which requires the extreme degree of the necessity of fighting, conquering, annihilating, self-assertion. The human race ranks highly because it belongs to the class of beasts of prey. Therefore we find in man the tactics of life proper to a bold, cunning beast of prey. He lives engaged in aggression, killing, annihilation. He wants to be master in as much as he exists.'[61]

Now you know the enemy. These are appalling words. They speak of life without love, without compassion, without joy. Actually, they are not describing life at all, they are portraying a kind of living death (which is precisely how most modern food animals are reared). Words such as these will serve to excuse any atrocity, any barbarism.

But they are not true. Man is demonstrably not a 'beast of prey'. The greatest achievements of human history – horticulture, for example – came about through co-operation, not lethal domination.

Eventually, the truth will emerge. Connections will be made, as you are doing now, and the forbidden knowledge will no longer be secret.

I hope we are not too late.

NOTES

1 *Dorland's Medical Dictionary*, 26th edition, W.B. Saunders & Co., 1985.

2 Phillips, R.L. et al,'Coronary heart disease mortality among Seventh-day Adventists with differing dietary habits: a preliminary report', *American Journal of Clinical Nutrition*, Oct 1978, 31, pages S191–8.

3 Snowdon, D.A. et al, 'Meat consumption and fatal ischemic heart disease', *Preventive Medicine*, Sep 1984, 13 (5), pages 490–500.

4 Hirayama, T., 'Mortality in Japanese with life-styles similar to Seventh-day Adventists: strategy for risk reduction by life-style modification', *National Cancer Institute Monograph*, Dec 1985, 69, pages 143–53.

5 Frentzel-Beyme, R. et al, 'Mortality among German vegetarians: first results after five years of follow-up', *Nutrition and Cancer*, 1988, 11 (2), pages 117–26.

6 Burr, M.L. and Butland, B.K., 'Heart disease in British vegetarians', *American Journal of Clinical Nutrition*, Sep 1988, 48 (3 Suppl), pages 830–2.

7 *The New York Times*, 8 May 1990.

8 *Meat Trades Journal*, 28 Jun 1984.

9 *Meat Trades Journal*, 20 Mar 1986.

10 The *Guardian*, 30 Nov 1984.

11 *Meat Trades Journal*, 25 Jun 1987.

12 *Meat Trades Journal*, 25 Aug 1988.

13 The *Independent On Sunday*, 10 Mar 1991.

14 Shaper, A.G. et al, 'Milk, butter and heart disease', *British Medical Journal*, 30 Mar 1991, 302, pages 785–6.

15 Elwood, P.C., 'Milk, butter and heart disease', *British Medical Journal*, 13 Apr 1991, 302, page 913.

16 *Newsweek*, 27 Jun 1988.

17 *US News & World Report*, 7 Dec 1987, 103 (23), page 57 (1).

18 UPI newswire, 22 Apr 1988.

19 UPI newswire, 16 Jun 1988.

20 *Newsweek*, 27 Jun 1988.

21 Advertising Standards Authority, Nov 1991.

22 *Marketing*, 13 Jun 1991.

23 'Who needs WHO?', Research report no. 16, Social Affairs Unit, 1992.

24 The *Independent*, 17 Aug 1992.

25 Ramsay, L.E. et al, 'Dietary reduction of serum cholesterol: time to think again', *British Medical Journal*, 19 Oct 1991, 303, pages 953–7.

26 Marckmann, P. and Sandstrom, B., 'Dietary reduction of serum cholesterol', *British Medical Journal*, 23 Nov 1991, 303, page 1331.

27 Evans, D.W., 'Dietary reduction of serum cholesterol', *British Medical Journal*, 23 Nov 1991, 303, page 1332.

28 Thompson, G.R., 'Dietary reduction of serum cholesterol', *British Medical Journal*, 23 Nov 1991, 303, page 1332.

29 Skrabanek, P, 'Dietary reduction of serum cholesterol', *British Medical Journal*, 23 Nov 1991, 303, page 1332.

30 Ramsay, L.E. et al, 'Dietary reduction of serum cholesterol concentration', *British Medical Journal*, 14 Dec 1991, 303, page 1551.

31 *Better Nutrition*, Feb 1990, 52 (2), page 24 (2).

32 Nutrition Research Newsletter, May 1989, 8 (5), page 56 (1).

33 Reis, G.J. et al, 'Randomised trial of fish oil for prevention of restenosis after coronary angioplasty', *The Lancet*, 22 Jul 1989, 2 (8656), pages 177–81.

34 Middaugh, J.P., 'Cardiovascular deaths among Alaskan natives, 1980–86', *The American Journal of Public Health*, Mar 1990, 80 (3), page 282 (4).

35 Cliff, W.J., 'Coronary heart disease: Animal fat on trial', *Pathology*, 1987, 19, pages 325–8.

36 Clarke, J.T.R. et al, 'Increased incidence of epistaxis in adolescents with familial hypercholesterolemia treated with fish oil', *Journal of Pediatrics*, Jan 1990, 116 (1), page 139 (3).

37 *Men's Health*, July 1989, 5 (7), page 4 (3).

38 Coleman, V., *The Health Scandal*, Sidgwick & Jackson, 1988.

39 Medawar, C., *The Wrong Kind Of Medicine?*, Consumers' Association, 1984.

40 Collier, J., *The Health Conspiracy*, Century, 1989.

41 Campbell, T.C., 'A study on diet, nutrition and disease in the People's Republic of China, Part I', *Asocicion Medica de Puerto Rico Boletin*, Mar 1990, 82 (3), pages 132–4.

42 Turpeinen, O., 'Effect of cholesterol-lowering diet on mortality from coronary heart disease and other causes', *Circulation*, Jan 1979, 59 (1), pages 1–7.

43 Collens, W.S., 'Atherosclerotic disease: an anthropologic theory', *Medical Counterpoint*, Dec 1969, 1, pages 53–7.

44 The *Guardian*, 23 Nov 1988.

45 *Meat and Food History: Background papers on meat related topics*, Meat and Livestock Commission, 1992.

46 *Meat Messenger*, Issue No. 12, Summer 1992, Meat and Livestock Commission.

47 *Academic American Encyclopedia*, 1992.

48 Morris, D., *The Naked Ape*, Jonathan Cape, 1967.

49 Fornaciari, G and Mallegni, F., 'Palaenutritional studies on skeletal remains of ancient populations from the Mediterranean area: an attempt to interpretation', *Anthropologischer Anzeiger*, Dec 1987, 45 (4), pages 361–70.

50 *Meat In The News*, Meat and Livestock Commission, February 1992.

51 Eaton, S.B. and Konner, M.N., 'Paleolithic Nutrition: A consideration of its nature and current implications', *New England Journal of Medicine*, 312 (5), pages 283–9.

52 Zihlman, A. in Dahlberg, F. (ed.), *Woman the gatherer*, Yale University Press, 1981.

53 Truswell and Hanson, 'Medical Research amongst the Kung', in Lee, D. (ed.), *Kalahari hunter-gatherers*, Harvard University Press, 1976.

54 Coon, C.S., *The Hunting Peoples*, Jonathan Cape, 1972.

55 Benney, N., in *Reclaim the Earth*, The Women's Press, 1983.

56 Howard, J., *Darwin*, Oxford University Press, 1982.

57 Zihlman, A., op. cit.

58 *Meat In The News*, op. cit.

59 Associated Press, 8 Sep 1992.

60 Budker, P., *Whales and whaling*, Harrap, 1958.

61 Spengler, O., *Der Mensch Und Die Technike*, Munich, 1931.

CHAPTER TWO
APOCALYPSE COW!

When people eat the flesh of an animal, they consume a substance which has been literally and metaphorically deconstructed. The pink, shrink-wrapped cuts of meat on the supermarket shelves don't look as if they've been hacked from anything that was once alive; city children often find it difficult to believe that 'meat' and 'animals' are at all connected. Metaphorically, we're sold the idea that this substance consists of protein, vitamins and other good things – rather like taking a vitamin pill. When you awaken from this fantasy, you're in for a nasty shock. Meat comes from once-living animals, and animals – like us – are creatures that are subject to sickness, disease and sometimes pestilence. It's all part of our common bond.

Every year millions of people become ill (and sometimes die) because they catch a disease from another animal species. Food poisoning is one of the most common complaints, caused by bacteria from the salmonella group of organisms which live in the intestinal tract of animals. This is just one example of a 'zoonosis', the scientific term for any disease which originates in animals and can be passed on to humans (sometimes in a much more virulent form). Other zoonoses include anthrax, rabies, leptospirosis, listeriosis, toxoplasmosis, brucellosis, tuberculosis and trichinosis – all serious, often fatal, diseases which are transmitted from animals to humans across the species barrier. But outside the research laboratory, very few people realise just what a grave health threat zoonoses may pose to all of us.

Zoonoses behave in strange, often unpredictable, ways. The process of human-animal disease transmission is going on all the time; new diseases are continually being created, transformed, mutated and activated. Some diseases may lie dormant for hundreds of years, just waiting for suitable conditions to appear before they re-emerge and decimate a population that has little or no immunity to them. The stark reality is that, today, three quarters of the world's rural population suffer from one or more diseases that have been passed on to them from a reservoir of infection in the animal population. But don't make the mistake of believing that it's only people who live in Third World countries who are at risk. Apart from a few widely publicised diseases (such as rabies and salmonella), most people, including a surprising number of doctors and scientists, are hugely ignorant of the legacy of disease which humans and animals jointly share. For example, very few people know that:

- *The common cold came from our ancestors' contact with horses. As a species, humans first succumbed to rhinoviruses (the group of viruses which produce the common cold) from their association with horses. The cold is a recent disease in humans; we have only suffered from it since we became urbanised, about 10,000 years ago. At that time, the rhinoviruses present in horses mutated and crossed over into the human population, where they now number more than 80.*[1]

- *Measles originated in the wolf population. Measles emerged as a new disease in humans about 6000 years ago. Evidence shows that wolves first passed on the distemper virus to dogs, where it then mutated and became the rinderpest virus which infected cattle, which once again mutated and established itself in the human population as the disease we now know as measles.*[2]

- *Syphilis first arose from contact with monkeys. Syphilis originated in Stone Age populations between 25,000 and 18,000 BC from a reservoir of infection existing in monkey populations. Originally a disease disseminated by bodily contact, it evolved to become a sexually transmitted disease as the wearing of clothes increasingly restricted skin-to-skin intimacy solely to the act of copulation.*[3]

- *Cholera originated from sheep and cows. Cholera is one of the newest of all human pandemics, first making its appearance in Calcutta in 1817, from where it quickly spread all around the world. The cholera organism almost certainly mutated from similar infections present in sheep and cows, and its rapid (and opportunistic) transmission is frightening evidence that, whether we realise it or not, zoonoses are our constant companions.*[4]

Where will the next zoonosis come from? As we keep our food animals in ever more intensive conditions, and feed them ever more unnatural diets, we are increasingly tempting fate. One day, a new disease may spring up, as others have done in the past, which will prove incurable and lethal to its human host.

Perhaps it already has.

Have you ever seen an old black-and-white disaster movie called *The Day The Earth Caught Fire*? It's still occasionally shown on television, during the grave-yard shift. Here's the plot: Russian and American teams of scientists have, unbeknown to each other, triggered a series of simultaneous nuclear explosions, which together throw the Earth off its path, spinning towards the Sun. As the temperature rises, the truth can no longer be concealed from the public, and pandemonium breaks out. At the eleventh hour a desperate rescue plan is attempted; yet more nuclear explosions are detonated, this time intended to correct the Earth's wayward orbit. The bombs go off, and the world waits in trepidation to hear its fate. The last shot of the movie slowly pans over a sweltering newspaper office, where tomorrow's paper is ready to run with one of two alternative headlines. 'Earth Saved!' reads one version. 'Earth Doomed!' reads the other. Which one will be used? We never discover. Fade-out and credits.[5]

Today, that is just about the situation that applies to the 'Mad Cow Disease' catastrophe. For the best part of a decade, unknown numbers of the meat-eating population have been exposed to a potentially lethal agent, whose sinister characteristics seem to have come straight out of a science fiction movie. This is how expert microbiologist Professor Richard Lacey puts it:

> 'If an evil force could devise an agent capable of damaging the human race, he would make it indestructible, distribute it as widely as possible in animal feed so that it would pass to man, and programme it to cause disease slowly so that everyone would have been exposed to it before there was any awareness of its presence.'[6]

Of course this is *not* science fiction, because just such a 'lethal agent' started to emerge in the cow population in the early 1980s. A few years later, it had become a grim plague.

JUST WHEN YOU THINK IT'S SAFE . . .

Our species has a very short attention span. With the new decade, our concentration drifted elsewhere. Cows were still succumbing to the disease, of course, their brains riddled with holes, but this was no longer news. Some people even thought the problem had been solved.

They were wrong.

We're waiting.

Waiting to see how many people will eventually die from the human equivalent of Mad Cow Disease. The British government has set up a special team to monitor the figures. They should know any decade now.

Today, cases have been identified in America, France and Switzerland, and no doubt they exist, unrecognised so far, in other countries, too. Certainly, the sheep pestilence from which Mad Cow Disease originated exists plentifully in America and Canada, most of Europe, and in many other countries. But Britain is the first country where it has crossed the species barrier and reached epidemic proportions. Who knows where it will strike next?

Perhaps there was 'an evil force' controlling this disease, after all, because it could hardly have chosen a better country than Britain for a beach-head. Where America has its Freedom of Information Act, Britain has its Official Secrets Act, spawned from a society which has no written constitution to safeguard the rights of the individual, and which values the commercial confidences of food manufacturers as if they were state secrets.[7] As a result, the British people were never told the full story of Mad Cow Disease. In the heat of the crisis, some officials behaved as if their prime duty was to suppress public concern, and thereby minimise economic loss to the meat industry. Both government and the meat industry were, in effect, saying to the public, 'Keep on eating the beef until we've figured out what's wrong with it.'

At all costs, 'panic' had to be avoided. But panic, or alarm, is a natural human survival mechanism. It protects us from exposing ourselves to foolish risks. And with Mad Cow Disease, so many of the risks still remain extremely unclear. What connection, for example, might it have with Alzheimer's Disease? By the age of 85, one in four people will suffer from this dreadful condition.[8] In the laboratory, there seem to be some ominous similarities.[9]

So here is the extraordinary history of Mad Cow Disease – almost certainly more truthful, and more complete, than anything you have yet seen. There are clear lessons to be learnt here, which we fail to learn only at our extreme peril.

COUNTDOWN TO PLAGUE

The story of Mad Cow Disease has three parallel threads, which eventually converged to tie a knot of Gordian complexity in April 1985, when it was first observed on a British farm.[10]

Thread One begins on the 15th January 1755, precisely. It was on that day that the British Parliament was petitioned by sheep farmers to impose severe restrictions on those who dealt in sheep purchased from breeders. The reason was the emergence, in epidemic proportions, of a disease they termed 'rickets' (also rather quaintly known as 'goggles'), an invariably fatal affliction which was wiping out entire herds. From contemporary accounts of the symptoms, it is clear that this disease was what we now know as 'Scrapie'. It is a horrible disease – one of its distinctive characteristics is an uncontrollable urge for the animal to rub or scrape itself until the wool is entirely worn away and the bleeding skin is exposed – hence the name. Scrapie has been present in many countries for hundreds of years. In Britain, it was responsible for the virtual extermination of at least two entire breeds of sheep, the Wiltshire Horn and the Norfolk Horn.[11]

The British Parliament responded to this early animal health request with characteristic and precedent-setting decisiveness.

It did nothing.

Thereafter, Scrapie waxed and waned, as epidemics do. Between 1750 and 1820, there were severe outbreaks in East Anglia, Wessex, France (around Rambouillet) and Germany (around Frankenfelde and Stolpen). In the Bath area of Britain, a contemporary agricultural writer recorded that the disease 'within these few years has destroyed some in every flock around the County and made great havock in many'.[12]

Then, between the years 1820 and 1910, outbreaks of Scrapie declined, until, by the turn of the twentieth century, it had virtually ceased to exist in Europe. This is the way of epidemics – they run their course. Slowly, more resistant sheep are bred. But only a fool would have claimed that the disease had been conquered.

From 1910 onwards, Scrapie began to re-emerge, in East Anglia, southern Scotland, many areas of France, eastern Germany, Hungary and Bulgaria. Sheep once considered to be resistant started to succumb.

A very slow fuse had started to burn.

Pulling the Wool

Let's take a moment to consider the symptoms of Scrapie. It has an incubation period which ranges from 1½–5 years, during which time there are no recognisable symptoms. All this time, the 'infectious agent' is replicating in the animal, finally reaching its brain. The first outward sign that something is wrong is a general restlessness, and a fixed, fearful expression in the animal's eyes. Its pupils dilate; it hangs its head; its movements become aimless and its legs stiff and unbending. Then it starts to grind its teeth; its lips start to twitch, which soon spreads to the muscles around the shoulders and thighs. If suddenly startled, the animal may fall into an epileptic fit. Then the intense itching begins. Finally, the animal becomes completely uncoordinated, paralysis sets in, and it dies.[13] A post-mortem will reveal characteristic spongy, hole-riddled areas of brain where the infectious agent has destroyed cells. This sponge-like quality gives rise to the name 'spongiform', which scientists use to describe this kind of distinctive pathological feature.

How many sheep have Scrapie? It is a simple question to ask, but an impossible one to answer with any certainty. Because, since that first petition to Parliament in 1755 until the time of writing this book, the British government never considered Scrapie to be sufficiently important to make it officially notifiable. So here we have an incurable disease, caused by a mystery 'infectious agent', capable of great devastation of sheep flocks, strongly implicated in the development of Mad Cow Disease, and very similar to certain dementia-producing diseases in humans. And we don't know how many sheep are carrying it, because it has never been made officially notifiable. As the British Veterinary Association mildly put it to the House of Commons in a memorandum: 'We can only guess at the incidence of the disease. That has been an omission.'[14]

Actually, successive governments cannot share all the blame for this state of affairs. As one expert, Dr. K.L. Morgan of Bristol University, explains:

> The potentially disastrous economic effect of its identification in flocks producing pedigree and breeding stock has resulted in a reticence to acknowledge the presence of Scrapie. The concealment of clinical cases is such that once the first case is diagnosed and the signs recognised, other cases may be disposed of without the knowledge of the attending veterinarian.'[15]

Scrapie is bad news for everyone. And like all bad news, no-one wants to know about it. Until it's too late.

A Big, Bad Bug

And what actually causes Scrapie? You will notice that I have been careful to describe it as the infectious agent. That's because, so far, no-one can be certain. Whatever it is that causes Scrapie lies at the very limits of present-day science. If you look it up in a medical or veterinary dictionary, you may find Scrapie described as a 'slow virus' disease. That definition is inaccurate. Scrapie is not caused by anything remotely similar to other recognised viruses. Yet for decades, scientists were happy to classify it as a 'viral' disease, for the simple reason that it was inconceivable that it could be anything else. As recently as 1989, the Academic American Encyclopaedia wrote:

> 'Slow viruses are disease agents *not yet identified but assumed to exist*, because the diseases resemble virus diseases in their epidemiology' (my emphasis).

The Scrapie 'virus', however has never behaved like a virus should behave. No viral particles could be identified from infected tissues. No viral antibodies could be recovered in the laboratory.[16] The Scrapie 'virus' also violated one of the three golden rules of biology known as 'Koch's Postulates', which were established a century ago by the German physician Robert Koch. Koch's third postulate states that, in order to prove that a given infection is caused by a particular agent, the agent must not only be isolated from the patient but must also then be capable of being grown in a culture.

But that's not all. Heroically, Scrapie resisted the most prodigious efforts to kill it, such as being bombarded with radiation, being cooked at high temperatures and being doused with strong disinfectant chemicals. None of these lethal assaults could kill the thing that causes Scrapie. As one expert commented with justified exasperation:

> 'The fourth decade of my association with Scrapie ended in 1978, with the causal agent still obscure, and virologists as adamant as ever that theirs was the only worthwhile point of view. To explain findings that did not fit in with a virus hypothesis, they re-christened the causal agent an "unconventional virus". Use of this ingenious cover-up for uncertainty made "virus" meaningless – for is not a cottage an unconventional castle?'[17]

Thread Two: The Laughing Death

The place is Papua New Guinea, mostly unexplored by Westerners until the second half of the twentieth century. Before then, nothing but the occasional gold prospector, the odd missionary, motivated by greed or creed, had risked death by malaria to penetrate its secret interior. And there were rumours of cannibalism amongst the indigenous population . . .

It is in the distant interior of this island that the Fore tribe make their home. The Lutherans were the first to reach them, in 1949, and the Australians followed two years later with a patrol outpost at a place they called Okapa. The temperature here in the hills is a comfortable 20°C all the year round, although the humidity can sometimes be disagreeable for a Westerner. The hills and the valleys, once extensively wooded, are now a mixture of trees and grasslands, the result of 11,000 years of continuous human habitation. It may not be paradise, but on first inspection it seems pretty close.

How deceptive appearances can be. If you had journeyed here some 30 years ago, you would have noticed some men, but mostly women, standing and sitting in a distinctive way, their feet spread wide to give them a broad base, a stout wooden pole or spade tightly grasped between both hands, planted firmly in front of them, never relaxing their grip. They sit and stand like this because they must. Without a physical support, they will simply keel over, like an uprooted tree. You see, their spongy brains are no longer to be trusted to keep them upright.

These people are dying, and what is killing them is remarkably similar to the infectious agent that causes Scrapie in sheep.

The True Meat of Women

The Fore tribe cannot be described as living in a state of natural bliss. Sadly, this is no Garden of Eden. If we discount, for a moment, the Scrapie-like disease that killed up to 80 per cent of all women in some villages,[18] we find a society which has several strikingly miserable parallels with our own.

Over-population has rarely been a significant problem for the Fore, because of frequent tribal wars. In addition, a taboo against copulation whilst the tribe was engaged in warfare ensured that the birth rate was often low or declining. But the really evident similarity between us and the Fore is the universal malevolence among males towards their womenfolk.

Fore males live together in houses which are strictly segregated from the women and children. Male children are taught from an early age to be disdainful towards females. Adolescent boys periodically go into seclusion to cleanse themselves from the polluting effects that their mothers and sisters radiate. Because the act of copulation is perceived as being fraught with danger, only a married man risks indulgence, for he alone has the power to ward off the evil consequences of such intimate female contact. Worst of all is male contact with menstrual blood, which may sicken a man, cause vomiting, turn his blood black, corrupt his vital juices, cause his flesh to waste away, dull his wits, and so precipitate his death.[19]

It is this deep-rooted hatred of Fore women that, anthropologists believe, led to the outbreak of 'Kuru', as this form of Scrapie was called. Cannibalism – in particular, the eating of human brains – appears to have surfaced in the Fore tribe for two main reasons. First, the threat of population decline may have resulted in an association being made between cannibalism and increased fertility. Therefore, the more human flesh a woman consumed, the more likely she would be to give birth again, and so replenish the population stock which the belligerent males had thoughtlessly decimated.

And secondly, as the forests and their animal populations disappeared and hunting became less and less successful, there was a corresponding increase in the domestication of pigs, and the consumption of pig flesh – but only by males. The women, on the other hand, were strongly discouraged from eating highly prized pig flesh, and they therefore resorted to flesh of an altogether different type, which the men would never seek to expropriate. This is why, among Fore males, the human corpse is disparagingly referred to as 'the true meat of women'.[20]

The Unnatural History of Kuru

'I break the bones of your legs, I break the bones of your feet, I break the bones of your arms, I break the bones of your hands, and finally I make you die.'[21]

This is the curse which, the Fore believe, when recited by a sorcerer with appropriate gestures and artefacts, will inflict Kuru upon his enemy. As a clinical analysis of the course of the disease, it demonstrates an intimate knowledge of its progressively degenerative nature, starting first with increasing difficulty in maintaining balance ('I break the bones of your legs . . .') and resulting in complete incapacitation ('I break the bones of your hands . . .') before death intervenes.

The first Westerners to encounter Kuru were mystified, and made copious clinical records and case studies. Very often, the earliest dreaded sign of the impending tragedy would not even be noticed by the sufferer herself; a friend or family member would remark upon her shaky balance while crossing a narrow log bridge or climbing over a palisade fence separating agricultural plots. This stage may last for six to 12 months, during which time the woman's general physical and mental health gradually deteriorates. Eventually, it becomes obvious that walking is difficult and clumsy; the rhythmic and confident swing of her plaited bark skirt is replaced by an unsteady swaying. As the disease progresses, something called the 'Kuru tremor' takes hold, a rapidly repeating contraction of opposing muscles resembling shivering, sometimes of the whole body, sometimes just the muscles of the face – hence the label 'the laughing death'. Twitching and shaking make it all but impossible to speak, and the victim becomes effectively mute. At this stage even sitting upright becomes impossible, and friends may drive a stake into the ground in front of her to grasp, or suspend a rope from the ceiling of her hut so that she may pull herself upright with it. Soon, paralysis and incontinence set in, food cannot be swallowed, and death comes.

Because of its occurrence within families, and its predilection for women, it was first thought that Kuru was an inherited genetic disease. However, clever scientific detective work proved beyond doubt that Kuru was clearly infectious. How Kuru first arose among the Fore tribe is an unanswerable question. What is certain, however, is that the disease was transmitted by eating meat – in this case, human meat.

Even at the height of the devastating Kuru epidemic, the proportion of people infected with the disease in the population was *five times less* than the calculated incidence of Mad Cow Disease in the British adult cow population.[22] That tells you something about the breathtaking dimension of the plague amongst our animals.

Of course, it would be easy to dismiss Kuru as an isolated, freak disease, of no possible consequence to anyone in the modern world, if it were not for one fact.

We in the West also have our own form of Kuru.

Thread Three: Brain Death

There is a disease which is so feared by some members of the medical profession that pathologists have refused to perform autopsies on patients who are suspected of dying from it. Operating theatre technicians have refused to be present when these patients are operated on; recently, the director of a pathology laboratory was so worried about the possible risk of contagion that he ordered the destruction of histology slides taken from infected patients. Astonishingly, some hospitals have even refused to admit patients suffering from it.[23]

What possible disease could cause so much terror amongst doctors? It is, of course, a disease caused by that familiar 'infectious agent' – an 'agent' which cannot be destroyed by boiling, which is immune to ultraviolet and ionising radiation, which resists most common forms of disinfectant, and which can survive for long periods in apparently hostile conditions. Tissue samples taken from humans, fixed in formalin (a powerful disinfectant and preservative) and then embedded in paraffin have still been found to be capable of causing fatal infection.[24]

The name of this dreadful affliction is Creutzfeldt-Jakob disease. 'Creutzfeldt' (pronounced 'kroytz-felled') after the scientist who described the first case in a 22-year-old woman in 1920; 'Jakob' (pronounced 'yack-ob') after the physician who diagnosed the next three cases in the following year. It is often abbreviated to just its initials, CJD.

The disease shares many familiar symptoms with those already described. The time between infection and commencement of the first symptom can be very long indeed – up to 35 years has been recorded.[25] On the other hand, in cases where infected material has been placed in direct contact with a patient's exposed brain (for example, during brain surgery with improperly sterilised instruments), the disease can manifest itself within two years.

Forgetful periods are common at first. Poor concentration, difficulty in finding the right words, depression, inexplicable feelings of fear, and aggressiveness are all frequent initial symptoms. Patients complain that objects look 'strange', attacks of vertigo and dizziness occur, so does ringing in the ears. There is widespread tingling or numbness. It becomes more difficult to walk, an effort to climb stairs, fine movements such as writing or sewing become difficult or impossible. The patient may fall down while turning around. And all this is simply the first stage of the disease.

The second stage is characterised by a lack of control over bodily movements. Trembling, writhing and uncontrollable jerking spasms occur. At the same time, the body, or parts of it, may become very rigid. Visual disturbances and hallucinations occur.

The final stage of CJD consists of an appalling decline into a vegetative state. Patients become mute and unresponsive, incontinent and unable to feed. In medical language, they appear 'decerebrate' – as if the brain stem has been cut, so as to eliminate brain function.

It's not difficult to see why some health workers are so fearful of this terrifying and incurable disease.

Tying the Threads Together

So there we have it: an unholy trinity of three very closely related diseases, two of them present in humans, one in sheep; all fatal and all caused by an unknown agent which challenges the most basic concepts of modern biology. After all, just how can a disease be transmitted from one person to another, apparently without the help of DNA or RNA (chemical substances involved in the manufacture of proteins and essential to the genetic transmission of characteristics from parent to offspring)? The first scientist to answer this particularly thorny question will achieve lasting fame. 'It's the stuff that Nobel prizes are made of,' says Charles Weissmann, a leading researcher in infectious brain diseases.[26]

For decades, however, there was comparatively little mainstream scientific interest into this baffling group of diseases. That which cannot be smoothly explained is all too often ignored, even by scientists who should know better. And there was no overriding urgency to the problem: The incidence of Creutzfeldt-Jakob disease in the population was considered to be very low, Kuru had been slowly dying out ever since cannibalism had been outlawed amongst the Fore tribe, and Scrapie . . . well, Scrapie had always been with us. Why should increasingly hard-pressed scientific resources be allocated to this peripheral area of interest? A strict cost-benefit analysis, much beloved of those officials who control today's science, would not justify the investment.

In November 1986, all that changed for ever. In that month, brain tissues sent to Britain's Central Veterinary Laboratory, by puzzled veterinary surgeons, were scrutinised by experienced neuropathologists. Under the microscope, the distinctive hole-riddled areas of brain could be clearly distinguished and photographed. They had seen it before, of course, in specimens taken from diseased sheep. But this was something very new indeed – the samples under the microscope weren't from sheep, they were from cows.

Something rather strange seemed to have happened. Scrapie, present in the sheep population for hundreds of years, appeared to have crossed the species barrier and had mysteriously infected cows. Suddenly, decades of work by a few dedicated scientists into Kuru, Creutzfeldt-Jakob disease and Scrapie acquired a new and urgent relevance. Because, as Dr Tony Andrews, a senior lecturer at the Royal Veterinary College, put it:

'Now we know that Scrapie has jumped from sheep to cattle there is nothing to suggest it may not, in future, wind up in people.'[27]

Jumping the Species Barrier

It has been known for decades that Scrapie, Kuru and Creutzfeldt-Jakob disease are all extremely similar – so similar, in fact, that scientists term them all 'spongiform encephalopathies'. An encephalopathy is a word used to describe any degenerative illness of the brain: in this, case, one which causes parts of the brain to resemble a sponge riddled with holes. It has also been established beyond any doubt that many spongiform encephalopathies possess the ability to cross the species barrier. For example, if you take tissue from a human suffering from Creutzfeldt-Jakob disease and infect goats with it, they will die from Scrapie.[89] Now consider this evidence:

- The agent that causes Creutzfeldt-Jakob disease in humans has been experimentally inoculated into chimpanzees, capuchin monkeys, marmosets, spider monkeys, squirrel monkeys, woolly monkeys, managabey monkeys, pig-tailed monkeys, African green monkeys, baboons, bush babies, patas monkeys, talapoin monkeys, goats, cats, mice, hamsters, gerbils and guinea pigs. Subsequently, all these animals succumbed to spongiform encephalopathy.
- The agent that causes Scrapie in sheep has been experimentally inoculated into mink, spider monkeys, squirrel monkeys, cynomolgus monkeys, goats, mice, rats, hamsters and voles. Subsequently, all these animals succumbed to spongiform encephalopathy.
- The agent that causes Kuru in humans has been experimentally inoculated into chimpanzees, gibbons, capuchin monkeys, marmosets, spider monkeys, squirrel monkeys, woolly monkeys, rhesus monkeys, pig-tailed monkeys and bonnet monkeys. Subsequently, all these animals succumbed to spongiform encephalopathy.[29]

I want to make it clear that I don't approve of experiments such as these. They are often needlessly repetitive and horribly cruel to the poor animals concerned. The traditional justification for vivisection is that it advances human knowledge, but in the case of these experiments, that excuse is less valid than ever. Because in the public debate that arose after Mad Cow Disease – later known as bovine spongiform encephalopathy (BSE) – was diagnosed, officials went to great pains to stress the extreme implausibility of BSE being passed from cows to human beings, even though they must have known that scientific findings such as those above proved nothing of the sort.

Concealing the truth about spongiform encephalopathy was nothing new for British officials, however. They'd done it before.

A Breach of Trust

In the United Kingdom, during the period 1959 to 1985, several thousand children were injected with human growth hormone, a treatment for dwarfism. At that time, human growth hormone was extracted from pituitary glands removed from the brains of corpses. A considerable number was needed, about 100 glands to treat one child per year, and in order to achieve this quantity, mortuary technicians were offered a cash incentive. One technician explained:

'We were given 10 pence per pituitary, and you never sent away less than 20, or in some cases 40, depending on how quickly you could gather them.'[30]

While some technicians were careful not to take glands from patients who had died from infectious or dementing diseases, others were not so painstaking. The same technician said:

'All they were interested in was the cheque when it came in. The more you sent off, the more money came in. You could pick up 25–30 pituitaries and not argue about it, just take the whole lot and send them off, and that was £3.'

During the course of research for this book, I interviewed Dr. Helen Grant, one of Britain's most experienced neuropathologists. She explained to me in graphic terms what would happen:

'I was one of the pathologists who did post-mortems in those days, and I remember when I was about to drop the pituitary I had just removed into formalin, the mortuary technician would say, "Just a minute, doctor, do you want that pituitary?" And I would say, "No, I don't think I do, why?"

'"Oh well, can I have it?" he would ask.

'"You can have it," I'd say, "but what are you going to do with it?"

'"Well," he'd reply, "we've got to collect them for research."

'Well, of course I'd let him have it. It was for "research", you see. Never for one instant did I suppose that it was going to finish up being used in a therapeutic way, as a treatment on children. As far as I knew, these pituitaries were going for research, and research only.'

Because the supervising authorities failed to implement sufficiently stringent procedures for the collection of pituitary glands, an unknown number of children became infected with Creutzfeldt-Jakob disease.

To date, seven patients who were given growth hormone have died from CJD.[31] Not all recipients of human growth hormone treatment were (or will be) affected, and no recipients of the treatment after autumn 1985 are at risk (since which date the manufacturing process of human growth hormone has changed; an entirely pure form is now genetically engineered from the bacterium *Escherichia coli*).

But now comes the most outrageous part of this sad history. When evidence of the disaster began to emerge, the British Department of Health decided not to tell the patients that their lives might have been put at risk. They decided that to inform the patients who might have been injected with Creutzfeldt-Jakob disease would only cause panic. So there was no public enquiry into this unfolding tragedy. Just silence.

In the United States, the patients and families affected were notified as soon as the full significance of the situation was recognised. But not in Britain. Dr. Grant explains:

> 'There was pressure to inform the families at the time the problem was first recognised. But it was decided not to. Why did they sit on it for seven years? Why did they sit on it at all? I suppose because it looks bad. And the effect of it was, you see, that some of those children may later have become blood donors. How about that?'

In justification of its policy of secrecy, the Department of Health stated:

> 'The right to know was balanced against the anxiety that would be caused to these patients about a condition which is invariably fatal.'[32]

A blood-curdling statement indeed, and one which makes you wonder what other unpleasant facts have been concealed from us in an attempt to protect us from 'anxiety'.

After mounting pressure, it was eventually conceded in late 1991 that recipients of the injections should, after all, be contacted and informed of the risks. And what of compensation? There simply are not words to describe the profound personal suffering involved in this dreadful business. Surely, there is no possible excuse, and no honourable pretext, for not generously compensating both patients and their families? 'Any legal action would be defended on the grounds that as regards clinical factors at the time it was administered the treatment conformed with knowledge then available about good clinical practice,' said the Department in a statement to journalists working on a television programme.[33]

'We Thought We Were Safe But We Were Wrong'

So what can we learn from this tragedy? First, it has to be accepted that official concealment has already been practised with regard to spongiform encephalopathy. Concealment justified on the grounds of preventing 'anxiety', but concealment which also served to obscure lamentably lax procedures and any contingent liability. It also demonstrates, once more, the diabolically uncertain nature of the 'infectious agent' we're dealing with, and the very real limitations of expert knowledge in this field.

In the 1970s, the British Medical Research Council was concerned to ensure the safety of extracting human growth hormone from the pituitary glands of corpses. Professor Ivor Mills, emeritus professor of medicine at Cambridge University, was a member of the Endocrine Committee whose task it was to advise on safety.

'I was on the Endocrine Committee of the Medical Research Council in the 1970s when we had to consider whether it was safe to extract human growth hormone from human pituitaries, because you know the pituitary is attached to the brain and it seemed to us that there might be some possibility that Creutzfeldt-Jakob disease would be in a form that was in the pituitary and we might transfer that to the children who had to be injected with growth hormone. We took advice from many experts, including Scrapie experts, because we thought there might be some relationship between the two diseases and we were advised at that time the technique was safe. We have since been proved to have been wrong and I feel rather guilty myself that this happened and two children in this country, and rather more in the rest of the world, have been inflicted with dementing fatal illness. I am anxious since there is a relationship between Creutzfeldt-Jakob disease, I think, and Scrapie that the same sort of mistake should not be made a second time because, as Professor Southwood pointed out, the results could be very serious indeed.'[34]

As a result of the human growth hormone tragedy, Professor Mills became so concerned that the mistakes of the past should not be repeated that he personally testified in front of the British parliamentary committee inquiring into BSE. His evidence makes arresting reading, due in part to his distinguished scientific credentials, and in part to his obvious depth of concern. In a memorandum to the parliamentary committee, Professor Mills described how the experts at the time of the human growth hormone disaster had judged the procedure to be safe:

'We took advice from several experts, including Scrapie experts, and thought the technique was safe. Yet two children in this country got Creutzfeldt-Jakob disease. It is now known that this disease is very similar to Scrapie.

'We thought we were safe but we were wrong. We have made a mistake once and as a result two children got a fatal disease. We cannot afford to make the same sort of mistake again because the result would be much more disastrous. We now know much more about the agent which causes Scrapie and CJD and similar neurological diseases in man. The agent is unique and, in my opinion, highly dangerous to spread widely.'[35]

In his evidence, Professor Mills made four important recommendations. First, and most importantly, he wanted to prevent potentially infected material (lymphoid tissue, brains and spinal cords from cows, sheep and goats) from being fed back to food animals. Secondly, he wanted to see controls extended to include calves of six months and younger. Thirdly, he wanted to make sure that potentially infected material was not allowed to contaminate other meat in the slaughterhouse and afterwards, and lastly, he proposed that calves born to cows known to be infected with BSE should not be used for breeding. All eminently sensible precautions, in view of the devastating and enigmatic nature of the infectious agent. The human growth hormone tragedy had already proved that even the country's top experts could be terribly wrong about this disease. As Professor Mills had written previously in a letter to *The Times*:

> 'What I think we should learn from this is that it is not good enough to say the chances of harm, we think, are very small.'[36]

That is a sensible, cautious attitude, based on hard-learned experience. But it was not the position of the British government, who constantly sought to reassure the public that there was no risk. As John Gummer, the Minister for Agriculture, bluntly put it when testifying before the parliamentary committee,

> 'The plain fact is that there is no evidence that BSE poses any risk.'[37]

That, of course, was a politician's statement.

A JOURNAL OF THE PLAGUE YEARS

No evidence? Here is the chronology of BSE. Read it, and make up your own mind.

On the 13th December 1985, a portentous report appeared in the British press which, in hindsight, could be considered a curtain-raiser to the whole BSE saga. 'It's dog eat dog on Swedish farms,' exclaimed the headline.[38] 'Many of the Christmas hams now on sale here have come from pigs fed on the minced carcasses of sick animals,' wrote a journalist from Stockholm. The story had particularly revolting aspects. For several years, the rotten carcasses of diseased cows and pigs, as well as formerly loved pets, had been covertly processed to make food for cows, pigs, poultry and domestic pets. Dairy farmers began to suspect the wholesomeness of their animals' feedstuff when their cows started to fall ill and decrease milk production. Despite attempts to conceal the size of the scandal, news eventually reached Swedish consumers, who were predictably outraged that their pets were being recycled with quite so much ruthless efficiency. With great prescience, the report also queried the wisdom of turning 'the traditionally vegetarian cow' into a carnivore, not to say a cannibal.

For the average British reader, it was a relatively trivial story. What the Swedes got up to on their own farms was their own business. And in any case, it couldn't happen here, could it? There were probably laws to prevent that sort of thing. Yes, indeed. Ignorance can be so blissfully comforting. By the time this report appeared, the first cows were already dying from BSE on British farms. But as yet, very few people realised what was happening.

Nine months later, another strangely prophetic article appeared in the science pages of a British newspaper.[39] Headlined 'The disease that bugs biochemists and sheep', it described how the 'exotic, utterly mysterious agent' which causes Scrapie mystified scientists. How could a lethally infectious agent exist which appeared to possess neither DNA nor RNA? 'Future work in this field is sure to be immensely interesting . . .' the piece enthusiastically concluded, with massive understatement. Quite so.

The Renderers' Surrender

Meanwhile, dire things had been happening to an obscure part of Britain's meat industry. Rendering is a little-known but essential element of the strange economic equation that holds together all the diverse sections of the flesh trade. Renderers take all the bits and pieces of animals from slaughterhouses that no-one else wants, boil them up, and produce fat and protein. The fat is then turned into products such as margarine and soap, and the protein makes animal feed. Thus, they serve two essential functions: they act as a garbage disposal service for 1½ million tonnes of mangled corpses every year, and they act as a cheap source of feedstuff for the next generation of food animals, who in due course are fed to the following generation, and so on. As the chairman of their trade association put it, 'If there is no rendering industry, there is no meat industry.'[40]

No-one really seems to have considered whether it was such a good idea to force naturally vegetarian animals such as cows to become carnivores and cannibals. It made sound economic sense; so they did it. They still do it in America and many other countries, and they obviously see nothing wrong with the practice. But as former British Minister of Health Dr. Sir Gerard Vaughan explains, when you monkey around with nature, it is wise to expect some nasty surprises:

> 'One of the main areas of fault is the processing of animal food using parts of the same animals. It's not a natural instinct for one animal to eat its own species, in fact it's totally foreign to it. That seems to be nature's understanding of the bacteriological and biological dangers, because if one animal starts to eat its own stock, then the dangers of infection increasing are very great indeed.'[41]

By the mid-1980s many renderers were themselves close to the brink of extinction. The low price of vegetable oil on the world markets had made the renderers' own animal fat product uncompetitive, and the drive towards healthier eating had made edible animal fats increasingly unpopular amongst food manufacturers. Denied income from this vital market, the economic equation just wouldn't hang together any more. In 1986 alone, 10 per cent of the industry went bankrupt.[42] The industry appealed to the government for help, but in vain. Then they tried charging slaughterhouses a fee for the disposal of animal waste, to the considerable ire of the slaughterhouses, some of whom illegally took waste disposal into their own hands. 'I know of abattoirs in the North-West of England and the Midlands,' the then chairman of their trade association was quoted as saying, 'that minced up offal and started spreading it on fields rather than pay for it to be disposed of by renderers.'[43]

Survival in a harsh economic climate is largely a question of efficiency, and the rendering industry had already started to take steps to modernise its methods and reduce its costs. The old procedure for rendering animal flesh and bone was a two-stage operation: first, the foul brew would be cooked up in a huge vessel and the fat separated from the solid (known as 'greaves'), then the greaves would be further processed, often with a solvent such as benzene or petroleum spirit, to draw off more fat and finally leave a meat and bone meal product. The new procedure differed in several ways: it was a continuous process, not a batch system, and the use of solvents to extract fat was discontinued in favour of mechanical pressing and centrifuging. It would later be speculated that these changes in the rendering process were the root cause of the BSE epidemic.

Putting the Lid on

Sources indicate that another case of BSE was seen in January 1986, and this triggered the involvement of the government's own veterinary experts. We shall never know precisely when BSE was recognised as a Scrapie-like disease, although the chief veterinary officer for the Ministry of Agriculture said on television that it was diagnosed 'within a few weeks'.[44] The visible effects of spongiform encephalopathies are quite distinctive – no other disease leaves such a dramatic, hole-riddled brain as evidence of its infection. And the government's veterinary experts would certainly have seen this type of disease in sheep, under the name of 'Scrapie'. Now, the key question is: Why did it take so long – over two years – for the government to begin to take any effective action? After all, here we have an entirely new disease of cattle, very similar to the lethal and incurable Scrapie in sheep. If a disease of this severity and lethality suddenly starts crossing the species barrier, shouldn't that be taken as an extremely disquieting development? Surely the alarm bells should have started ringing as soon as those government experts saw the warning signs?

Well, perhaps they did. Neuropathologist and fellow of the Royal College of Physicians, Dr. Helen Grant, believes that there was a policy of official silence on the matter. And since she has spoken to many of the key scientists involved, she ought to know. She told me:

'A lid was put on it. As soon as they figured out what the disease was likely to be due to, they should have stopped feeding cattle with contaminated feedstuff. But the government didn't. They let it go on until the Southwood committee finally stopped it.'

Professor Richard Lacey agrees. Much reviled in official circles ('he seemed to lose touch completely with the real world,' sniffed the 1990 parliamentary enquiry into BSE),[45] Professor Lacey was one of the most outspoken critics of government policy during this period. And his views could not lightly be dismissed; as a clinical microbiologist with a world-wide reputation, and a Fellow of the Royal Society of Pathologists, he was well placed to comment. Professor Lacey believes that money was at the root of official inaction:

'The available evidence suggests that there has been a carefully orchestrated manipulation of public opinion by the Government in order to avoid taking action. The main reason for this is the sheer scale of the action that would be needed. The cost of compensation for replacing, say, 6 million infected cattle could run into billions of pounds. Moreover, the adverse international publicity this would generate might effectively put the UK into quarantine with loss of food exports, tourism, and even a substantial part of our industrial base.'[46]

The Story Breaks

In November 1986, the official record shows that the government's Central Veterinary Laboratory formally identified bovine spongiform encephalopathy, but ministers within the government were not informed until June the following year.[47] And even then, it took a further 10 months – until April 1988 – for the government to decide to appoint a working party into the disease (known as the Southwood committee). Why all these delays?

Again, we can only speculate about the real reasons. The official justification is that 'transmission experiments' were needed, in which infected tissue from cows would be injected into mice. You might think that it had already been painfully established that the new disease had almost certainly been transmitted – from sheep to cows. More cynical observers might conclude that, in reality, a gamble was being taken that the outbreak was small and containable, and that it could be dealt with quietly before it blew up into a major 'food scare'. If this was the case, the bet failed miserably.

In October 1987, BSE finally went public – but in a very demure and modest way. A brief paper, barely covering two sides, appeared in the professional journal the *Veterinary Record* in the section entitled 'Short Communications', just above an advertisement for magazine binders.[48] It described the disease in clinical terms, showed some photographs of diseased brains, and proposed the official name for the disease: bovine spongiform encephalopathy. The paper concluded with a careful warning – despite BSE's striking similarity to other spongiform encephalopathies (such as Scrapie and Creutzfeldt-Jakob disease) its cause was unknown, and 'no connection with encephalopathies in other species has been established'.

The response of other media was appropriately low-key, most treating it as something of a scientific oddity. 'There have been suggestions,' wrote the agriculture correspondent of *The Times*, a couple of months later, 'that it could be linked to a sheep disease called Scrapie', but the short article concluded by quoting another expert as not yet seeing BSE as a serious threat to cattle health.[49]

By April 1988, it must have become excruciatingly obvious to those in authority that BSE was not going to go away peacefully. By now over 400 cases had been reported nationally, even though the disease was still not officially notifiable. Clearly, we were on the verge of an epidemic of unknown magnitude, and something had to be done. The shrewd political response to this sort of tricky situation is to appoint a committee. Committees give politicians breathing space; their advice is not binding; and they provide an effective shield with which to deflect criticism. And that is what the government did – on the 21st April, the Southwood committee was announced to the world. 'A working party headed by Sir Richard Southwood, professor of zoology at Oxford, has been set up by the Ministry of Agriculture and the Department of Health,' reported the *Sunday Telegraph*, 'after complaints from vets that the Government has been dragging its feet.'[50] Sir Richard had the advantage of also being chairman of the National Radiological Protection Board, and was therefore accustomed to a high-profile, controversial position. Tagged on to the official announcement was some typical public relations baloney. 'The Ministry of Agriculture said there was no evidence of the disease being transmitted between animals and no evidence of it being passed on to people through meat and milk.'

The Subtle Art of Deception

The propaganda battle had now begun in earnest. At stake was a market for beef and veal worth £2 billion and, as in all battles, truth became the first casualty. It became impossible, for example, to establish just how hard the market for beef had been knocked. At the worst of the crisis, the head of Britain's largest chain of retail butchers was quoted in their trade journal as saying there had been no reduction in beef sales.[51] However, the parliamentary committee said the market had dropped by 25 per cent, and newspaper reports indicated that it might have plummeted by as much as 45 per cent.[52][53]

Official statements began to be peppered with the sort of evasive language normally only used in times of war. Defensive phrases such as 'no evidence' and 'no proof' recurred time and time again in official proclamations. In particular, the defence of 'no evidence' would be used repeatedly to quell rising public concern. When the Ministry of Agriculture claimed that there was 'no evidence of the disease being transmitted between animals' they were, of course, being economical with the truth. The evidence already strongly suggested that there had indeed been 'transmission between animals', inasmuch as cows had almost certainly contracted BSE from Scrapie in sheep, and extensive experimental work had already established that Scrapie could be transmitted to many other mammals.

As far as transmission between cows was concerned, it was simply too early to make any kind of prediction. As the government's own vets rather embarrassingly pointed out within a few days of the Ministry's nonsensical proclamation quoted above: 'Until it is known whether or not the cow is an end host, it is not possible to say if the offspring of affected animals will themselves be infected.'[54] The two announcements were, of course, directed towards totally different audiences – the government's vets were speaking to other professionals, and the Ministry of Agriculture was addressing the public at large. Thus do our rulers deceive us.

By now, intelligent people were starting to ask some penetrating questions. A thoughtful paper appeared in the *Veterinary Record* under the title 'Bovine spongiform encephalopathy: time to take Scrapie seriously'.[55] It was written by K.L. Morgan, a lecturer at Bristol University, who pointed out that tissue taken from patients dying from Creutzfeldt-Jakob disease could infect goats with Scrapie, and also cause a similar disease in cats. And Scrapie could also be transmitted from sheep to monkeys. Further, it had been experimentally demonstrated that passage through animals could alter the 'host range' of the Scrapie agent – in other words, if Scrapie had jumped from sheep to cattle, it might now become more directly infectious to human beings. This, of course, would be the ultimate 'nightmare scenario'.

A couple of weeks later, the government's own veterinary service also published an article in the same journal, which included the following passage:

> 'BSE must be seen in perspective. The number of confirmed cases (455) is very small compared with the total cattle population of 13 million. The number of cases is expected to increase but if, as is anticipated, it behaves like similar diseases in other species only small numbers of incidents relative to the total number of cattle disease incidents are likely to occur.'[56]

Did you follow that? The logic is, to say the least, convoluted. What they appeared to be saying was this: There are currently only 455 cases of BSE, so don't panic. And even if that number increases, it will probably be a small fraction of the total number of sick cows around, so there's still no need to panic. Four years later, we are well on the way to 100,000 cases of BSE, but presumably there are still lots of sick cows who *don't* have BSE. Apparently, we are supposed to find this reassuring.

Brain Food

To their credit, the Southwood committee worked quickly, although there was criticism that none of its members had experience of spongiform diseases.[57] In June 1988 BSE was at last made a notifiable disease, and a six-month ban was imposed (effective from mid-July) on the feeding of 'animal protein' to cows and sheep. For six months, at least, these animals were to be allowed to live like vegetarians again – cannibalism was no longer compulsory – although poultry and pigs still continued to be flesh feeders. By August, it was announced that cattle known to be infected with BSE (i.e. already in the terminal stages of the disease) were to be slaughtered and their carcasses destroyed. While it was clear that the Southwood committee was spurring the government into some action, it was unfortunate that the committee's full report would not be made public until February 1989.

Nevertheless, the action taken was far from adequate. Following an article in the *British Medical Journal*, *The Times* pointed out in June that there was still no legislation to stop manufacturers adding cows' brains to meat products, and no requirement for appropriate labelling.[58] 'It seems odd,' said Dr. Tim Holt of St.

James' Hospital in London, 'that they have banned cattle from eating these cattle brains, but they have not banned humans from eating cattle brains.'[59] Odd? Or scandalous?

So here was the position: because the disease could not be detected by any test in the living cow, only those cows who exhibited symptoms – and who were therefore in the last stages of the disease – had to be destroyed. Other cows – and in particular, meat from the most suspect organs – could still enter the food chain. It was now more than 18 long months since the official identification of the disease by the government's Central Veterinary Laboratory.

Whistling in the Dark

In August 1988, the Ministry of Agriculture announced that it would pay farmers 50 per cent of the market value for cows which had to be slaughtered following infection with BSE.[60] This penny-pinching compromise pleased no-one. Farmers were outraged that they were being denied full compensation; consumers were worried that farmers would be tempted to sell infected cows into the food chain rather than destroy them for half their value. About this time, too, concerned voices began to be heard on the subject of Scrapie. Dr. Tony Andrews of the Royal Veterinary College, for one, was quoted in a newspaper interview as saying:

'There is evidence to suggest a link between Scrapie and Creutzfeldt-Jacob disease, which causes premature senility in people. There have been experiments where tissue from the brains of dead victims of this disease has been put into goats which then contracted Scrapie.'[61]

Dr. Andrews wanted the government to introduce an eradication policy for Scrapie, starting by establishing a register of Scrapie-free sheep. The Ministry of Agriculture was characteristically cool about the idea, saying that 'in the light of known medical evidence' it could not be justified. An anonymous ministry spokesman then outlined the official line:

'Scrapie has been known about for 400 years and there is no evidence that it has ever spread to people.'

The Department of Health also got in on the act, echoing the belief that there was 'no proof' of a link between Scrapie and human diseases. This hypothesis would later be given ministerial weight when John Gummer testified before the parliamentary committee enquiring into BSE in 1990.

The ministry's propaganda machine seriously lost credibility a few days later, however, when Dr. James Hope, head of a government-funded but independent research unit studying the disease, seemed to contradict their unctuous reassurances. While agreeing with the government's position that there was 'no evidence' that humans could catch the infection from eating beef (indeed, how could there be such a slow-developing disease?), he went on to say:

'Of course there is alarm because it's potentially a great threat to the livestock industry as well as to human health. Because it jumped from sheep to cow, it might be better fitted to jump from cow to human.'[62]

Yes, indeed it might. And that was a possibility which no amount of official whistling in the dark could exclude.

By now, several countries had decided to ban the import of British cattle. And in October, the results of the government's 'transmission experiments' to mice showed that BSE could indeed infect other species – extremely bad news for the public relations blowhards. Then, in December, legislation came into force prohibiting the sale of milk from 'suspected' cattle, and the ban on recycling animal protein back to cows in their feed was extended for 12 months. Also in December, scientists from the government's Central Veterinary Laboratory published damning evidence showing that the source of BSE was contaminated cattle feed.[63] 'The results of the study,' the scientists wrote, 'do, however, lead inevitably to the conclusion that cattle have been exposed to a transmissible agent via cattle feedstuffs.' In other words, cows had been eating Scrapie-infected sheep.

Apocalypse Cow

So began a strangely apocalyptic period in the history of British farming, dramatically illustrated by some extraordinary contemporary photographs. As I write, I am looking at some of the most surreal pictures I have ever seen in my life; pictures of the farming folk of Merrie England, busily burning hundreds, eventually thousands, of their own cows.

Here is a photograph which is both ludicrous and chilling. It shows a secret Ministry of Defence location, where cows are being burnt. Operatives dressed in nightmarish chemical warfare suits are clambering over earth mounds, digging ditches, manoeuvring heavy-duty Army cranes from which dead cows swing. They look like doomsday Lilliputians swarming over a herd of upturned bovine Gullivers. The very notion that scores of British cows should receive secret military funerals is beyond farce, beyond satire. While some cows go to make meat pies, other cows receive state funerals. For services unrendered, perhaps.

Another widely reproduced photograph of the time starkly conveys the surreal quality of it all. It is, simply, a picture of Armageddon. There they lie, like vanquished warriors, a herd of supine cows, legs splayed, carcasses bloated with gas, while the flames of hell lick around them and ghostly clouds enshroud them. If you were to photograph the end of the world, it would probably look something like this. It was a disturbing, archetypal image which millions of people saw in the national press; maybe it reminded them of the cryptic forces which modern agriculture had unleashed on the world, and how very close we all might be to bio-cataclysm.

But perhaps the most widely seen image of all was that of four-year-old Cordelia, daughter of Britain's Minister of Agriculture, John Selwyn Gummer. No history of BSE is complete without mention of Cordelia, the little girl who, for a few awkward minutes in 1990, was conscripted into service for the Ministry of Agriculture, and posed with daddy before the world's media, cow burger in hand – a spectacle which one seasoned journalist movingly described as a 'deeply distressing sight'.[64] Today, as politicians increasingly demand that the intrusive media leave their personal lives unexamined, and threaten oppressive legislation to enforce their 'right to privacy', it is appropriate to remember Cordelia.

What possessed the Minister of Agriculture to involve his little girl in such a public relations exercise is hard to fathom. Perhaps it was intended to reassure us all that, if the Minister was willing to expose his own family to British beef, then all must be well. But to many, it must have seemed a deeply suspect PR gimmick. This 'televisual pantomime' – as the science editor of the *Independent* newspaper called it – 'of the Minister of Agriculture attempting to force feed his daughter with a beefburger' was all too easy to see through. 'This is the man,' wrote the science editor, 'or to be charitable, the successor to the men, who acquiesced in turning cattle into carnivores and chickens into cannibals. And he is surprised that the public does not take his word on food safety.'[65]

It was also a particularly capricious hostage to fortune. If, in later life, Mr. Gummer's daughter should ever fall ill with any meat-related disease (and I sincerely hope she does not), then you can be sure that those press photographs will reappear to haunt her. In the apt words of Shakespeare: 'Upon such sacrifices, my Cordelia, the gods themselves throw incense.'[66]

An Offal Year

Things were looking decidedly bleak for the meat industry, and in 1989 they got even worse. The year started with a lambasting for the government from a very surprising source – Lord Montagu of Beaulieu, one of England's most prominent landowners. One of his tenant farmers reported having a cow with BSE in 1987, which, Lord Montagu learnt with astonishment, could legally be sent to market. 'I am amazed at the slow reaction of the ministry and the complacent attitude it had at the beginning,' stormed his lordship, who also wrote to John MacGregor, then Minister of Agriculture. With an inevitable turn of phrase, a ministry official once again answered the charge of complacency by repeating the official mantra:

'There is no evidence to suggest that BSE can be transmitted to humans through meat.'[67]

Of course there was no evidence. With an incubation time of upwards of 20 years, the ministry will still be waiting for 'evidence' in 2010. Said Professor Richard Lacey:

'They are guessing, and hoping. They have a public voice which is trying to reassure everyone, but an inner fear. I think they know there's a real problem much worse than they're letting on.'[68]

'The Implications Would Be Extremely Serious'

Public and professional disquiet was steadily mounting. In early February, the *Guardian* newspaper ran a front-page report headlined 'Meat risks report "held back"', which alleged that a report into the risks of BSE to human transmission was being officially expurgated; this was subsequently officially denied.[69] A few days later, in the same paper, a letter from expert neuropathologist Dr. Helen Grant of London's Charing Cross Hospital was published, which pointed out:

'There are no laboratory tests to identify such [BSE-infected] animals: the only way to establish the diagnosis is to examine the brain. Such animals, thought to be healthy, will be slaughtered and enter the food chain . . . there is no doubt that animals harbouring the virus but seeming healthy have finished up as beef.'[70]

The next day, *The Times'* medical correspondent, Dr. Thomas Stuttaford, echoed rising medical concern:

'Neither Mrs. Thatcher nor her scientific advisers can be sure that these organisms [BSE] . . . have not been already picked up by people as they enjoyed a piece of marrow in an Irish stew, or ate a meat pie which had contained brains or meat from an infected, but not yet stricken, animal.'[71]

The Southwood report was published at the end of February, and its main conclusion was:

'From present evidence, it is likely that cattle will prove to be a "dead-end host" for the disease agent and most unlikely that BSE will have any implications for human health. Nevertheless, if our assessments of these likelihoods are incorrect, the implications would be extremely serious.'[72]

Officials responded warmly to the first part of the conclusion. The Southwood committee had also described the risk to humans as 'remote', and this now became the official buzzword, largely replacing the 'no evidence' slogan used up until now. But quite soon, it would be demonstrated that cattle were not, in fact, the 'dead-end hosts' which the committee had proposed, and that the disease could be further transmitted to other species (cats, for example). Nevertheless, the Southwood report would now be used to give additional substance to the assertion that 'beef is safe'. As Professor Lacey pointed out:

'Even after the cat deaths, the only official action seems to be the parrot-like claim from ministers that our beef is completely safe.'[73]

The key concern now was this: it was known that the 'infectious agent' was concentrated in certain organs, notably the brain, spleen and thymus glands. Cattle could be infected with BSE, but not show obvious signs, and there was nothing to stop their organs ending up in the food chain as offal. The Southwood committee had wondered whether meat products containing brain and spleen should be labelled as such, but 'did not consider that the risks justified such a measure'.[74]

However, they did suggest that offal – brain, spinal cord, spleen and intestines – should not be used in the manufacture of baby food – a rather contradictory recommendation in view of their basic postulate in favour of the safety of beef.[75] Further, it was announced that the government's chief medical officer advised mothers not to feed infants under 18 months on this material.[76] This contradiction was spotted by one member of parliament, who promptly asked the Prime Minister (Margaret Thatcher):

> 'If, as appears likely, BSE is a threat to humanity, why not ban it [offal] for all human food – or, if it is not a danger, as it is not, according to the Minister of Agriculture, why ban it for babies?'

The Prime Minister dodged the question, saying there was no point setting up a committee and then not taking their advice.[77]

Estimate of the likelihood of a meat-eater consuming BSE infective agent[78]	
Period	Chance of eating infected beef
1986 – 1988	1 in 10,000
1988 – 1989	1 in 1000
1990 – 1993	1 in 200
1993 – 2000	1 in 1000

Questions in the House

Meanwhile, some startling revelations were coming to light. On the 13th March, a question was asked in the House of Commons to establish whether the Ministry of Agriculture had commissioned research to find out whether BSE would infect human cells. Donald Thompson, the parliamentary secretary to the Ministry of Agriculture, answered:

> 'No, but trials are under way using marmosets, which are primates.'[79]

This was a staggering admission. The official line had always been that there was 'no evidence' that BSE posed any risk to human health. Well, of course there was 'no evidence'. If you don't commission the research, you don't have the evidence!

This Alice-In-Wonderland logic had surfaced in parliament a few days earlier, when Mr. Thompson was asked how frequently the Ministry of Agriculture had tested samples of cattle feed, to check that the ban on cows and sheep in cattle food was actually working. He replied:

'Ministry officials are empowered to take and test samples of ruminant feedstuffs if they have reason to believe the ban on the use of ruminant-derived protein is being broken. To date, there has been no reason to believe the law has been broken and such action has not been necessary.'[80]

In other words, the Ministry had never tested cattle feed because there was no evidence of wrongdoing. And if you don't look, you don't find . . .

On the 16th March, the Minister of Agriculture was asked an all-too-explicit question in parliament by MP Ron Davies. Would he now ban the sale of those organs from all cows and sheep which are known to harbour the infectious agent? The parliamentary secretary to the Ministry of Agriculture made it clear that they had no intention of taking any such action. In justification, he presented two arguments. Firstly, carcasses of BSE-suspected cows were already being destroyed. Secondly, Scrapie had been present for 200 years 'without any evidence of a risk to humans'. Therefore, it would not be 'appropriate' to ban these organs from sale. But, he was asked, the Southwood committee recognised that there was a danger to human health from the consumption of infected organs. And as far as Scrapie was concerned, now that it had suddenly demonstrated that it can leap across the species barrier (implying a dangerous new mutation) surely this should mean that all organs which act as a reservoir of infection should now be banned from sale? Mr. Thompson disagreed, reiterating that 'the Southwood report concluded that it was most unlikely that BSE would have any implications for human health'.[81]

This was yet another hostage to fortune, when the government abruptly decided, just four months later, to reverse its policy and ban cows' offal from sale. In retrospect, it seems obvious that policy was being made 'on the hoof'. As one policy position after another became untenable, so it was unceremoniously dumped.

On the 13th April, Mr. Thompson was asked whether he would introduce restrictions on the movement of calves born to cattle infected with BSE. Mr. Thompson said he had no such plans. This hygiene measure was important, because as long as there was a possibility of 'maternal transmission' of BSE (i.e. from cow to calf) the transport of BSE-infected calves around the country might spread the disease. Again, this reveals an extraordinary inconsistency in the government's policy. One of their key policy justifications was the similarity of BSE to Scrapie. Since Scrapie hadn't infected humans – they argued – BSE wouldn't either. But Scrapie was clearly transmissible from mother sheep to lamb – there was no doubt at all about this. As Dr. James Hope explained:

'In a flock of sheep, the disease is transmitted principally from mother to offspring, that is, from ewe to lamb. That's not to say that it is a genetic disease. We believe infection either occurs in utero before birth, or immediately after birth via the placenta. The placenta is highly infectious, and poses a threat to the newly-born lamb and other members of the flock.'[82]

The following day, evidence emerged that diseased cattle were being sent to slaughterhouses, when Mr. Thompson stated in reply to a question from MP Ron Davies that there had been 40 cases of BSE-infected cows being detected in slaughterhouses.[83] No one could say, however, how many cows had slipped through undetected.

The Fix

In May, the Women's Farming Union added their voice to rising public demands for a complete ban on the use in any food products of all brain and spinal cord material from cows and sheep.[84] The government must take steps, they said, to ensure that BSE and Scrapie could not be spread through the food chain.

The government's position had now become virtually untenable. An opinion poll for *Marketing* magazine revealed that only 2 per cent of the population believed the government completely on matters of food.[85] The 'no evidence' pretext was now well past its sell-by date, and the Southwood committee's report had not provided the absolute, cast-iron assurances which the government and the meat industry now desperately needed to reassure the public. Clearly, something drastic had to happen.

The Southwood committee had spawned another committee, under the chairmanship of Dr. David Tyrell, a retired virologist. This time, its members included scientists with experience of spongiform diseases. Again, they worked with commendable speed, and presented a report to the government in June. Disgracefully, it was not made public for seven further months.[86] However, a few days after receiving the (still secret) Tyrell report, the government abruptly reversed their position on the sale of offal, and a total ban was announced on the sale for human consumption of all cows' brain, spinal cord, thymus, spleen and tonsils.

It was a victory, of sorts. One of the problems was that the ban would not come into effect for a further five months in England and Wales, and not in Scotland for seven months. Said Dr. Hugh Fraser, a neuropathologist at the Institute of Animal Health in Edinburgh:

'They could have introduced a ban six months ago. They ought to have a ban as soon as possible. It doesn't seem right to delay it.'[87]

While welcoming the ban, MP Ron Davies, who had asked so many penetrating questions in the House of Commons, demanded more drastic and immediate action – such as random testing on cows' brains in slaughterhouses to determine the true size of the epidemic. This eminently sensible measure would be steadfastly opposed by the government.

There was no denying that it was a fix. Just three months earlier, they had told the House of Commons that a ban on the sale of offal would not be 'appropriate'. Now, it looked very 'appropriate' indeed. But would it be sufficient to reassure an increasingly leery public?

The Cows Come Home

1990 was the year that the cows came home to roost, or whatever it is that cows of ill omen do. Professor Richard Lacey wrote:

> 'During the last weeks of 1989 and early in 1990, findings of spongiform encephalopathy in, first, zoo animals such as antelopes, and then domestic cats, were published, completely invalidating the Southwood committee's hope that BSE was a "dead-end host", that is, that it would not spread beyond cattle.'[88]

So the key question was no longer 'Can BSE spread to other species?' but rather, 'How many other species can it infect – and is *homo sapiens* one of them?'

A few days into the new year, a report from trading standards officers revealed that cattle infected with BSE were still being sent by farmers to market – hardly surprising, in view of the low level of compensation being offered by the Ministry of Agriculture. Flying in the face of common sense, a Ministry official commented that compensation was 'not an issue' in safeguarding the public from BSE-infected animals. 'We have no evidence,' the official all-too-predictably commented, 'to suggest that farmers are dishonestly sending animals to market knowing they are infected.'[89]

1990 was also the year of the spin-doctor. From now onwards, the disquieting results of animal 'transmission' experiments would start to emerge. Yet, with sufficient ingenuity, even the worst results could be made to seem encouraging. For example, in early February results were published showing that BSE was capable of being transmitted from one cow to another.[90] Gloomy though this might at first seem, a positive 'spin' could be made to appear by pointing out that the cattle concerned were injected with infected material, and this artificial technique would never occur naturally. When a further experiment showed that mice (a different species) could contract BSE simply by *eating* infected cow brains, it was pointed out that the amount given to the mice (9 grams) was proportionately far higher than the amount likely to be eaten by a human being. Well, it was supposed to *sound* like good news.

'So What?'

In February 1990, the investigative television programme *World In Action* examined BSE, and included a pugnacious interview with Britain's food minister, David Maclean.[91] It was a sprightly enough performance for a debating society, but it must have done little to reassure the public that their food was in safe hands. 'Your critics say that meat inspectors simply aren't as qualified as vets to spot BSE-suspect cattle at abattoirs,' commented the interviewer.

'Well, maybe they aren't,' declared Mr. Maclean. 'I wouldn't expect them to be as qualified as vets, vets after all do a five-year training course. I wouldn't expect them to spot them. So what?'

'Well,' said the interviewer, 'they're missing a good many BSE-suspect cattle, it is suggested.'

'Well, so what?' snapped Mr. Maclean.

The thrust of Mr. Maclean's argument was that, since the most suspect organs from all cows were now being removed at slaughterhouses, it didn't matter if some BSE-infected cattle were reaching the slaughterhouses undetected. 'We're cutting the offals out of every cow, not just the BSE-suspect ones, every single cow,' he said. 'And that's the final preventative measure.'

The programme also included evidence from an experienced environmental health officer which graphically revealed that this 'final preventative measure' was by no means the absolute guarantee of safety the government evidently hoped it would be.

'When you split down the carcass,' he said, 'there will be bits of the central nervous system tissue that get scattered all over the rest of the meat. And when the carcass is sawn down, what they do is to hose that off. But again, that in itself is a compromise, because how do we know we get rid of it all? And how do we know what we produce is satisfactory? The whole animal is full of nerves, it's impossible to remove it all. It is the job of my meat inspectors to make sure that none of the banned offal gets through. But there will be some central nervous system that is left behind that is not covered by the banned offal anyway.'

'So suspect tissue is going in to the human food chain?' asked the interviewer.

'Yes, certainly,' was the unequivocal reply.

Heavy Petting

In April, as the number of detected cases of 'mad cows' passed the 10,000 mark, the government announced the commissioning of a study to examine the connection between BSE and Creutzfeldt-Jakob disease in humans.[92] This action was taken at the behest of the Tyrell committee's report, which stated:

'Many extensive epidemiological studies around the world have contributed to the current consensus view that Scrapie is not causally related to CJD. It is urgent that the same reassurance can be given about the lack of effect of BSE on human health. The best way of doing this is to monitor all UK cases of CJD over the next two decades.'[93]

Professor Lacey was scathing. 'In two sentences, the government's intent is revealed in absolute clarity. Its action is intended somehow to reassure, rather than to take any curative action.'[94]

What happened next was totally unforeseen. If an evil alien intelligence had indeed been plotting the next move of the infectious agent, it could not have done better than what followed. The British, as is widely known, are besotted with their pets. Although we are content to allow our food animals to live mean and miserable lives – out of sight – we will not tolerate any insult or injury to our beloved companion animals. So when the first pet cat died from a uniquely distinctive BSE/Scrapie type disease, the nation was appalled and outraged.

With hindsight, it was entirely logical that, if the infectious agent was present in cattle feed, the same infectious material could also be present in pet food. However, the reality of the pets actually dying, and all the negative public relations implications, doesn't seem to have been considered – there hadn't even been a routine 'no evidence' statement from the government. But once diagnosed, officials acted quickly to put this right, saying there was no evidence 'at this stage' of a link with pet food or, indeed, with BSE.[95]

But remarks made by the president of the British Veterinary Association raised the possibility that many more cats might be infected, when he was quoted as saying: 'Vets are presented with cats showing nervous disorders like this one every day. Some can be treated, some can't and have to be destroyed. But in 90 per cent of cases when they do have to be put to sleep owners don't want us to carry out a post mortem.'[96] Wisely, the Pet Food Manufacturers' Association had already advised its members not to include cattle offal in their products, but in view of the long incubation time of spongiform diseases, there could be no guarantee that many more cats would not subsequently be discovered to have 'mad cat disease'.

There was now something perilously close to a state of panic. Within days, beef had been removed from the menus of more than 2000 schools across the nation. The parliamentary opposition called upon the beleaguered Minister of Agriculture to take immediate further action or to resign. In an amazing public admonishment, a former chief veterinary officer, Alex Brown, broke the customary silence imposed on civil servants to lambaste successive governments' policies concerning the recycling of sheep and cows in cattle feed:

> 'No one was more alive to the potential risk involved in tampering with the eco-system than I was. I continually drummed it into everyone around me that we should never, never forget that nature has a right to do funny things to man. You should also never dismiss the unknown, because it is unknown.'[97]

That, of course, is precisely what officials had been doing when they continually asserted that there was 'no evidence' of any risk.

Clearly, the government and the meat trade were losing the propaganda war, and they had to counterattack. The Meat and Livestock Commission decided to launch a £1 million advertising campaign. 'It is not a response to the latest scare over BSE,' said their marketing director, 'it reflects our concern about the general pressure to eat less meat.'[98]

Colin Cullimore, managing director of the Dewhurst chain of high street butchers, laid into Professor Lacey. 'Professor Lacey is being alarmist,' declared Mr. Cullimore to *The Times*. 'He is a scientist, but he is making statements without any evidence.'[99]

The Minister of Agriculture, John Gummer, broadcast to the nation to condemn 'scaremongers'. 'The public has absolute confidence that I am not going to be pushed off what is the right action merely to curry favour with one or two people,' he said.[100]

In parliament, David Maclean, the food minister, lashed out at 'so-called experts' who failed to submit their evidence, and another backbencher complained of 'a bogus professor'.[101] (While speaking in the House, members of parliament are protected by parliamentary privilege against the laws of libel.)

Improper Suggestions

On Wednesday the 23rd May 1990, the Agriculture Committee of the House of Commons opened its proceedings into BSE. For the Minister of Agriculture, it was to be a fateful day. As an astute politician, John Gummer must have realised the crucial importance of a favourable verdict – if the committee vindicated his handling of the crisis, it would provide him with some sorely needed political backing. But if, on the other hand, it censured him, then who knows what might happen. The uncomfortably recent 'salmonella and eggs' crisis was a stark reminder that ministerial heads could and did roll over matters such as this . . .

There was always the possibility that events could take a disastrous turn, but as he prepared to testify that afternoon, he must have felt a certain degree of quiet confidence. He was not, after all, alone. On his left sat Keith Meldrum, chief veterinary officer at the Ministry of Agriculture. Next to him sat Elizabeth Attridge, head of the animal health division of the ministry. And on the minister's right was Dr. Hilary Pickles from the Department of Health, joint secretary to the Tyrell committee. All in all, a high-powered team, combining political acumen with scientific erudition. It would be difficult for things to go too far wrong.

The minister kicked off with a long introductory statement, expressing his pleasure with the committee's decision to hold an enquiry, outlining the course of the disease since its detection, and summarising the government's response. It was, as one would expect, executed with proficiency, and the formal nature of the proceedings precluded any awkward interruptions or cross-examination until the minister had finished speaking.

He started well. Although Mr. Gummer could never be accused of Churchillian oratory, his mind was sharp, and the structure of his speech was logical – stressing the government's deep concern, its swift response to the crisis, and the firm grasp his ministry had over the problem. It was a good beginning. Perhaps he should have left it there. He certainly could have done, because he had already said enough to create a favourable impression.

But he didn't.

He was well into his stride when something altogether astounding happened.

The official line had always been that, since there was 'no evidence' that Scrapie could infect humans, it therefore followed that BSE couldn't infect humans. This was a central tenet of the government's policy position. But on the afternoon of Wednesday 23rd May 1990, before the House of Commons Agriculture Committee, John Selwyn Gummer, Minister for Agriculture, went far, far further than that. This is what he said:

> 'The plain fact is that there is no evidence that BSE poses any risk. Some may argue that BSE is a new disease so how can we be so sure. Well, there is good historical evidence because BSE is very similar to sheep Scrapie which has been in the sheep population for over 250 years without any suggestion that it poses a risk to humans. Neither have extensive studies shown a link between Scrapie and the human disease CJD.'[102]

To the assembled members of the parliamentary committee, it must have sounded very persuasive. As Mr. Gummer spoke, flanked by experts, he must have appeared both impressive and credible.

There was just one problem.

The evidence was against him.

Scrapping over Scrapie

Whatever possessed Mr. Gummer to make such a breathtaking assertion we may never know for certain. He could just as easily have used the formulaic weasel-words so beloved of politicians – 'no conclusive evidence', 'no proof', and so on – which would have adequately conveyed his message without putting his neck on the line. But he didn't.

He'd now gone on record, before a committee of the House of Commons, claiming that Scrapie had existed in the sheep population 'for over 250 years without any suggestion that it poses a risk to humans'. 'Suggestion' is defined by the Oxford English Dictionary as 'the putting into the mind of an idea . . . an idea or thought suggested, a proposal'.[103] In effect, he seemed to be implying that the very notion that Scrapie might pose a risk to humans was so inconceivable that no scientist would even propose the idea.

But this was rubbish. For at least 15 years, there had indeed been 'suggestions' from scientists that Scrapie might play a part in the development of CJD, Creutzfeldt-Jakob disease. It was unthinkable that the minister's experts were not aware of this. But that afternoon, the experts were on Mr. Gummer's team. They were there to support him, not to cross-examine him. It is, of course, conceivable that Mr. Gummer had been misinformed by his expert advisors. This is highly unlikely, however, as a close examination of his words reveals. For immediately after claiming that there hadn't been 'any suggestion' that Scrapie posed a risk to humans, he alluded to 'extensive studies' examining the link between Scrapie and CJD. The obvious question that arises from this is: If there hadn't been 'any suggestion' that Scrapie might pose a risk, why had 'extensive studies' been performed? There is a conspicuous error of logic here.

What would have happened that day, if the Agriculture Committee had taken steps to widen their enquiry, and examine this new area in detail? We can only speculate, of course. They might have come to the same conclusions, in any case.

There again, they might not.

From the government's point of view, the worst possible outcome of the committee's enquiry would have been a failure to exonerate their conduct during the BSE disaster, coupled with a widening of the enquiry into the related area of Scrapie and sheep. Given the existing high level of anxiety among the population, it was conceivable that such a chain of events could have precipitated a governmental crisis of uncontrollable dimensions.

Perhaps in his desire to avoid opening this particular can of worms, Mr. Gummer simply went over the top, and abandoned the careful language of politicians. If so, it was an astounding mistake, and he was indeed fortunate not to have been challenged about it.

Until now.

The First 'Suggestion'

The first major 'suggestion' that sheep Scrapie might be linked to Creutzfeldt-Jakob disease in humans was presented to thousands of the world's scientists on the 29th November 1974.[104] That day, an issue of the widely read journal *Science* was published, carrying a letter signed by six distinguished scientists, including D. Carleton Gajdusek, the Kuru expert and later Nobel prize winner.

The letter was in response to a research paper published in the same journal earlier in the year.[105] The authors of the earlier paper were intrigued by the preponderance of CJD among certain population groups within Israel. Jews and their families who had emigrated from Libya were particularly susceptible – up to 78 times more likely to suffer from CJD than the general population. In response to this strange finding, the six scientists wrote:

'This finding may be related to the dietary habit of eating sheep's eyeballs, which are a gastronomic delicacy among Bedouin and Moroccan Arabs and also Libyans. A disease of sheep, Scrapie, has clinical and histopathological features similar to those of CJD ... If the CJD agent is found in the cornea, retina or optic nerve, the ingestion of eyeballs of sheep harbouring the Scrapie agent might possibly lead to the development of CJD in susceptible individuals and thus account for the high incidence of the disease in Libyan Jews.'

This 'suggestion' wasn't simply idle speculation. It had recently been tragically proven that CJD could be transmitted from one person to another when the recipient of a corneal transplant, unwittingly taken from a donor suffering from CJD, subsequently contracted CJD and died from it. Therefore, if the CJD agent was present in human eyeballs, it might also be present in sheep's eyeballs.

One of the authors of the original study replied to this suggestion with some interesting evidence:

> 'We knew that brain and spinal cord, mainly from sheep, was a delicacy among Libyan Jews. Inquiries even revealed that a favourite method of preparation is light grilling, which could conceivably leave an infectious agent viable.'[106]

However, he went on to say that having considered the idea, they then rejected it, on the grounds that the consumption of sheep's eyeballs was not limited to just the Libyan Jewish population, 'so we deleted reference to it in our final manuscript,' he explained, concluding that 'brain is a more likely source of the putative CJD agent than eyeballs.'

And so the ongoing debate began – not in public; but amongst scientists, and in the rarefied pages of professional journals. Evidence would be produced in favour of the theory, and evidence would be produced against it. But no-one could now claim, with any truthfulness, that there had not been 'any suggestion' that Scrapie posed a risk to humans.

On the Trail

Let's stay with the Scrapie/CJD story for a little – not to discomfit the unfortunate Mr. Gummer further, but so that we can understand some aspects of these enigmatic spongiform diseases.

After the publication of the initial report in *Science*, more research was conducted into the Libyan Jewish population. It produced more tantalising evidence, but not clear proof. One piece of research, for example, showed that the vast majority of CJD patients had indeed been known to consume sheep's brains – but so did other 'controls', without apparently succumbing to CJD.[107] What did this mean?

It simply meant that a clean-cut cause-and-effect relationship could not be easily established. While it was notable that the CJD sufferers were more often exposed to animals than the control group – and, significantly, they ate brains that were far more lightly cooked – this was not in itself strong enough evidence. Another study summarised it like this: 'The results suggest either a common source of exposure or a genetic influence on susceptibility to the virus.'[108]

The science of epidemiology – which is really detective work by numbers – is at its strongest when a clear cause-and-effect relationship can be proven. In order for the Scrapie-CJD theory to be proven beyond doubt, it would have to be shown that people suffering from CJD differed significantly from the general population in their exposure to the Scrapie agent in sheep meat. As long as there were people in the population who didn't contract CJD, but who were similarly exposed to sheep meat, it could not be conclusively demonstrated that Scrapie caused CJD. So, although the evidence so far didn't *prove* the connection between Scrapie and CJD, it didn't *disprove* it either.

Let's take a moment to consider the six links in the chain of disease transmission shown in the table below:

1	Characteristics of the agent
2	A Reservoir
3	Portal of exit
4	Mode of transmission
5	Portal of entry
6	Susceptibility of host

In a way, it looks rather like a game of Russian Roulette – you have to be rather unlucky to lose and become infected. Before anything can happen, the infectious agent itself must be one of a strain capable of causing disease (there are several different Scrapie strains). Then, there has to be something that acts as a reservoir of infection. This in itself is a powerfully suggestive argument in favour of a connection between Scrapie and CJD, because CJD would have died out by now if it was purely confined to human beings – there is almost certainly a natural reservoir of it outside our own species, which periodically re-infects us when conditions are right.

Next, there must be a way of getting the disease out of the natural reservoir – in the case of Scrapie, the most infectious parts of the sheep are the brain, placenta, spleen, liver and lymph nodes. Then, there has to be a method of carrying the infection to the new host. Well, in the case of sheep, that's easy enough – we eat them. So far, so good – or bad, as the case may be. But all this still isn't enough to infect the host. Two more essential steps are necessary.

The first is the route into the host itself. Now, we know that the effectiveness of different routes of entry to the host can vary extremely. At one end of the spectrum, we know that Scrapie can sometimes be transferred very easily from one sheep to another simply by allowing the sheep to graze on pasture previously grazed on by infected sheep – no other contact is needed.[109] And at the other end of the spectrum, it has been demonstrated that sometimes only direct inoculation into the brain with infected material will succeed in transferring infection. So between these two extremes, there is a huge variety of routes, some far more successful than others. This is a significant point, because there is evidence to suggest that eating Scrapie-infected meat may not, in itself, be sufficient to produce an infection – there may also have to be some kind of accidental inoculation, such as biting the skin of the mouth at the same time, or lesions of the lips, gums or intestines.

Finally, there has to be an existing susceptibility to the disease in the new victim. Some breeds of sheep are far more susceptible to Scrapie than others. And by implication, some humans may be more susceptible, too. As we will see later, this is the 'joker in the pack', because Scrapie/CJD is peculiar in having both a genetic and an infectious component. Tricky stuff, indeed.

You can see that there are many, many possible factors which can affect the transmission of disease – and its subsequent detection. Because of this, it is not always possible to tease out a clear cause-and-effect relationship from the numbers. For example, in one study of 38 American CJD patients, it was established that at least 10 of them had eaten brains within the previous five years – apparently, a very significant finding.[110] However, nearly as many people in the 'control group' had also eaten brains, and didn't get CJD. As one scientist commented, while reviewing the results:

> 'The chance of a person's getting the disease depends on a complex sequence of events . . . it is important to remember that exposure to a suspected mode of transmission may not be enough to result in disease, and some ingenuity in the method of inquiry will have to be introduced. For example, in this study, a high but equal proportion of both patients ate brains. What could be critical is that the patients may have experienced some coincidental events, such as concurrent trauma or acute respiratory infection which caused a break in the skin or mucosal lining thus allowing the CJD agent a portal of entry.'[111]

The fact is, even with the best team of scientists available, it could be next to impossible ever to provide the sort of conclusive epidemiological proof that would convince everyone that Scrapie can cause CJD. One major stumbling block is the sheer length of time between infection and onset of disease – how many people can accurately remember what they had to eat 20 years ago? Also, bear in mind that many CJD patients are not properly diagnosed until after death, so scientists have to question their next of kin, which makes it even more difficult to get accurate responses.

There are problems, too, simply recognising CJD. Until 1979, the International Classification of Diseases (a system used to codify causes of death) didn't even include a specific category for Creutzfeldt-Jakob disease.[112] In Britain, approximately 75,000 people die every year from 'dementing' diseases. Yet the official statistics show that only 30 to 40 people die from CJD. There is good evidence to believe that the true figure is far, far higher, probably in the region of 9000 cases.[113]

And here's yet another problem. In America, it has been found that areas with the largest number of reported outbreaks of Scrapie (Illinois, Texas, Indiana, Ohio and California) have no more cases of CJD than the national average. Is this reassuring evidence? By no means. It actually tells us very little at all. As one reviewer commented:

'Such a comparison is of limited value, since Scrapie-infected material may have been widely disseminated throughout the country in processed meat.'[114]

Today, most of the food we eat has been transported hundreds, sometimes thousands, of miles. Therefore, a local outbreak of Scrapie might result in a cluster of CJD cases far away in another continent!

A further report reveals that we can't even be certain that sheep with Scrapie will be accurately diagnosed. Examining the marketing of sheep in Pennsylvania, scientists concluded that:

'Sheep were usually marketed before central nervous system signs of Scrapie were expected to appear . . . opportunities to detect the disease were limited . . . sheep producers in the area knew little about Scrapie despite the fact that the disease has been reported in the area.'[115]

All these difficulties present formidable obstacles to epidemiological surveys. In France, a 12-year study of Scrapie in sheep revealed that the disease had been diagnosed 'in virtually every region where sheep are raised.'[116] It also found that lamb consumption among nation-wide categories of increasing population density correlated with an increasing frequency of CJD. A year later, a continuation of the same study still found that:

'There is a correlation between lamb consumption and CJD mortality rates in different nation-wide population categories.'[117]

However, five years later, the scientists had identified a total of 329 patients dying of Creutzfeldt-Jakob disease, but were unable to conclude there was a clear connection with lamb consumption, or with any other single factor.[118]

Such equivocal evidence is hard for scientists to come to grips with. Therefore, when something more substantial comes along, it is eagerly seized upon, and previous theories are forgotten.

And that is precisely what happened next.

Bad Genes?

'Clusters of CJD have long been known,' declared *The Economist* magazine two months after Mr. Gummer testified to the House of Commons Agriculture Committee. 'The most famous was among some Libyan Jews in whom CJD was almost 40 times more common than normal. Since they ate sheep, it was thought that Scrapie might be to blame. Further research showed that the sufferers were related.'[119]

'Although it may be worrying that such clusters of CJD exist,' the writer explained, 'the good news is that they seem to have been caused by bad genes, not bad mutton.' Well, maybe it wasn't such good news, after all. Initially, it had been proposed that there was a simple family connection between the Libyan Jews who suffered from CJD – in other words, it was a hereditary disease.

Subsequent work, however, failed to confirm this.[120] What *was* subsequently established by genetic detective work was that the Jewish CJD patients displayed a specific genetic mutation.[121] So were 'bad genes' the cause of CJD? The answer would come from the largest – and for us the most worrying – cluster of CJD cases yet discovered; right in the middle of Europe.

A Plague in Slow-Motion

Cases of Creutzfeldt-Jakob disease among Libyan Jews were 40 times more common than normal – and that was considered to be extraordinary. Today, in Slovakia, an epidemic of CJD is developing. I use the word 'epidemic' deliberately because, in certain areas, the incidence of CJD is more than *three thousand times* the ordinary level.[122]

It seems strange to think of an epidemic which has an incubation time measured in decades. When people drop like flies – from cholera, for example – then the drama momentarily hits the headlines, and we are all horrified, until we forget about it. But with CJD, there is no instant three-minute tragedy, conveniently pre-packaged for the evening news bulletins. There is no news angle for a plague which is running in slow-motion.

Whatever is developing in Slovakia is a matter of intense interest, and deep concern, to many scientists. Some experts believe that we are now seeing the beginning of a world-wide epidemic of 'Kuru virus', encompassing the sudden appearance of BSE, an upsurge in Scrapie in sheep, and CJD in humans. 'We have a major problem in human disease,' grimly warns one authority.[123]

When a conventional epidemic strikes, time is the enemy. You need time to identify the causative agent, time to study it, and time to develop countermeasures. When the period between infection and death may be just a few days, you never have enough time. But that's not the case with CJD. Which means that the Slovakian epidemic is the best-studied, most-investigated outbreak of CJD ever. In the past couple of years, we have learnt more about the cause of CJD than we've ever known before.

Compelling New Evidence

This chapter began with a film plot in which Russian and American scientists battled to save the world from annihilation. A real-life parallel has been going on in Slovakia, as both Americans and scientists who were formerly under communist jurisdiction now co-operate to comprehend the nature of the epidemic now in progress. Here is a summary of the findings so far:

• The epidemic has two centres. One is located in the rural Lucenec area of south central Slovakia, with some cases being reported from across the Hungarian border. The other is based further towards the north, in the Orava area, to the west of the High Tatra mountains on the Polish border.

The two areas differ significantly in some key respects. In the south, the disease progresses steadily, continuing to claim about the same number of people every year. In the north, however, it suddenly erupted in the late 1980s; two small villages, with a combined population of less than 2000, have had more than 20 cases of CJD in the last three years alone.[124]

- Once again, initial research first suggested that the disease had a genetic origin.[125] Nine CJD victims from the north, and six from the southern cluster had their DNA sequenced, and it was found that they all had a similar mutation. This discovery led some scientists to claim that CJD was 'caused' by a genetic mutation – back to the 'bad genes' theory described earlier. However, subsequent evidence has shown that as a comprehensive explanation, it simply isn't tenable, for the following reasons:

- Genetic screening has established that the mutation in question was present in people living in the northern Orava region at least as far back as 1902, and probably much earlier. Yet it was only recently, in 1987, that CJD suddenly exploded in frequency there. Obviously, if 'bad genes' was the root cause of CJD, there would have been cases of CJD as long as people had been carrying the genetic mutation. This clearly points to another triggering factor in the environment, such as the emergence of Scrapie.

- When scientists studied families in which CJD had claimed more than one victim, they found that CJD occurred more or less at the same time – but not at the same age. If the disease was purely genetic, it would be more likely to occur after a certain number of years. This evidence also suggests that, suddenly, an environmental source of infection appeared, with tragic consequences.

- After extensive genetic screening, it was established that many people could carry the genetic mutation but remain perfectly healthy.[126] Further research work with CJD outbreaks in Chile has now established that, among one identified group of people with the mutation, only half the expected number actually developed CJD.[127] This is very convincing evidence that an environmental factor triggers the disease in those susceptible to it.

- A case history illustrates this with great clarity. Three children were all found to be carrying the genetic mutation. Two of the children grew up in their birthplaces, within the southern cluster of CJD. Both these children subsequently contracted CJD and died. The third child, however, didn't contract CJD, even though she carried the mutation. The difference was that she was taken away from the area while still an infant, and lived and grew up in Bratislava, well outside the danger area.[128] But why should there be 'danger areas', in any case? The answer to this lies in recent agricultural history. In an attempt to stimulate the Slovak sheep farming industry, sheep were imported from 1970 onwards from England and France, and the breeds chosen (Ile de France and Suffolk) are both highly susceptible to Scrapie.[129] Furthermore, careful research work has revealed that most of these sheep went into regions which are now suffering from CJD.

- The evidence becomes more incriminating still when you examine the jobs that the CJD patients had. Well over half of them worked in livestock farming or meat processing.[130]
- Further laboratory work has now confirmed that Scrapie definitely exists in these flocks of sheep, and worryingly, Scrapie infection has now been identified there in sheep not manifesting any clinical symptoms of the disease.[131]

To summarise, this evidence strongly supports the theory that the most recent epidemic of CJD is the lethal result of genetically susceptible people being exposed to the Scrapie agent in sheep.

News From Wonderland

Early in 1992, it seemed as if the 'all-clear' had sounded . . .

'Beef given a clean bill of health,' proclaimed the headline in the *Meat Trades Journal*.[132] 'The results of the latest batch of tests on BSE suggest the disease cannot be passed from cattle to humans.' The report continued: 'British beef has been given a clean bill of health by a government scientist claiming tests on monkeys may have proved BSE cannot be transmitted from cattle to man.' 'I am absolutely convinced BSE can't be transmitted easily from cows to humans,' the government scientist was quoted as saying. 'I don't believe the meat of any cow is a risk to man and am certain that the meat arriving at any butchers always has been and still is fit to eat.'

Reassuring words, indeed, they were based on an experiment which involved transmitting BSE to marmosets, small monkeys belonging to the same biological family as humans. Two marmosets were injected with tissue taken from BSE-infected cattle, and another two were injected with material taken from Scrapie-infected sheep. The two marmosets infected with Scrapie both died, but the other two lived on. 'I feel certain that the monkeys have passed the danger period,' the scientist was quoted as saying. 'I would have no worries if butchers told any customers still refusing to eat beef that there is little or even no chance of them developing the disease.'

Just two months later, the *Meat Trades Journal* carried the following stark, doom-laden headline: 'Primates are affected by BSE'.[133]

What had happened? Why, one of the two BSE-infected marmosets had died, and the other one was only expected to live for a few more weeks. Yes, they'd both got Mad Cow Disease. So did this change everything? Did the government scientist quoted above now consider that BSE was more of a threat to human health? Not at all. The article quoted the scientist as now saying: 'We now know that BSE is even less of a risk.' And the Ministry of Agriculture commented . . . (Do I really need to write this for you? I mean, by now, you know what's coming, don't you?) . . . that there was no cause for concern about human health.[134] So that was all right then.

The End?

Clearly, the most important question facing us now is this: What effect is BSE going to have on our own health? It is very, very difficult to say. So much is still unknown – such as, for example, the precise nature of the infective agent itself. Various experts have differing opinions. Professor Lacey, for example, puts forward a persuasive hypothesis suggesting that beef products, rather than sheep, are a possible source of CJD in humans. He writes:

> 'The recent description of the experimental transmission of the agent causing BSE from a cattle source to a pig suggests that the chances of a spread to man may be somewhat more than 50 per cent. Pig tissues are well known to resemble those of humans. Because it is not possible to identify which British herds are free of the BSE agent, the advice must stand that British beef poses a risk and should be avoided. The danger is greatest in young children and pregnant women on account of the anticipated long incubation period of a BSE-like illness in man.'[135]

It takes a considerable degree of integrity, and courage, to speak out in public, as Professor Lacey, Dr. Grant, and other concerned experts have done. It is far easier, and far less professionally risky, simply to remain silent. I put more store by these people, who are prepared to risk official censure sometimes bordering on public mockery, than I do by any number of official spokespeople who obediently do their masters' bidding.

The final words belong to Dr. Helen Grant. She is one of those rare people whose open honesty is obvious to all, and whose sharp mind and ready humour are a delight to witness. I trust her, and her judgement. 'Tell me what's wrong with this,' I asked her. 'The official line is that, since humans have lived with sheep Scrapie for at least 200 years, BSE cannot be a threat to human health.' Her eyes twinkled. 'Yes, and do you know what the fault is with that argument? If Scrapie can be so easily transferred from sheep to cattle, it must mean that the infectious bits and pieces are present in the sheep's carcasses. And what are the infected bits and pieces? The brain and spinal cord. So the brains are still inside the sheep's carcasses, are they? Interesting! That means they haven't gone into our food chain, doesn't it? They never thought that out.'

I thought about it, and it made sense. 'So because sheep's brain and spinal cord mainly went into cattle feed, they couldn't have got into human food?'

'Yes. Do you know, my phone bill for that year was horrendous, because I rang round the abattoirs to ask them what was the routine – what did they normally do in the abattoir with sheep's brains? "Oh nothing," they told me. "Sheep's brains? Labour-intensive, not worth digging them out." Now cattle brains, that's a different matter. They are always removed, because they're bulkier, and there's lots of "meat", so cattle brains were always removed. So the reason we never got any bother with Scrapie previously was simply because we didn't swallow the perishing brains! And that still hasn't been admitted by anyone.'

'Do you think there's been an official cover-up?' I asked.

'Oh yes. I've seen it in action myself. I went to an official briefing on BSE for scientific journalists, where five scientists were going to present their papers. Before the briefing started, I heard the chairman say to the scientists, "I want you to come along to room XYZ to discuss policy." I remember thinking "Policy – what policy? These scientists are giving papers on their work. Where's the policy?" Later, something else struck me as very strange. In one of the papers being presented, all of the references cited other work which concluded that "there is no relationship between Scrapie and CJD". Now that was very odd, because there are several papers which say there may be a relationship. But those were left out. So that was the "Policy" – being careful what they say. I was a bit shattered, because the scientist concerned was very distinguished, not someone I would have described as "malleable".'

'Let's just speculate about BSE's possible impact on the human population,' I said. 'The official line is that there's no risk.'

'That's foolish, of course. They should never say "never". Then they trot out the erroneous argument that we never had any bother with Scrapie. Now even if that were true – and I don't think it is – we never swallowed the sheep's brains. But we did swallow the cattle's brains. So it's not the same.'

'Nevertheless,' I said, 'they imply that there's no risk. Presumably, in view of the long incubation period involved, by the time any increased incidence is discovered, the politicians will be well out of the way.'

She twinkled again. 'They'll be in the House of Lords!'

'So from your own expertise in this area, what's your best possible estimate of what might happen?'

'Let's examine the factors. First of all, it's going to remain a rare disease, I think. It's been rare because very few people are genetically susceptible, as with sheep. Australia and New Zealand have seen to it that they don't have Scrapie, because they have never imported genetically susceptible sheep. In the second place, it will only be those people who swallowed a fair amount of infected material between approximately 1981 and November 1989. You've got to have a fair amount of bad luck, and must have eaten a lot of meat pies, and pâté, and stuff like that. Or you had the misfortune to have had a slice of steak that was covered with a coating of brain material because of the circular saw that they used to use in the abattoirs. So there won't be a huge number of cases. I doubt if it will increase by more than a factor of three, that sort of order.'

'A gentle increase like that wouldn't be too difficult to massage away in the figures, would it?' I asked.

'It wouldn't be difficult at all, In fact, they will probably cover themselves by saying "We now have better criteria for establishing a diagnosis". So it will get lost in that statistical fog – the goalposts may be moved.'

So there the matter rests, at least for the time being. If we have indeed been saved from a plague of inconceivable dimensions, it is probably due to an accident of human genetics, a lucky break for our species. Lucky for most people, that is. A minority may not be so fortunate.

Meanwhile, we'll just have to wait and see. But I will make one prediction: as long as humans continue to consume the flesh of animals, kept unnaturally, fed unwholesomely, then the spectre of another new, BSE-like disease will always be with us.

And next time, we may not be quite so lucky.

NOTES

1 T-W-Fiennes, R.N., *Zoonoses and the Origins and Ecology of Human Disease*, Academic Press, 1978.

2 Ibid.

3 Ibid.

4 Ibid.

5 Interestingly, this scenario may not be as unlikely as it sounds. An article in *War and Peace Digest* (Vol. 2, No. 3, Aug 1992) states: 'On June 19, 1992, the United States conducted an underground nuclear bomb test in Nevada. Another test was conducted only four days afterwards. Three days later, a series of heavy earthquakes as high as 7.6 on the Richter scale rocked the Mojave desert 176 miles to the south. They were the biggest earthquakes to hit California this century. Only 22 hours later, an "unrelated" earthquake of 5.6 struck less than 20 miles from the Nevada test site itself. It was the biggest earthquake ever recorded near the test site and caused one-million dollars of damage to buildings in an area designated for permanent disposal of highly radioactive nuclear wastes only fifteen miles from the epicenter of the earthquake.'

6 Lacey, R., *Unfit for Human Consumption*, Souvenir Press, 1991.

7 See Cannon, G. *The Politics of Food*, Century, 1987.

8 Prusiner, S.B., 'Molecular biology of prions causing infectious and genetic encephalopathies of humans as well as scrapie of sheep and BSE of cattle', *Developments in Biological Standardization*, 1991, 75, pages 55–74.

9 Carp, R.I. et al, 'The nature of the unconventional slow infection agents remains a puzzle', *Alzheimer Disease and Associated Disorders*, Spring–Summer 1989, 3 (1–2), pages 79–99.

10 Wells, G.A.H. et al, 'A novel progressive spongiform encephalopathy in cattle', the *Veterinary Record*, 31 Oct 1987, page 419.

11 Parry, H.B., *Scrapie Disease in Sheep*, Academic Press, 1983.

12 Anon, 'On the disease called goggles in sheep; by a gentleman in Wiltshire', 1788, *Bath Papers*, 1, pages 42–4.

13 Jones, T.C. and Hunt, R.D., *Veterinary Pathology*, Lea & Feibiger, 1983.

14 'Bovine spongiform encephalopathy', the *Veterinary Record*, 23 Jun 1990, pages 626–7.

15 Morgan, K.L., 'Bovine spongiform encephalopathy: time to take scrapie seriously', the *Veterinary Record*, 30 Apr 1988, 122 (18), pages 445–6.

16 Lantos, P.L., 'From slow virus to prion: a review of transmissible spongiform encephalopathies', *Histopathology*, Jan 1992, 20 (1).

17 Quoted in *Developments in Biological Standardization*, 1991, 75, page 56.

18 Seale, J.R., 'Kuru, AIDS and aberrant social behaviour', *Journal of the Royal Society of Medicine*, Apr 1987, 80 (4), pages 200–2.

19 Hornabrook, R.W. (ed.), *Essays on Kuru*, Institute of Human Biology Papua New Guinea Monograph series, No. 3, 1976.

20 Lindenbaum, quoted in Hornabrook, R.W. (ed.), ibid.

21 Hornabrook, R.W. (ed.), op. cit.

22 Calculated from Ministry of Agriculture data, and data in *Developments in Biological Standardization*, 1991, Vol. 75.

23 Bastian, F.O. (ed.), 'Creutzfeldt-Jakob disease and other transmissible spongiform encephalopathies',

Mosby Year Book, 1991.
24 Ibid.
25 Ibid.
26 The *Independent on Sunday*, 14 Jun 1992.
27 The *Sunday Telegraph*, 21 Aug 1988.
28 Morgan, K.L., op. cit.
29 From data cited in Dealler, S. and Lacey, R., 'Beef and bovine spongiform encephalopathy: the risk persists', *Nutrition and Health*, 1991, 7 (3), pages 117–33.
30 *CheckOut '92*, Channel 4 Television, 8 Jul 1992.
31 The *Guardian*, 7 Jul 1992.
32 *CheckOut '92*, op. cit.
33 *CheckOut '92*, op. cit.
34 Agriculture Committee Fifth Report 89–90 : Bovine Spongiform Encephalopathy (BSE) Report and Proceedings of the Committee, HCP 449 89/90.
35 Memorandum submitted by Professor Ivor H. Mills, Agriculture Committee Fifth Report 89–90, op. cit.
36 Mills, Professor Ivor H., letter to the Editor, *The Times*, 15 May 1990.
37 Agriculture Committee Fifth Report 89–90, op. cit.
38 The *Guardian*, 13 Dec 1985.
39 The *Guardian*, 8 Aug 1986.
40 Agriculture Committee Fifth Report 89–90, op. cit.
41 BBC Television, 1989.
42 *Meat Trades Journal*, 28 Aug 1986.
43 Ibid.
44 *Country File*, BBC Television, Jul 1989.
45 Agriculture Committee Fifth Report 89–90, op. cit.
46 Lacey, R.W. et al, 'The BSE time-bomb? The causes, the risks and the solutions to the BSE epidemic', *The Ecologist*, 1991, 21 (3), pages 117–22.
47 Agriculture Committee Fifth Report 89–90, op. cit.
48 Wells, G.A.H. et al, op. cit.
49 *The Times*, 29 Dec 1987.
50 The *Sunday Telegraph*, 24 Apr 1988.
51 *Meat Trades Journal*, 5 Apr 1990.
52 Agriculture Committee Fifth Report 89–90, op. cit.
53 Kew, B., *The Pocketbook of Animal Facts and Figures*, Green Print, 1991.
54 'Disease update: bovine spongiform encephalopathy', The *Veterinary Record*, 14 May 1988, 122 (20), pages 477–8.
55 Morgan, K.L., op. cit.
56 'Disease update: bovine spongiform encephalopathy', op. cit.
57 Lacey, R., op. cit.
58 *The Times*, 9 Jun 1988.
59 *Meat Trades Journal*, 9 Jun 1988.
60 *Meat Trades Journal*, 4 Aug 1988.
61 The *Sunday Telegraph*, 21 Aug 1988.
62 *Associated Press*, 2 Sep 1988.
63 Wilesmith, J.W. et al, 'Bovine spongiform encephalopathy: epidemiological studies', *The Veterinary Record*, 17 Dec 1988, 123 (25), pages 638–44.
64 The *Independent*, 14 May 1990.
65 Ibid.
66 *King Lear*, V. iii. 20.
67 *The Times*, 30 Jan 1989.
68 Granada Television, 1990.
69 The *Guardian*, 11 Feb 1989.
70 The *Guardian*, 15 Feb 1989.
71 *The Times*, 16 Feb 1989.
72 Report on the working party on bovine spongiform encephalopathy, HMSO, 1989.
73 Agriculture Committee Fifth Report 89–90, op. cit.
74 Agriculture Committee Fifth Report 89–90, op. cit.
75 *Hansard*, 27 Feb 1989.
76 Ibid.
77 *The Times*, 1 Mar 1989.

78 From data cited in Dealler, S. and Lacey, R., op. cit.
79 *Hansard*, 13 Mar 1989.
80 *Hansard*, 9 Mar 1989.
81 *Hansard*, 16 Mar 1989.
82 BBC Television, 1989.
83 *Meat Trades Journal*, 27 Apr 1989.
84 The *Daily Telegraph*, 26 May 1989.
85 Cited in Agriculture Committee Fifth Report 89–90, op. cit.
86 Agriculture Committee Fifth Report 89–90, op. cit.
87 *The Sunday Times*, 9 Jul 1989.
88 Lacey, R., op. cit.
89 The *Independent*, 4 Jan 1990.
90 Dawson, M. et al, 'Preliminary evidence of the experimental transmissibility of bovine spongiform encephalopathy to cattle', *The Veterinary Record*, 3 Feb 1990, 126 (5), pages 112–13.
91 Granada Television, 1990.
92 The *Independent*, 14 Apr 1990.
93 Agriculture Committee Fifth Report 89–90, op. cit.
94 Agriculture Committee Fifth Report 89–90, op. cit.
95 *Daily Mail*, 11 May 1990.
96 The *Independent*, 11 May 1990.
97 The *Evening Standard*, 25 May 1990.
98 *The Times*, 15 May 1990.
99 Ibid.
100 The *Independent*, 16 May 1990.
101 The *Independent*, 18 May 1990.
102 Agriculture Committee Fifth Report 89–90, op. cit.
103 *The Shorter Oxford English Dictionary*, Third Edition.
104 Herzberg, L. et al, 'Letter: Creutzfeldt-Jakob disease: hypothesis for high incidence in Libyan Jews in Israel', *Science*, 29 Nov 1974, 186 (4166), page 848.
105 Kahana, E. et al, 'Creutzfeldt-Jakob disease: focus among Libyan Jews in Israel', *Science*, 11 Jan 1974, 183 (120), pages 90–1.
106 *Science*, 29 Nov 1974, 186 (4166), page 848.
107 Goldberg, H. et al, 'The Libyan Jewish focus of Creutzfeldt-Jakob disease: A search for the mode of natural transmission', in Prusiner and Hadlow (eds), *Slow Transmissible Diseases of the Nervous System*, Academic Press, 1979.
108 Neugut, R.H. et al, 'Creutzfeldt-Jakob disease: familial clustering among Libyan-born Israelis', *Neurology*, Feb 1979, 29 (2), pages 225–31.
109 Palsson, P. and Sigurdsson, B., in 'Proceedings of the 8th Nordiska veterinary congress', Helsinki, 1958.
110 Bobowick, A. et al, *American Journal of Epidemiology*, 1973, No. 98, page 381.
111 Malmgren, R. et al, 'The epidemiology of Creutzfeldt-Jakob disease', in Prusiner and Hadlow (eds), op. cit.
112 *International Classifications of Diseases*, Eighth Revision, National Center for Health Statistics, US Government Printing Office.
113 *The Lancet*, 1990, 336, pages 21–2.
114 Masters, C.L. et al, 'Creutzfeldt-Jakob disease: patterns of worldwide occurrence', in Prusiner and Hadlow (eds), op. cit.
115 Davanipour, Z. et al, 'Sheep consumption: a possible source of spongiform encephalopathy in humans', *Neuroepidemiology*, 1985, 4(4), pages 240–9.
116 Chatelain, J. et al, 'Epidemiologic comparisons between Creutzfeldt-Jakob disease and scrapie in France during the 12-year period 1968–1979', *Journal of Neurological Science*, Sep 1981, 51 (3), pages 329–37.
117 Cathala, F. et al, 'Maladie de Creutzfeldt-Jakob en France. Contribution à une recherche épidémiologique', *Rev Neurol* (Paris), 1982, 138 (1) pages 39–51.
118 Brown, P. et al, 'The epidemiology of Creutzfeldt-Jakob disease: conclusion of a 15-year investigation in France and review of the world literature', *Neurology*, Jun 1987, 37 (6), pages 895–904.
119 *The Economist*, 28 Jul 1990.
120 Korczyn, A.D., 'Creutzfeldt-Jakob disease among Libyan Jews', *European Journal of Epidemiology*, Sep 1991, 7 (5), pages 490–3.
121 Goldfarb, L.G. et al, 'Mutation in codon 200 of scrapie amyloid precursor gene linked to Creutzfeldt-Jakob disease in Sephardic Jews of Libyan and non-Libyan origin', *The Lancet*, 8 Sep 1990, 336 (8715), pages 637–8.

122 Gajdusek, D.C., 'The transmissible amyloidoses: genetical control of spontaneous generation of infectious amyloid proteins by nucleation of configurational change in host precursors: kuru-CJD-GSS-scrapie-BSE', *Journal of Epidemiology*, Sep 1991, 7 (5), pages 567–77.

123 *Scientific American*, Aug 1990.

124 Gajdusek, D.C., op. cit.

125 Goldfarb, L.G. et al, 'Mutation in codon 200 of scrapie amyloid protein gene in two clusters of Creutzfeldt-Jakob disease in Slovakia', *The Lancet*, 25 Aug 1990, 336 (8713), pages 514–15.

126 Mitrova, E. et al, '"Clusters" of CJD in Slovakia: the first laboratory evidence of scrapie', *European Journal of Epidemiology*, Sep 1991, 7 (5), pages 520–3.

127 Goldfarb, L.G. et al, 'Creutzfeldt-Jacob disease associated with the PRNP Codon 200 Lys mutation: an analysis of 45 families', *European Journal of Epidemiology*, Sep 1991, 7 (5), page 477–86.

128 Mitrova, E. et al, 'Focal accumulation of CJD in Slovakia: Retrospective investigation of a new rural family cluster', *European Journal of Epidemiology*, Sep 1991, 7 (5), pages 487–9.

129 Mitrova, E. et al, '"Clusters" of CJD in Slovakia: the first laboratory evidence of scrapie', op. cit.

130 Mitrova, E., 'Some new aspects of CJD epidemiology in Slovakia', *European Journal of Epidemiology*, Sep 1991, 7 (5), pages 439–49.

131 Mitrova, E. et al, '"Clusters" of CJD in Slovakia: the first laboratory evidence of scrapie', op. cit.

132 *Meat Trades Journal*, 9 Jan 1992.

133 *Meat Trades Journal*, 12 Mar 1992.

134 The *Independent*, 5 Mar 1992.

135 Lacey, R., op. cit.

CHAPTER THREE
PIGTALES

Schizophrenia literally means 'split mind', and it is the most common form of psychosis in our society. The word is used to describe an abnormal splitting of psychic functions; ideas and feelings are often rigidly isolated from each other. For example, a sufferer may express frightening or sad ideas in a happy manner.

Meat-eaters demonstrate a kind of 'split mind'. The whole point of eating meat is to obtain pleasure (there is no other valid justification). Listen to the conversation of gourmets – folk who take their pleasures with frightening seriousness – and you will hear people engaged in nothing else but the earnest pursuit of indulgence, as they debate, with great feeling, the comparative delights of such delicacies as veal, goose liver pâté, frogs' legs, and an endless agenda of even more recondite morsels.

Well, there's nothing wrong with taking pleasure in what you eat. Food is, after all, one of humanity's greatest delights and sources of comfort, is it not? But what puzzles me is this. When savouring a tender mouthful of veal, or deliberating over those oh-so-succulent cuisses de grenouille, how do you stop yourself thinking about the misery and pain which the animal experienced? I mean, doesn't the thought of a baby calf, crying in fear to be reunited with its mother, upset you – just a little? Doesn't it take the edge off your appetite? Evidently not. Meat-eaters do not allow such unruly thoughts to interfere with the weighty processes of ingestion and digestion.

How wonderful is the meat-eater's mind! It has an infinite capacity to relish the very finest degrees of pleasure, fused with a limitless ability to ignore the cruellest obscenities of suffering. A split mind.

But actions have consequences. When someone eats veal, the consequence is that the market for veal increases, and more baby calves will be born and live sad and wretched lives. Yet in the divided mind of the meat-eater, no connection between his action and the inevitable consequence has been made, because unpleasant thoughts like that are simply not permitted. And so he learns to live in a kind of dream-world, where actions don't have consequences, and self-gratification takes precedence over everything. Now I wonder, is this the sort of ethic that is going to make our world a better place to live in? Or rather, does it embody the obsolete morality which has been responsible for creating so many of the world's most desperate problems?

This chapter is no judgement-ridden lecture, intended to make you feel purgatorially guilty or terminally depressed. Here, you'll encounter people, and ideas, which you can explore, accept, or reject. Make of them what you will.

'Don't you eat meat?'

The sneering question was the one I had been gloomily expecting, but nonetheless dreading. A blushing, tongue-tied youth of barely eleven years, I had become wearily accustomed to the humiliating rituals and seasonal embarrassments of Christmas dinner with my extended family. As the stench of a turkey with third-degree burns fought with me for possession of the meagre contents of my stomach, I turned to seek out my interrogator. Whose turn was it to be this time – one of my precocious girl-cousins? Or a primly-disapproving uncle or aunt? This year, I was particularly unlucky, for it was my grandfather himself, sternly venting his displeasure with his only grandson. It was clear, he informed the gathering, that I was a spoilt child; the bare plate in front of me proved as much. He had told my mother not to feed me fruit when I was young, but she hadn't listened; now she had to live with the consequences. He wouldn't be surprised if I eventually turned out to be a clergyman, or even a ballet-dancer; maybe even both. Evidently, with me as the sole male heir, the blood-line was doomed to extinction.

I listened glumly. It was a seasonal sport, but about as amusing for me as the Boxing Day meet is for the fox. I had learnt to be silent, sometimes to apologise, never to try and explain. My parents understood, but they were the only ones. I knew there was something wrong with me, or rather with my strange dietary preferences, but it was as impossible for me to change as it was for me to stop being left-handed; although the teachers had tried to correct that, too. Then one of my know-it-all cousins spoke.

'He's a vegetarian!' she declared, accusingly.

I hung my head still further, and scowled. I had never heard the word before, but it sounded horribly similar to 'vegetable'. All eyes turned to me, looking in curiosity at this new-found species, half-human, half-plant. Under the table, the vegetable's knuckles blanched.

And so the awful ceremony continued, with pauses only for the occasional dissection of turkey or grandson. It was hard to understand my grandfather, I reflected. A local politician of some stature (being both Lord Mayor and Sheriff of Nottingham) he had involved himself in campaigns to end the misery of performing animals in circuses, and was in many ways a humane and enlightened person. That he so utterly failed to understand my motives for not wishing to devour animal flesh was forever a mystery to me. Alas, I never had the courage to discuss it with him.

Years later, my work meant I had to eat endless business lunches. This reached its nadir one day when my secretary informed me that I had not one, not two, but three lunches booked with clients or prospective clients, followed by a business dinner the same evening. I started eating at 11.45 in the morning, and didn't really stop until midnight. And although the company had changed since those awkward, angst-ridden Christmases, the eternal question remained:

'Don't you eat meat?'

By now, I knew how to brush it aside with a casual remark or two. Depending on the company, it was either something I did for 'health reasons', or because I simply didn't feel like eating meat. Still the hint of apology, still the defensive gesture to ward off disapproval or ridicule. I knew no other vegetarians, and didn't care to know any. In common with most inherently conservative businessmen, perhaps I sensed there might be something vaguely subversive about them, and I didn't want to be contaminated. Although I could live with, and where necessary, cunningly conceal my own personal idiosyncrasy, I couldn't cope with other people's 'weirdness'. When Tom Regan wrote that 'merely to content oneself with personal abstention is to become part of the problem rather than part of the solution', he could have been writing about me.

But things change. My father died from a heart attack, my mother died a long and painful death from cancer. I was divorced, my business was sold. Suddenly, I found myself with time to think about the questions which most of us spend our lives trying to avoid: What was I going to do next? I didn't know. What was really important to me? Again, I'd never truly considered the matter. Laboriously, I figured out a few fundamentals, such as the principle that how we live our lives on a day-to-day basis is of far more practical significance than any grand goal or personal ambition we may believe we are struggling to attain ('*il faut cultiver son jardin*', as Voltaire economically expressed it). It dawned on me that vegetarianism was an uncommonly sensible way of living – it harmed no-one, and had no part in the systematic brutality towards animals which I had witnessed so often in my country childhood. My enthusiasm had been kindled, and I set about learning all I could about it. Although it is a trade secret rarely acknowledged by teachers and professors, the best way to learn about a subject is actually to teach it. So I did – evening classes, weekend courses, wherever I could pack a few people into a room and harangue them on the subjects of nutrition, ethics, cookery, economics, and ecology. I became the equivalent of a human sponge, soaking up information during the day, spurting it out in the evenings.

What I learnt fascinated and absorbed me. As a child, I had come to the conclusion that my obstinacy was a solitary and unique phenomenon, a personal abnormality or deficiency in a world where carnivores were the norm. I had grown up in close contact with many kinds of animals, and knew how to communicate with them. I knew, for example, the language that chickens used to tell each other where food was found, and the 'words' that mother geese used to summon their goslings. Eating creatures that were, in effect, my friends seemed to me to display all the moral superiority of the cannibal. When, many years later, I discovered that mine was not an isolated and freakish persuasion, that countless other people shared my qualms about slaughter, and that vegetarianism had a long and mightily distinguished history, I was frankly astonished. I no longer considered myself to be simply 'squeamish'; I now walked in the company of Da Vinci, Empedocles, Gandhi, Lincoln, Paine, Plutarch, Pythagoras, Schweitzer, Shaw, Tolstoy, and Voltaire. That vegetarianism was not merely a 'food fad', but had a rock-solid ethical basis was indeed wondrous news to me. With this revelation, I began to wonder how any educational system could allow children to emerge without at least some basic exposure to the ideas of the great ethical thinkers of history. When the only moral education that children receive comes from the inarticulacies of Rambo and the antics of Mutant Turtles, then we will indeed be a society in terminal decay.

IDEAS TO CHANGE THE WORLD

Ethical theory can be divided into two main schools of thought, each with different ideas about 'rightness' and 'wrongness'. The first school argues that an action's 'rightness' or 'wrongness' depends on the consequences of the action. For example, if you borrow your friend's watch without telling him, thus causing him to miss an important meeting, then the act of borrowing is wrong, because the consequences are detrimental to your friend. However, if your friend doesn't miss his watch, then it could be argued that since no negative consequences ensued, what you did was not wrong.

An extension of this line of thought is that an action's 'rightness' or 'wrongness' can only be judged by its overall impact on the balance sheet of happiness. This is commonly known as Utilitarianism. There is no such thing as an absolute 'right' or 'wrong' action. If an action creates more happiness than suffering, it is good; but if it creates more pain than pleasure, then it is wrong. Vegetarianism takes on a very different complexion when seen in this light: surely the pleasures of eating meat are totally outweighed by the huge amount of suffering inflicted on the animal population? Peter Singer, author of *Animal Liberation*, strongly condemns eating meat for this reason, but true to the coherent and flexible basis of Utilitarian philosophy, can still foresee certain restricted circumstances in which even vivisection would be right. As he says, 'if one or even a dozen animals had to suffer experiments in order to save thousands, I would think it right and in accordance with equal consideration of interests that they should do so.'[1]

The other school of thought argues strongly and unequivocally in favour of animal and human rights. Rights are absolute and not subject to this kind of cost-benefit analysis. Tom Regan, author of *The Case for Animal Rights*, is a chief advocate of this theory. He supports vegetarianism on the principle that the basic moral right of all beings, including animals, is the right to respectful treatment. He also holds that animals, like humans, have inherent value in themselves – they have the potential to lead fulfilling lives, and should be allowed to do so. Where the inherent value of an animal is debased – as, for example, in the case of the degrading conditions in which battery chickens are kept – then their rights have also been violated. The obvious attraction of this philosophy of animal rights is that it provides clear and unambiguous guidelines about the way we should treat animals. Anyone who accepts the philosophy of human rights should, logically, also accept the validity of animal rights. If you do not, you could be accused of 'speciesism', a prejudice akin to racism and sexism.

Plainly, these are revolutionary ideas, capable of bettering the lives of humans as well as animals. As they emerge and develop, a new kind of ethic is evolving, a universally appropriate morality which, in the words of Einstein, will 'widen our circle of compassion to embrace all living creatures and the whole of nature'. By exposing ourselves to these revolutionary ideas, and by taking practical steps to incorporate them into our daily lives – the first step is surely to become vegetarian – we can actively extend Einstein's 'circle of compassion'. And ultimately, that is the only way in which the world's dire problems can be resolved once and for all.

As a small boy of eleven I knew nothing of these vast ideas. For me then, and for countless numbers of other little boys and girls who are upset at the thought of killing animals, not eating their carcasses seemed to be the very least I could do. Adults, of course, habitually try to deceive children – grinding up animal flesh until it's in tiny indistinguishable pieces, or stuffing it inside a banana-shaped casing and calling it a 'banger'. None of this duplicity succeeds, because children are far smarter than adults often imagine. And I had already seen cruelties which haunted me, and still do. The modern countryside is not a place where compassion is in evidence. Rather, it is a production unit, a factory, a place where everything must earn its keep, and when it does not, is summarily executed. I saw unwanted kittens beheaded with a blunt spade; perhaps kinder than drowning, but not much. I witnessed a cat, guilty of some nameless crime, nailed to a gatepost and left to a lingering crucifixion. Hedgehogs had their throats cut by farm labourers who uncovered them; I once asked one why he did it and he told me it was because his father had done it before him. I nursed a mallard with lead shot in its wing back to health, but mercifully, it could never fly again. Greedy men would ride to hounds two or three times a week; if no fox or hare could be persuaded to break cover, then they would chase pets instead. Images of death abounded: the mole-catcher with his row of forty, fifty, sixty small strychnined bodies suspended on barbed wire; the eerie tree of dead crows, flapping a lifeless warning to their comrades. And then there was the slaughterhouse itself, a place so ghastly that my mind long refused to believe what my eyes had witnessed . . .

THE MYTH OF CRUELTY-FREE MEAT

It had been a year or two since I last visited a farm, and I wanted to see for myself what was new. When *Why You Don't Need Meat* first appeared, the meat trade accused me of picking on the worst examples of their business, and they tried to reassure a sceptical public that the cases of cruelty I cited were not typical of their trade. In response to this, I set out to find the very best they could offer . . .

It is a dull, windswept morning in the heart of the English countryside, and I am looking for the most humanely produced meat that exists. The meat and farming industries accused me of being too emotional (such an ugly crime for a man), and argued that most farmers are kindly, caring folk who weep night and day when their cherished pets go off to the slaughterhouse. A meat producer wrote to me, inviting me to learn more about his ideas. The animals on his farm are spared the worst excesses of modern intensive agriculture, and for thousands, perhaps millions of people, this concept has certainly proved appealing. These ideas are now starting to spread throughout the meat industry, as more farmers and retailers realise the significant profit potential of 'cruelty-free meat'.

Cruelty-free?

Actually, the industry hasn't quite had the effrontery yet to call it that, but no doubt it will come. For the moment, 'welfare' is the chic word to use. Apparently, any day now we will see labels on meat products written to assuage the increasingly squeamish consciences of the meat-eating public – perhaps with the endorsement of a major animal charity. I can see it now: Meat Without Misgivings, Bacon Without Bother, Flesh Without Fret. So here I am, on a farm somewhere in England, chatting to one of the younger generation of farmers who try to produce their meat in a kinder and more ethical way. First, I raise the question of terminology.

'You describe your meat as "high welfare",' I say. *'Is that the same as "cruelty-free"?'*

'I suppose so,' the farmer answers. 'It's terminology, isn't it? I think, all in all, our welfare standards are the highest in the country, if not in Europe, if not the world.'

'So are you saying there's actually no cruelty involved at all in your method of meat production?'

'I think that would be tricky, wouldn't it? I see what you mean, now. Interesting. I don't think you possibly could say that really. I mean, it depends what you mean by "cruel", doesn't it? What I'm seeking to do is to rear the animals in much the same way as any one of our customers would if they did it themselves. In other words, if I gave a customer ten porkers, I don't think he would build a mini-factory farm with gridded floors, cut their tails off, medicate them, or choose a growth promoter. He'd find an old coal shed or something and put some straw in it, and feed them scraps. That's what we're doing. That's a fascinating question. You remind me of a guy from BBC TV who asked much the same thing – if animals have rights or not. I think it's a difficult one to answer, it needs a long discussion, more than half an hour's continuous chat, and probably as long to think about it.'

I can't believe that he hasn't thought about this, and come to a clear conclusion in my own mind. *'But you must have thought about it.'*

'Yes, but not exactly in those terms. I think our methods are undoubtedly high welfare, but I agree with you that's not the same as cruelty-free. I mean, supposing you have a pig that doesn't want to get on the lorry. We will pick it up and carry it on. If you wanted to be utterly cruelty-free, I think you'd have to let it go, and hope that it wanted to go next week. So at the moment, I'd say we are definitely high welfare, but to say that we are absolutely cruelty-free, 100 per cent, would be difficult.'

'What about the rights argument? Do animals have rights?'

'You'd have to talk to a priest about that.'

'What do you think? Do you think animals have rights?'

'I think they deserve respect and kindness, particularly if you're using them as a source of food. I think they do anyway, but particularly . . . I say particularly because in a way you're then using them, as opposed to living with them. It's not as if they're performing some other function, such as a guide dog.'

'It could be said that what you're really doing is just being kinder to meat-eaters, rather than being kinder to the animals. Because it avoids the unpleasant thought in their minds that the animal they're eating has suffered.'

'I wish that thought was stronger in their minds in the first place. I mean, I actually don't think that thought lurks much in people's minds.'

We stop talking and tour the farm. The wind is piercing, and the driving rain is turning the chalky soil the same leaden grey as the sky. I am grateful for the shelter of the first farm building we are herded into; but I am not prepared for the sight that meets my eyes. It is dark, but in the gloom I can see three vast sows, confined by metal frames, their teats exposed and constantly available to the baby piglets that run and squeak as we enter. The sheer bulk of these sows is breathtaking, even majestic. But these are not just female pigs, they are mothers too. In the narrow farrowing crates in which they lie imprisoned, they are all but denied access to their own babies. They cannot even turn round. I catch one of the sow's eyes, and I understand the particular distress she is experiencing. When we resume our conversation, I tackle him about this.

'I wanted to talk to you about some of the things we've seen today. Now the first thing that we saw was the farrowing crate, which has been criticised by various organisations. Can you tell me why it's been criticised?'

Suddenly, he has become very distant.

'I presume because it restricts the freedom of the sow. For the period of giving birth to the piglets.'

'And what's your feeling about that?'

'I think if you don't, and the sow then treads on the piglets, or savages them, or they suffer in any other way, then it can create more problems than it solves. It remains a totally unsolved concept in this company. We allow all forms of farrowing, and both have advantages and disadvantages, and after six years we have no clear policy. And any honest welfare person wouldn't have either.'

I don't believe this. Compassion in World Farming (an animal welfare group, and certainly an 'honest' one) has a clear policy on farrowing crates – it wants them banned. And the government's own Farm Animal Welfare Council has gone on record as stating that farm animals should have 'the freedom to express most normal, socially acceptable forms of behaviour' – a freedom which the farrowing crate obviously violates.

'I must say I found it quite disturbing to see those sows like that. The first thing that hit me, when I went into the building to see those sows lying down, completely immobile, was what a common bond there is between the human animal and the pig animal that your industry abuses. It seems to me that the meat trade is founded upon the exploitation of the female reproductive qualities of animals. Have you ever thought about that?'

'Well', he replies, 'I think one of the saddest things in this country is that people are so far removed from all methods of production, and to anyone not used to keeping animals I'm sure it all seems terrible. I mean, those pigs are actually having a much easier time than my own wife had during the birth of our children.'

'How do you know?'

'Well, because they don't seem agitated, actually. I mean, they are not actually in the crates for very long, and they go in them very freely.

'But you are perverting natural maternal instincts, aren't you? Because you're confining that sow. You're stopping it from leading the full life of a mother, which all mothers, whether humans or pigs, are entitled to. Surely that is a fundamental right?'

'I don't think so. Life is more complicated than that. That's too naive. That's dangerous stuff.'

Dangerous stuff? Or simply a question to which the industry doesn't – as yet – have an anodyne answer? There's another question, too, which has been bothering me.

'What about veal? Isn't that inherently much more troubling? Killing a small baby animal is surely one of the most horrible things that anyone can do. Doesn't that upset you?'

'Well, it upsets a lot of people, and I have to remind myself that it's similar to lamb. Killing a calf aged six months to a year is similar to killing lambs, which doesn't give me a problem, really.'

'Why not?'

'If it's to die, and has lived a decent life, probably the length of time it's lived is irrelevant. It's a bit like one of the free-range chicken definitions, which actually gives a different echelon to an animal which has lived longer, in fact 30 days longer. I think that is absurd. I mean, you may accuse me of being an evil man for killing animals at all . . .'

No, I don't think he's evil. After all, he has recognised some of the cruelty inherent in meat production, and is trying to do something about it. But I do think that he – in common with most farmers and butchers – hasn't thought through all the moral issues involved in the business.

'*Your male cattle are castrated,*' I say. '*Are you quite happy to do that?*'

'As long as it's done with anaesthetics.'

'*But essentially, it's still a mutilation.*'

'Yes, of course', he replies.

'*But you think it's worth doing?*'

'Well it doesn't actually harm the animal. I mean it depends what you call a mutilation. I've got a pierced ear, that's mutilation. I don't see it as a big problem. I mean, it's part of keeping animals.'

'*How do they feel afterwards?*'

'That would be an interesting research project. I've no idea.'

'*You've got no idea?*'

'With beef animals, if you don't castrate them, you can't keep them outside. You're terribly stuck then. I would say that the choice between being able to get out in a field and castration weighs in favour of being outside. Very difficult decision.'

Difficult indeed. Now I want to raise another 'difficult' issue, the question of human and animal rights. In common with many people of his generation, this young farmer was, at one stage, involved in anti-apartheid demonstrations.

'*Why do you feel that's an important issue?*' I ask.

'Because there is discrimination in South Africa between two sorts of human beings, based entirely on colour.'

'*So it's wrong to discriminate between the* same *species on the grounds of colour or genetic constitution, but it's OK to discriminate between species? That's OK?*'

'I don't have a problem with it,' he replies.

'*What makes it all right?*'

'I go back to the simple fundamental fact that there is no moral problem, for me, with eating meat. It's every man's choice, and I'd love to spend hours arguing the philosophical points, and heaven knows, eventually one might be convinced. But my gut feeling tells me that eating meat's all right, as long as the animal isn't abused.'

'*But surely,*' I reply, '*there's no greater abuse than taking a life unnecessarily?*'

'I mean during its life.'

'*So it's OK to commit the ultimate abuse – taking a life – but it's not all right to commit lesser abuses along the way?*'

'I would say,' he answers, 'that living torture is probably worse than death. Isn't making an animal suffer during its life worse than killing?'

'*I think it's better if the animal had never been born.*'

I feel I have to push him now, to make him explain precisely how he justifies what he does.

'*But tell me, in your mind, is the pleasure you get from eating a lump of steak worth the amount of suffering to the animal?*'

'That's what you call a loaded question.'

'*No, it's not a loaded question. I'd like a clear answer.*'

'The clear answer is this, that I do not believe that the way we rear animals involves suffering during their lives. Nor even their death. So they don't suffer. So I'm not balancing the pleasure against the suffering, I am balancing it against the life of an animal. And I square that simply because I do not see the eating of meat, and therefore the killing of an animal that has lived properly, as fundamentally wrong. And that's where you and I differ.'

'*The only difference is that I view the taking of a life as murder,*' I say. '*But you don't, do you?*'

'Not of animals. Do you?'

'*Yes, of course.*'

There is an iciness in the air now.

'Are you prepared to do violence about it?' he asks me.

His eyes have narrowed, and his lips are pursed. His question takes me aback.

'*What do you mean?*' I exclaim.

'I mean,' he says, 'there are those who burn lorries and blow people up because of it. What about that, you answer that, you owe me that much.'

'*Of course not,*' I wearily retort, and I stop the conversation.

BRAIN BATTLES

Violence. Ultimately, it always comes back to violence. The concealed violence of the meat-trade claims some kind of moral superiority over the public violence of the animal terrorist. When reason fails, savagery triumphs. Let's abandon this sterile side-track and talk to someone who does his fighting with his brain. Meet Dr. Alan Long. This man has one of the best science brains in the country. With two first-class degrees in organic chemistry – one from London, the other from Cambridge – he is sharp, observant and analytical. Not the kind of person to take an overly sentimental view of animals.

'I wouldn't call myself a great animal lover,' he explains. 'I'm a great respecter of animals, and I think that when they die, they're entitled to be treated with respect, just as when human beings die.'

And Dr. Long certainly knows about death, and how it comes to our food animals. For several decades, he has visited livestock markets and slaughterhouses, witnessed and recorded events there, and used this information to become one of the most informed and respected critics of our process of meat production.

'*What made you get involved in this area?*' I ask him.

'My mother was involved with a campaign in the 1930s to abolish the pole-axe,' he says. 'Because of that, there was always an interest in animal matters in my family. I was very much influenced by the parable of the Good Shepherd, and I couldn't equate the allegory with the dead lamb on my plate – I decided that I preferred to see lambs in fields rather than mutton on the dinner plate. My parents were understanding, and said, "All right, you don't have to eat lamb if you don't want to," and so we thought it through, and gradually all of us became vegetarian.

'My mother was still concerned with animal welfare, and I went to see livestock markets with her. I suppose I was an enquiring little boy, and I asked more and more questions and eventually decided that something ought to be done about the things I saw.

'So I started to collect facts, and I became a campaigner for what was then generally perceived as a really nutty cause. I decided that the best way of dealing with this perception was to get the facts, and use them in an unemotional way.'

'At the same time, you were also pursuing your scientific career, weren't you?'

'Yes, I took a degree in London, and then went on to Cambridge with several scholarships, because one wasn't enough. I had quite a hard time at university to remain a vegetarian, but I'm pleased to say that before I left there was a vegetarian table, and it's now very easy to eat a good vegetarian diet there.'

'Now, your scientific background is quite significant, because many within the meat trade and the farming industry accuse people who feel concerned about the plight of food animals of being over-emotional.'

'Yes, I thought that the animals should speak for themselves. And I don't believe in accumulating facts without making them work. I don't believe in collecting facts like stamps, just to look at and admire. I was, if you like, a self-appointed shop steward for the animals, constantly putting out all the facts I could uncover.'

'Some farmers have said to me that the animals in their care are really quite happy for most of their lives, and that people like you and me are just being over-emotional. How do you react to that?'

'I would say that they have no real grounds to make that assertion, in fact they're being emotional by saying that. You have to ask, "How do they know? Where is their evidence?" On the other hand, we certainly have scientific evidence that many animals suffer very considerably – you can do lots of experiments on stressor hormones, you can look at dehydration, that sort of thing. And I think that many farmers would privately agree that animals sent to market suffer very greatly.'

'But if a farmer said to you, "Prove that my animals suffer pain", what would you say to him?'

'Well, a lot of work has been done on this. The first thing you can point to is the general condition of the animal – many of them look quite poorly, their coat isn't in good order, that sort of thing. Then there's the environment – they suffer from the cold and the wet, just as we do. You can see cows in hot weather that are so thirsty that they try to lick the urine off the concrete, and that to my mind is definitely a sign that they are suffering. You can analyse the concentration of stressor hormones, the concentration of sugar levels in blood – all of these are signs of stress. There are 30 cases of mastitis for every 100 cows a year in the British dairy herd. Mastitis is an inflammation – it doesn't take much imagination to think that an animal doesn't like mastitis any more than a human being does – inflammations hurt. Lameness is another prevalent problem. And you can survey the health of animals arriving at the slaughterhouse. One study has shown that 30 per cent of all chickens have sustained broken bones being transported to slaughter. There's a skip at one market I visit where they throw the bodies of pigs that have actually died on their way to market. This sort of evidence is really incontestable.'

'Pigs are particularly susceptible?'

'Yes. With modern breeding, you can see that they have been bred to be fleshy – they're really travesties, these animals. And that puts a strain on their hearts, because their organs haven't been adapted to meet the demands of intensification, they are ill-equipped to deal with the effects of stress.'

'But farmers say that it's in their own commercial interests to keep the animals happy, otherwise they don't put on weight, or produce enough eggs.'

'That's an old, discredited argument. Of course, they do put on weight – unhappy people will put on weight. Unfortunately, cows will continue to give milk, even though they're in a very bad way. This is understandable, because nature's way of ensuring survival is that certain processes will go on, right to the end. The other point you have to remember is that these animals don't have to live very long. Nearly all the animals that are in production in this country are killed off in one way or another before they reach puberty. Sadly, what farmers do instead of looking after their animals with "tender loving care" is to shovel in loads of drugs to keep them going – growth boosters and that sort of thing.'

'Now, if we lived in the best of all possible worlds, where all the little animals were happy, and happily trotted off to the slaughterhouse, where they were painlessly "put to sleep", would you then eat meat?'

'I think that your own choice of words indicates that this is really mimsy-whimsy; I'm afraid that's simply a fantasy. Look, a half-tonne bullock is not going to succumb that easily in the slaughterhouse. It's rather like the dentist telling you "this won't hurt", and you know damn well it's going to hurt!'

'Tell me about your first slaughterhouse experiences.'

'Well, the first I went into was a little poky place. To some extent, I'd been prepared for it, because my mother had been into slaughterhouses, of course, and she'd told me. She didn't encourage me to go in, but when I was a big boy, and ready for the facts of life and death, I went in. And I was horrified, and I am still horrified. They were slaughtering horses. They weren't stunning them first, they were just cutting their throats.'

'How on earth can you cut a horse's throat without somehow immobilising it?'

'They hoist it up by its back legs.'

'So you've got this enormous horse, just struggling on the end of a chain?'

'Yes. You will see an animal in that sort of circumstance, struggling about for a minute or so, rather like a huge, wriggling puppy.'

'Now, people say small slaughterhouses are better?'

'Well, that varies a great deal. Today, most slaughterhouses are very large, they're essentially killing factories. You can go to a slaughterhouse where they're killing chickens, and the enormity of the whole scale is really quite appalling. These chickens being shackled upside-down on conveyor belts, and being mechanically eviscerated, and mechanically de-feathered. The sheer scale of the massacre – which is what it is – is rather horrifying. And the whole smell and stench of death pervades the place. The trouble with a killing factory is, if anything does go wrong, then it can go very wrong indeed.

'Smaller slaughterhouses are run on a "parish pump" basis, and supervised by local authorities – which is something that should change. It's quite wrong that this sort of operation should be supervised by local authorities who haven't got the proper means to do it. Also, you get local politics coming into it – there's often a reluctance in the local community to do anything much about the local slaughterhouse, unless it causes a nuisance to the neighbours.'

'You've actually seen religious slaughter, which very few people have. What happens?'

'At the first one I went to, the foreman said to me, "I can tell that you find this horrifying – I do too." Well, he was an honest man, I suppose. The Jewish law requires that you use a very big knife to cut the throat, and one stroke. In order to do this ritual, the animal has to be prevented from struggling about too much. The biggest problem that presents itself with large animals is to get them down. So in the early days, they used to tie the animal, hobble it, and throw it on the floor. When the animal was hobbled, it couldn't struggle very much, so several men pushed it over, and it lay on the ground, and they cut its throat. It got exhausted, and eventually bled to death. Well that process was very unpleasant, and there were hygienic as well as welfare objections to that. So they invented the casting pen – with the help of animal welfarists, which I find incredible.

'It was a pen into which the animal was driven, and the sides were moved up so that it was squashed in. Then the pen was capsized, so the animal was turned upside down. Those are very unpleasant circumstances, particularly for animals with a big rumen, like bullocks and cows, because the weight of their stomach actually prevents them from breathing.

'Then there begins an appalling time of battering about and screaming and wriggling – as I say, a half-tonne bullock, wriggling like a little puppy. Sometimes, it will get a leg out, sometimes both legs. It would be battering its head up and down, I saw these dreadful, blood-covered casting pens where the animals were bleeding from where they'd bashed their heads on the floor, because they were so terrified. I used to time this, and it was always over a minute before the animal's struggles died down. Then they'd put some water onto its neck, which started it struggling again. Then you'd have another bout of this awful struggling. And by that time, it was more or less exhausted, and couldn't do much more. Then, an ordinary slaughterman held the animal's head, and the Jewish slaughterman, a rabbi, shochet, took the big knife, called a chalof, and cut the animal's throat from side to side. In some slaughterhouses, after the cut has been made, an ordinary slaughterman may shove a knife into the chest, through the wound, to make it bleed more plentifully. And then it's hoisted up by a back leg, and left there to bleed for a while, before it's finally butchered – "dressed" as they say.

'One ritual slaughterhouse I went to was interesting because they had some West Indians there, they had white non-Jewish, and they had Jewish. When it came to the tea-break, all three groups went into separate huddles. The white ones invited me first, and offered me a cup of tea, and I said, "Don't you mix with the others?" No. They are racists, almost fascist, in fact, in their attitudes. The same goes for Smithfield meat porters. You can hardly find a more fascist bunch of fellows. Now, I find that an example of the sort of brutalisation that the whole thing inspires.'

'Is halal significantly different?'

'With halal, quite a number will now accept prior stunning of the animal. And there are moves afoot for both types to accept slaughter in an upright position, so that the cut has to come from below. Sheep and goats are the main halal animals. I spoke to one Imam who was slaughtering, and he was complaining that he was really a holy man, a learned man, and he really shouldn't be doing things like this. The procedures in that place were appalling – they were not using any pre-stunning, and the sheep and goats were twitching about for minutes on a cratch while they bled to death. Some of them were actually falling off onto the floor. And the animals were being slaughtered within sight of one another, which shouldn't have been happening. The whole thing was dreadful. The foreman of that particular place told me that he tried to do something about it, but he had trouble with race relations. He told me that he dare not stir things up, and the meat inspector told me the same. All they had succeeded in doing was to try to improve things by severing the spinal cord immediately after the cut had been made. Well that is a very dubious procedure, and that whole process was ghastly.

'Another problem with ritual slaughter taking place in a "mixed" slaughterhouse is that it tends to corrupt the other system. You have to remember that slaughtermen are paid on a per-kill basis, so the quicker they can get the animals through, the more it pleases them, the more money they get. If they can cut out a "nicety" – and I use the word rather cynically – like stunning, then that will speed things up. So the ordinary slaughtermen will look at the ritual slaughterers avoiding it, and they are tempted to do the same.'

'Have you come across any slaughtermen who have had qualms of conscience about what they do?'

'Yes, I often ask them, "Would you want to do this again?" Every foreman I've asked that has said "No." They've told me that they're glad their children aren't doing it. Who would want their daughter to marry a slaughterman? One told me he'd never go back to slaughtering horses, and he would never have anything to do with ritual slaughter. Yes, I find that there are men who think about it in more detail. But, of course, it attracts a number of people who are, unfortunately, brutal types. Every now and then you see a case of cruelty to animals, and it often involves a butcher or slaughterman. It's not really surprising.'

'Yes, I know of several unpleasant cases like that. There was a slaughterman who was recently sentenced to a maximum security hospital for strangling a woman and then drinking her blood – a real-life vampire. He said he used to do the same thing to the animals in the slaughterhouse.'

'Yes, and I've come across a bunch of slaughtermen who were tormenting a cat. They'd shoved a knife in its mouth – this is an illegal way of slaughtering animals, but it is used for poultry, to stab through the mouth into the back of the throat. It's a horrid, slow death.'

'Now, one more question about ritual slaughter. Many people would avoid ritually slaughtered meat if they knew the animal had been killed like that. But there's no way of telling, is there? The meat on your plate could easily come from an animal killed like this?'

'Yes. What happens is that the hindquarters of the animal can't be "porged" – which means taking out the blood vessels and the sciatic nerve – so butchers find it more profitable to sell just the forequarters to the Jewish trade, and to pass the hindquarters into the ordinary trade. The Farm Animal Welfare Council did take this up some years ago, and wanted the government to insist that meat be labelled with its method of slaughter. And the government turned it down. So today, you'll find that those people who criticise ritual slaughter may in reality be sitting down to eat a bit of meat killed by the very method they abhor.'

'Can you explain what the practice of "pithing" is?'

'Pithing is basically unhygienic. This was one of my most awful early experiences in slaughterhouses. There would be the animal, already stricken, and there would be some fellow, who would go up with this long metal rod, and shove it into the animal's brain, through the hole where the bolt had penetrated. They used to say that it scrambles their brains, but the animal then reacts very violently, and makes terrible groaning noises. After that, they kick much less. With a big animal, the kicking endangers the slaughterman, so they like to do this with the bigger animals. But it does give you appalling qualms – whether the animal has lost sentience at that time, or not.

'Newcomers to slaughterhouses are very much put off when they see the bits that have been butchered are still beating – you can see them still pulsing. You can say that, physiologically, that's just involuntary muscular movement, and there's no pain in that, but this area is very imprecise. I've seen a video of research work where a sheep was decapitated – its head was completely severed – and there was still evidence that there was a sense of feeling – you could still see the brain waves in this surviving brain. The whole question of sentience is a real problem.'

'But we're told, by the meat industry and the government, that adequate rules and regulations exist, and vets supervise the whole process. That it's really just like having a much-loved pet put to sleep.'

'One of my complaints is about the vets. Because they keep on producing reports, and commissions, and councils – and lists of recommendations are drawn up, and lists of abuses, but improvements are painfully slow in coming. The veterinary profession has not been as considerate as it should have been. After all, doctors are very, very reluctant to officiate at executions. But vets are really quite keen to be in at the kill. It's a massacre of their patients, really. They swear a vow to do the utmost for the welfare of the animals in their care – well, you do have to question what they're doing in slaughterhouses.'

'OK, once any compassionate person knows about these things, they're going to stop eating animal flesh. What else can they do?'

'In my view, quite a number of the dairy industry's nasty secrets haven't really been explored. The veal trade, for example – where do the calves come from? Most of them come from the dairy herd. Where does BSE come from? It comes from the dairy herd. Where does most of Britain's beef come from? From the dairy herd. Because clapped-out cows from the dairy herd are literally burgered. So really, most of Britain's beef industry is a by-product of the dairy industry. And that is what I'm trying to explain to people now. If they're cutting down on their consumption of animal products, they should cut down on dairy produce as well as meat.'

'What keeps you awake at night?'

'What horrifies me most is the thought, "Will I get hardened to all this?"'

EVERYDAY EVIL

I can well understand Dr. Long's concern. When murder – let's not beat about the bush, that's what it is – becomes routine, then we lose the ability to be shocked by it, and part of our humanity withers and dies. I've seen evidence of this so often in my encounters with the meat trade – their perplexed, bewildered expressions as they try to understand what it is that other people find so distasteful about their grisly business. So here I would like to present two straightforward and unsensational accounts by first-time observers of what they discovered inside these anonymous killing factories:

'Before we went in, our guide, the manager of the place, gave us a short description of what we would see. We then went in, and the things that immediately struck us were the noise (mainly mechanical) and the awful stench.

'The first thing we saw was the cattle being killed. They came one by one from the holding pens, up an alleyway into a high-sided metal pen. A man leaned over with a captive bolt pistol, and shot them between the eyes. This stunned them and they fell down. The side of the pen then lifted up and the cow rolled out on its side. The cow appeared to be completely rigid, as if every muscle in its body was tensed. The same man put a chain round the cow's hock and electrically winched it up into the air until just its head was resting on the floor. He then got a large piece of wire (we were told it was not electrified) and stuck it into the hole that the pistol had made. We were told that this killed the cow by severing the connection between its brain and its spinal cord. Every time the man inserted the wire into a cow's brain, the animal kicked out and struggled even though it was apparently unconscious. Several times while we were watching this operation, a cow fell from the pen kicking as if it had not been properly stunned, and the man had to use the pistol on it again.

'Once the cow was dead it was winched right up so that its head was about 2–3 feet off the floor. It was then moved round to a man who slit its throat. When he did this a torrent of blood poured out, splashing everywhere, including all over us. He also cut the forelegs off at the knee. Then another man cut off the head which was put on one side. Then the hide was removed by a man who was standing up high on a platform, and then the carcass moved on again to where the whole body was split open, and all the lungs, stomach, intestines, etc. came flopping out. We were horrified on a couple of occasions to see a fairly large, well-developed calf come out as well, as the cow had been in a late stage of pregnancy. Our guide told us that this was a regular occurrence.

'The carcass was then split into two by a man who sawed it down the spine with a sort of chain saw, and then the cow was ready for cold store. While we were there only cattle were being "processed", but there were sheep waiting in the holding pens. The animals that were waiting to be killed showed obvious signs of extreme fear – panting, staring eyes and frothing at the mouth. We were told that the sheep and pigs were killed by electrocution, but this method could not be used on cattle as it required such a high voltage to kill, that the blood would separate and the meat would look as if it were full of black dots.'[2]

Ghastly though it is, it is by no means unusual for a calf to be disembowelled from its mother's womb in the slaughterhouse, as the Ministry of Agriculture makes clear: 'It is normal husbandry practice to put cows in-calf during lactation, even if they are to be culled. In the normal course of events, there-fore, many cows going to slaughter would be in-calf.'[3] Now, here is another eye-witness account, witnessed this time by representatives of the media:

> 'In the holding pens the slaughtermen were trying to move a young but fully grown steer. The animal could sense and smell the death before him and didn't want to move. Using prods and spikes they drove him forwards and into a special restraint where he received a meat tenderising injection. A few minutes later the animal was driven reluctantly into the stunning box and the gate slammed shut. He was then "stunned" by a captive bolt pistol, his legs buckled beneath him, the gate was opened, and he sprawled out onto the floor. They pushed a wire into the half-inch diameter puncture made in his forehead by the stun bolt and twisted it. He kicked whilst the wire was in and then was still. A chain shackle was put around one of his hind legs, at which point he started struggling and kicking whilst he was hoisted over the blood bath. He was then still. A slaughterman then approached with a knife. Many people saw the steer's eyes focus on the slaughter-man and follow him as he approached. He struggled both before and as the knife went in. The universal opinion, including that of the press, was that this was not a reflex action – he was conscious and struggling. The knife was inserted twice and he bled into the bath.'

The 'tenderising injection' is one of the most obscene recent developments in slaughterhouse practice. Because many of the cows being sent to slaughter are, as Dr. Long observed, 'clapped out', their flesh has to be made tender so that it can be sold at a premium. An enzyme injection is therefore administered, while the animal is alive. This means the cow's jugular vein is opened up, and up to 2.5 litres of enzyme fluid is injected. In a few minutes, the animal's bloodstream literally begins to dissolve its own flesh. Then it is slaughtered. This process is so horrible that even the meat trade has qualms about its use, and it will, by all accounts, be ended soon. However, I dread to think what dreadful new process will be invented to replace it.

'IS IT KOSHER?'

Now there's one huge problem associated with ritual slaughter, and that is the racist dimension. Extreme right-wing elements have already tried to use this issue as a means of stirring up hatred against both the Jewish and Moslem communities. And both communities have reacted strongly to defend what they perceive as a racially motivated attack on an important aspect of their way of life. Because of this, most animal welfare organisations have been reluctant to confront the problem, and there is little likelihood of effective legislation being passed to ban it.

But that doesn't mean that we could, or should, ignore this dreadful cruelty – described to me by one official veterinary surgeon as the most revolting sight he had ever seen inside a slaughterhouse. He told me how it would sometimes take up to four slaughtermen to hold one sheep down as it struggled on the slaughtering table. The slaughterer then cuts the animal's neck open, and allows it to bleed to death. 'He didn't cut so much as saw,' the vet said to me. 'It took four attempts by the slaughterman before the animal's arteries were finally severed.'

You see, religious slaughter – or ritual slaughter, as it is more popularly known – is not purely the concern of those Moslems and Jews who believe it to be necessary. It is actually the responsibility of us all. Because every person who eats meat will, without knowing, have eaten ritually slaughtered food at some time. It is estimated that approximately 70 per cent of the meat that comes from animals which have been killed by ritual means actually ends up on the open market, finding its way into everyone's school meals, restaurants, and meat products of all descriptions. Britain is now a major exporter of ritually slaughtered meat to Europe and the Middle East.

It seems difficult to understand how two of the world's great religions can sanction such a plainly barbaric practice. In fact, it is only strictly orthodox Jews and Moslems who insist on it, there being an increasing number of modern Jews and Moslems who are prepared to see the practice end. Basically, there are two main reasons for the Jewish (schechita or kosher) and Islamic (dhabbih or halal) ways of slaughtering. First, the animal must be 'whole' if it is to be consumed by humans. This is taken to mean that it must not be sick or damaged in any way. It is therefore argued that pre-stunning, even if it occurs only a few seconds before death, is not acceptable since it results in a damaged animal. The second reason is to exsanguinate (bleed out) the animal, since consuming its blood is not permitted. 'Blood is unhealthy,' explained Dr. A.M. Katme, a representative of the Islamic Medical Association. 'It is full of toxins, urea, and organisms. The consumption of blood is forbidden for Moslems . . . It is arrogant for someone who is not a Moslem to presume that he can teach us the practice of our faith. God protect us from those who think that they know better than he.'[4]

I have no wish to teach anyone their faith, but I must respectfully point out a fundamental reality of ritual slaughter, which is that it is never possible to drain all of the animal's blood from its body, as any vet can testify. So if a Moslem is to observe the prohibition against the consumption of blood correctly, it therefore follows that he should not eat meat. This is an often-overlooked reality, and it should be urgently and honestly addressed.

The proponents of ritual slaughter claim that the loss of blood the animal suffers is so sudden that it induces rapid unconsciousness. This is how the Chairman of the Schechita Committee of the board of Deputies of British Jews explains it: 'The animal's throat is cut, and the whole operation can, if done properly, take less than half a minute. People imagine that because the animal has its throat cut while fully conscious it must be in pain. But what has been found as a result of experiments conducted in the late 1970s at the University of Hanover is that the animal becomes unconscious within two seconds of its throat being cut.'[5]

Rabbi Berkovits, registrar of the Court of the Chief Rabbi, has defended schechita on much the same grounds, claiming that it is preferable to the conventional stunning and slaughtering process.[6] Here, again, the supporters of ritual slaughter are on very shaky ground. The majority of vets I have spoken to strongly dislike ritual slaughter for the 'pain, suffering and distress' it causes the animals (to use the words of the government's Farm Animal Welfare Council). Research undertaken by the Institute for Food Research and the Institute for Tierzucht und Tierverhalten in Germany clearly shows that animals killed by ritual slaughter may remain conscious for up to two minutes after their throats are cut.[7] The researchers implanted electrodes in the cerebral cortex of animals to be slaughtered, and found that the animals' brains would respond to a stimulus up to 126 seconds after the cut.

Research carried out at a New Zealand university found that calves were making attempts to get up off the floor five or six minutes after their throats had been cut. One Birmingham vet (Birmingham has several major slaughterhouses that are devoted to ritual killing) believes that it can take up to 12 minutes for the animal to lose consciousness.[8] 'How would you feel about the same fate for your cat or dog?' he asks. 'There's no difference.' Another vet explained to me that all animals (including humans) have several arteries supplying the brain, and not just the carotid ones that are slashed in ritual slaughter. He explained that another major artery, the vertebral one, ran close to the spinal column, and it would be quite impossible to sever this (unless the whole head was cut off). Consequently, this artery goes on supplying blood to the brain even after the others have been cut, thus prolonging the animal's death agony.

Sadly, the government has already said that it does not intend to act to ban this monstrosity. When responding to the Farm Animal Welfare Council's recommendation that ritual slaughter should be banned within three years, agriculture minister John MacGregor said: 'The government has to recognise the serious implications for the religious communities if they were no longer required to prepare meat as their faiths require. We do not believe that we would be justified in imposing such a burden on these communities.'[9]

The whole question of ritual slaughter is not going to go away. Unless this matter is tackled now – preferably by compassionate people from within their own communities – I predict that it will become a severely divisive issue, exploited by extremists and racists for their own evil ends. Fortunately, there are people within both communities who understand the need for immediate reform.

Over the past few years, I have been fortunate enough to make a number of friends within the Jewish vegetarian movement. They have struck me, without exception, as being caring, concerned people, whose sincere approach to Judaism may provide others in their community with some serious food for thought. An editorial in the Jewish vegetarian magazine is well worth pondering, as it relates to our treatment of food animals today:

'The Sabbath day was granted to all, and Rashi comments that even domestic creatures, at least on that day, must not be enclosed but shall be free to graze and enjoy the work of creation. If now they are incarcerated in darkened containers seven days and seven nights in each week for the entire period of their lives; if they neither see the luminaries of the heavens nor experience the sweet smell and the taste of the pastures, has not the most sacred Sabbath Law been flagrantly violated, and can the flesh of their bodies be Kosher?

'Would the law that states "Thou shalt not muzzle the ox when he treads the corn" acquiesce in the computerised feeding of chemical fatteners, whilst the poor beast scents the dew and clover in the meadows beyond his darkened cell?

'When it is written "Thou shalt not yoke an ox with an ass" does it imply that a calf may spend its entire life standing on slats, never to lie down, and effectively chained to the sides by its neck to prevent it doing so? Is this a perversion of the Torah? And when the unfortunate victim is slaughtered, can its remains be considered Kosher?

'If the law forbids one to cause distress to a mother bird by removing eggs from the nest in her presence, would it concur that during its lifetime a hen could be shut up in a receptacle of twelve square inches, its beak removed and feathers clipped? And after its throat is cut, would its body be Kosher for food?

'If a cruelly treated animal shall be considered unfit for food, and if the measure of the cruelty is determined by its ability to walk, do the authorities inspect the incarcerated animals in the factory farms, and is there any record as to whether they are able to walk to their own slaughter on their own emaciated legs? And if not is their flesh Kosher?

'If the law forbids the mixing of species, even of plants, and confusion of sexes, would it condone the injection of female hormones into the male beast, even though it is acknowledged to be cancerous in practice? And when this distortion of blood is covered with earth, even as is human blood, is this respect for the Creator who saw that all he created was very good? Or is this confusion, is it defilement, is it sacrilege, and is the flesh still Kosher?

'Shall a certificate of Beth Din convey that Torah min hashamayim has been sincerely observed, or shall it become a licence for misinterpretation, evasion, and permission to the beholder to bow down each man to the God of his own stomach?

'Let all who are observant and devout remember that the responsibility is their own; no Jew can use an intermediary, whether in this case it be a Beth Din, a Board of Shechita, or just a Kosher butcher with a label on his window. If unaware of the facts, his is a sin of omission; if he is aware and chooses to ignore his personal responsibility, his is a sin of commission; he is eating Trefah.'[10]

These are good and wise questions, which raise fundamental issues of conscience that all of us, Jewish or not, must urgently consider.

THE SECRET LIFE OF PIGS

But perhaps I'm being over-emotional again? It's so very easy to do. We humans often confuse facts with feelings, especially when it comes to animals. Indeed, we're taught to do so; think about the insults that children learn from an early age – 'You greedy pig!', 'Your room is a pig-sty!', 'You're pig ugly!'. Odd, really, to think that a wild pig is one of nature's cleanest creatures – it will groom itself impeccably, when permitted to. And the pig is intelligent, too – at least as clever as a smart dog.

The problem with pigs is that they are uncomfortably similar to humans in many ways. Pigs, for example, know how to have fun. They will play with each other, and with humans, for hours on end. Some folk swear that pigs have a keen sense of humour, and that they've heard a pig laugh. That mother sows have a highly developed maternal instinct is a matter of record.

'Horizontal man' is how vivisectors describe the pig, because the arrangement of their internal organs is so similar to ours. Pig flesh even tastes like human flesh, according to those who've enjoyed both. Who knows, perhaps behind every bacon-eater there lurks an ancestral cannibal . . .

The problem with pigs is that they are really far too similar to us, too similar for their own good. Perhaps that is why we disparage and ridicule them so much – it puts some metaphorical distance between the two species, without which their ruthless exploitation would be far more distressing for us.

I'd like you to meet someone who knows about pigs. Andrew Tyler is a good journalist and a talented writer, and I am fortunate to know him as a friend. He has spent time – much time – in drafting a searingly exact account of the relationship between our two species, a task which has taken him to a couple of dozen farms, often working incognito. Here, he shares with us extracts from the diary he kept while working on a large pig farm; an establishment raising thousands of animals each year for both slaughter and breeding, and regularly attaining top marks from Ministry inspectors. It is almost certainly the most accurate account of the secret life of pigs that you will ever read.

'Woke up tired, burning eyes at 5.30. Got to the farrowing house to find a litter of 10 piglets had been born in the night. They were still shuffling clumsily around their mother when Ed, the farm manager, demonstrated the art of teeth and nail clipping – done with a pair of steel pliers. First the teeth – two on top, two on the bottom, both sides. The piglet is seized, his jaws forced open, and the little pointy teeth clipped off down to the gum. Such squealing! Then an inch of tail is removed and a squirt of purple antiseptic applied to the belly over the umbilical. The operation ends with the young one being chucked back in the pen. Always they are thrown, grabbed – by a back leg or ears, no matter how small . . .

'Next I watch 10 sows moved from the house where they'd been impregnated to the pens in which they'll wait out their pregnancy – all of them assumed now to be "in pig". After which I check inside the drug cabinet, a battered tin object with its doors permanently swung open, and see the staggering array of antibiotics, de-wormers, growth boosters, antiseptics used here.

'At 8.30 weaning begins – among the cruellest of the host of daily cruelties. Twelve sows with 113 piglets on them are removed from their farrowing crates – tight-fitting metal contraptions which allow the mother to stand and flop down but not turn around. She has been sharing it with her young since their birth 21 days earlier. Now the sow is removed with a shout and a slap – backward, down the steep stone step into a central aisle that is slippery with shit and piss; you'd have thought they might have cleaned it out first. They are slapped on the head, pulled by the tail, and kicked out of the joint, most of them struggling to remain with their young who stare bewilderedly at the tailboard of the crate and the direction in which their mothers are being taken. The men will be back for them soon.

'Most of the sows rebel and try to return to their young while being driven across the yard to their next stop; the service house. Here they are penned up and, incredibly, an hour later, checked to see if they are ripe for another servicing. Being a mixed batch, unused to each other, and disturbed through being torn from their young, the sows fight amongst themselves. Simultaneously, they are howling and screaming for their young. Ed tells me that fighting and complaining go on "for a day or two". Another worker, Mac, goes round them, pushing and actually riding on their backs, examining their vulvas. He decides one is ready to be served. He leads her out to a boar in a facing pen. When she gets there they discover her milk sac is bleeding and raw, possibly from a fight, possibly from being stood on. Mac applies the remedy – an antiseptic spray and antibiotic injection in the neck – even while she's in the boar's pen. The boar tries to mount. She screams and runs. They try her again but realise she wasn't on heat at all but "stood" for Mac's riding because she'd been trapped by the other sows in the crowded pen and couldn't go anywhere.

'As well as the boars penned in the service house, there are half a dozen in indi-vidual old-fashioned brick stys just across the yard. A sow who was "served" on successive days three weeks ago but failed to fall pregnant is taken to one of these. She's mounted and, as the penis is inserted, she howls and begins bleeding, quite a lot of blood. They believe at first it's the male's sheath but continue anyway once they realise it's coming from her, Mac assisting entry with his fingers. It seems the boar has struck her bladder, a common complaint. They persist in cornering the female so the impregnation can continue but the sow eventually breaks loose making it impossible. She gets an antibiotic jab, a splash of purple marker on her shoulders and, I'm told, she'll be served again this afternoon. "Raped" is probably a more suitable word than "served" . . .

'Disease problems they have had to cope with include viral pneumonia, scours (a diarrhoea that in the young is often lethal), meningitis – which the owner describes as "virtually similar to the human kind", salt poisoning (an often lethal and often agonising condition caused by them not getting enough drinking water), plus there is the memory of the Aujeszky's disease outbreak some seven years ago . . .

'I ask the owner what happens to the dead animals. He'd already acknowledged that the smaller ones at the other farm were dumped on the muck heap to be spread on the fields. But here, being bereft of straw, they have no muck heap. He says the corpses all go into a "death pit", but he looks seedy when offering this. Maybe because, as I witnessed, his death pits seem to breach the health and safety rules by not being enfenced; or maybe it's because the small ones are actually tossed straight into the lagoons. The death pit is an incredible sight; a hole about seven feet deep, about 10 feet square and clogged with the decomposing corpses of grown and half-grown animals, some beginning to go green, the skin and flesh bubbling vilely. They are in a variety of twisted positions, rear ends and snouts up but none fully submerged. Perhaps these represent just the top layers of animals, unable to sink for the bodies of their comrades.'

So this is how bacon is brought into the world. Later, Andrew describes the conclusion of the process:

'No sign declares the name or nature of the business, and it is a condition of my entry that I withhold the firm's identity. Suffice to say, it is one of the south of England's largest, disposing of 500 pigs weekly; EC licensed and therefore regarded as being superior to the majority of the other 1000-odd UK operations . . .

'All the animals start in the lairage: a large stone area divided by bar gates into a system of pens. Before the pigs' throats are opened, they receive what the plant manager calls "electrical stimulation" of the brain. The manager has just such a phrase for every aspect of the killing process.

'The "stimulation" is accomplished by a pair of hand-operated tongs, like giant pliers, that are clamped either side of the pig's head just in front of the ears. This takes place in a stunning box, a walled-in area about 15 feet square into which about a dozen animals at a time are corralled.

'As the first dozen is driven into the stunning pen, one urinates on the trot and makes a screeching noise I hadn't heard before. Blood and mucus fly from his snout. The eyes close, the front legs stiffen, and when the tongs are opened, he falls, like a log, on his side. He lies there, back legs kicking, as the stunner turns to the next animal. He tells me that the tongs should be held on for a minimum of seven seconds to ensure a proper stun before the throat is cut. But, urged on by his mates further along the slaughterline, he is giving them 1½ seconds or less ("If you were from the ministry, I'd do it longer").

'When he has stunned three or four, he shackles each of them with a chain around a back leg. They are then mechanically lifted and carried to an adjoining stone room where a colleague cuts deep into the neck and the still pumping heart gushes out blood. They are supposed to stun and shackle one animal at a time, since the delay involved in doing them in groups means they could go wide awake to the knife . . .

'Suddenly an electrocuted animal slips her shackle, drops five feet to the stone floor and crash lands on her head. The stunner continues jolting more creatures while her back legs paddle furiously. Without re-stunning her, he hooks her up again and sends her through to the knife. This crash-landing routine is to be repeated several more times in the next few minutes . . .

'One animal slams down twice. One man curses him as he lies paddling, blood seeping from anus and mouth. Another man, meanwhile, is ear-wrestling a would-be escapee that is leaping at a small opening in the metal gate. "You can have it another fucking way then, you idiot," he cries, as he helps slap the animal down.

'There is just one more waiting for the tongs, a small quiet creature which, from her position near the gate, looks me directly in the eye, breaking my heart. The stunner chases her a few steps. The tongs first ineptly clasp her neck; the eyes close in a strange blissful agony. The tongs are adjusted, and like a rock she falls.'

IN PRAISE OF PITY

So is it wrong to feel moved, to be horrified or even to shed a tear when you read accounts of everyday atrocities such as these? The farmers and the butchers say it is wrong to feel upset by these things. They say that we are being emotional, sentimental, even hysterical. They say that our hearts rule our heads. But what I say is this: If you do not shudder when you learn about these dreadful things, then you are no longer a whole human being. Let's examine this.

For a man, the charge of 'being emotional' is particularly stinging, because emotion is thought of as a female quality. Evidently, a man who is accused of being 'emotional' is also implicitly accused of being something less than male. In a world where testosterone sets the agenda, this is a grave accusation indeed.

Think how this parallels those appalling words of hatred we uncovered in the first chapter: 'The human race ranks highly because it belongs to the class of beasts of prey . . . We find in man the tactics of life proper to a bold, cunning beast of prey . . . He lives engaged in aggression, killing, annihilation.' This sort of human being does not shudder, does not feel empathy, does not feel joy or love. The emotions of kindness, pity, mercy and compassion are far beyond his limited experience; mere weaknesses to be eliminated. And this is the sort of human being the meat industry implies we should become. Their motives are obvious, of course. Writer Brigid Brophy exposes them with great precision:

'Whenever people say "We mustn't be sentimental," you can take it they are about to do something cruel. And if they add "We must be realistic," they mean they are going to make money out of it. These slogans have a long history. After being used to justify slave traders, ruthless industrialists, and contractors who had found that the most economically "realistic" method of cleaning a chimney was to force a small child to climb it, they have now been passed on, like an heirloom, to the factory farmers. "We mustn't be sentimental" tries to persuade us that factory farming isn't, in fact, cruel. It implies that the whole problem had been invented by our sloppy imaginations.'[11]

But it is cruel, and – instinctively – most of us know it. What we must now do is to have the confidence to trust the gut revulsion we feel when faced with these barbarities. Robert Bly, path-finder of the men's movement, well understands the importance of gaining access to these prohibited feelings:

'Children are able to shudder easily, and a child will often break into tears when he or she sees a wounded animal. But later the domination system enters, and some boys begin to torture and kill insects and animals to perfume their own insignificance . . .
   'Gaining the ability to shudder means feeling how frail human beings are . . . When one is shuddering, the shudder helps to take away the numbness we spoke of. When a man possesses empathy, it does not mean that he has developed the feminine feeling only; of course he has, and it is good to develop the feminine. But when he learns to shudder, he is developing a part of the masculine emotional body as well.'[12]

Do not be deceived by those who would forbid us access to these essential emotions. When you feel compassion for the innocent victims of our food production processes, then you are reaffirming our basic kinship with all creation. When you feel empathy with the mother animals who are so grievously exploited in the name of commerce, then you are reaffirming the sacredness of the bond between all children and their mothers, of all species.

And when you stop consuming the products of this reprehensible system – and when you encourage others to do so as well – then you are starting out on the long journey that will make things better again.

NOTES

1 Singer P., *Practical Ethics*, Cambridge University Press, 1979.
2 Observed by Sarah Hicks and Jackie Bain.
3 MAFF letter, 24 Nov 1984.
4 Letter to the *Guardian*, 20 Aug 1985.
5 Edmunds, M., 'What is ritual slaughter?', *SHE*, Jul 1985.
6 Letter to the *Independent*, 15 Mar 1989.
7 *New Scientist*, 7 Apr 1988.
8 *Birmingham Evening Mail*, 3 Jul 1985.
9 *Meat Trades Journal*, 5 Nov 1987.
10 Pick, P.L. (ed.), *Tree of Life*, A.S. Barnes & Co., 1977.
11 Brophy, B., 'Unlived life – a manifesto against factory farming', quoted in Wynne-Tyson, J. (ed.), *The Extended Circle*, Centaur Press, 1985.
12 Bly, R., *Iron John*, Addison-Wesley Publishing Co., 1990.

CHAPTER FOUR
THE A TO Z OF GOOD HEALTH

*'There is no evidence at all that eating meat
causes any illness
or is in any way unhealthy'*
The Director-General of the
Meat and Livestock Commission, 1986[1]

*'. . . If meat really was bad for you we could be sure the Government and the majority of
health and diet professionals (not to mention the leaders of most religious groups) would
have joined together to tell us so.' [A point for your fact file: the Meat and Livestock
Commission – the organisation speaking for the whole of the meat industry – is an official
organisation set up and monitored by the Government.]*
Meat and Livestock Commission,
publicity to schoolchildren, 1992[2]

*The word 'holistic' is rarely used in conventional medical circles. When studies are
published which demonstrate the superiority of the vegetarian diet – in either a preventative or curative capacity, doctors and scientists seem to respond (if they respond
at all) by searching for the one 'magic ingredient' that makes vegetarians healthier. Is
it the reduced animal fat in their diet? Or the larger amounts of vitamins A and C?
Or the trace minerals? Or the amino acid pattern in the protein? If they could just
put their fingers on it, then the problem would be solved. And meat-eaters could
make suitable adjustments – eat leaner meat, or take a few more vitamin pills – and
then they'd be as healthy as vegetarians, too.*

*The problem is, vegetarianism doesn't work like that. Although all of these factors
seem to contribute to some extent towards making the vegetarian lifestyle healthier,
none of them by itself is the magic bullet that will serve as a corrective for the meat-
centred diet.*

*The straightforward reality is that all the ingredients of a healthy vegetarian diet
work together to preserve health and combat disease. And some of these ingredients
are much better understood than others. By and large, the studies that follow are not
wholistic, and many of them were performed with the intention of identifying that
single 'magic ingredient'. However, something very interesting happens when you*

bring all this research together. A picture starts to emerge of a way of living that is better in virtually every respect. So here, for the first time, is . . . a glimpse of the Whole Picture.

ARRESTING ARTHRITIS

Conventional medicine has traditionally poured scorn on the idea that arthritis may be treated with a vegetarian diet. However, my own experience of listening to the histories of many arthritis sufferers on radio phone-in programmes tells another story entirely. Whenever the subject of arthritis is raised, I can virtually guarantee that someone will call in to say how much their symptoms have improved, usually after adopting a completely vegan (i.e. animal-free) diet. Since meat and dairy products contain arachidonic acid, which can promote inflammation when converted in the body to prostaglandins and leukotrienes, this is perhaps not really surprising. And at long last, studies are emerging to confirm what many sufferers seem to have found out for themselves.

- Recently, the results of a very impressive randomised, single-blind controlled trial have been published.[3] Twenty-seven patients were asked to follow a fast for 7–10 days, then they were put on a gluten-free vegan diet for three and a half months. The food was then gradually changed to a lactovegetarian diet for the remainder of the study. A control group ate an ordinary diet throughout the whole study period. After four weeks the diet group showed a significant improvement in number of tender joints, number of swollen joints, pain, duration of morning stiffness, grip strength, white blood cell count, and many other measurements of health. Best of all, the benefits in the diet group were still present after one year.
- Another important study of rheumatoid arthritics involved a week of fasting followed by three weeks of a vegan diet. At the end of this time, 60 per cent said they felt better, with 'less pain and increased functional ability'.[4] A vegan diet excludes all meat, fish, eggs and dairy products but includes an abundance of fruits, vegetables, grains, pulses, seeds and nuts.

And in a very welcome move, it seems that the medical specialists are starting to endorse dietary treatment, too. Speaking at the launch of the Arthritis and Rheumatism Council's booklet *Diet and Arthritis*, consultant rheumatologist Dr. John Kirwan commented:

'As far as we can tell at present, low-fat diets, cutting out red meat, full-fat milk, butter and confectionery made with butter – together with an increased intake of coldwater fish or vegetable oil – may enable people to take fewer pain-killers and anti-inflammatory drugs.'[5]

BATTLING BLOOD PRESSURE

Hypertension is the medical name for high blood pressure, one of the key risk factors in the development of heart and cerebrovascular disease. The United Kingdom government estimates that over 240,000 people die every year as a result of a hypertension-related disease.[6] Thirty-three per cent – one third – of all deaths that occur in people under 65 are attributable to hypertensive causes.

Blood pressure is measured by the height in millimetres of a column of mercury that can be raised inside a vacuum. The more pressure there is, the higher the column will rise. Since blood pressure varies with every heartbeat, two measures are taken – one to measure the pressure of the beat itself (called systolic blood pressure) and another to measure the pressure in between beats, when the heart is resting (this is called diastolic blood pressure, and is the 'background' level). These two figures are written with the systolic figure first followed by the diastolic figure, like this – 120:80.

When we're born, our systolic blood pressure is about 40, then it doubles to about 80 within the first month. Thereafter, the increase is slower, but inexorable, for the rest of our life. Many people do not realise they suffer from hypertension. There may be no symptoms, and it may only be discovered during a visit to the doctor's surgery for another complaint. In its later stages symptoms may include headache, dizziness, fatigue and insomnia.

A pressure of 150:90 would be considered above average in a young person, and 160:95 would be abnormally high. In older people, systolic pressure could be 140 at age 60, and 160 at age 80 years. Comparatively small changes in the pressure of those people who are in the 'at risk' category could have very worthwhile results. This was emphasised by a government report, which stated:

> 'It has been estimated that a relatively small reduction (2–3mm) in mean blood pressure in the population, if the distribution were to remain similar to the present distribution of blood pressures, would result in a major benefit in terms of mortality, and that a shift of this magnitude would be comparable to the benefit currently achieved by antihypertensive therapy. This estimated benefit seems applicable to mild as well as severe hypertension.'[7]

If a small change in the population's blood pressure could be as beneficial as all the drugs that people are now taking, then *what are we waiting for?*

Scientists have known for a long time that some populations are apparently 'immune' from hypertension, and do not display the rise in blood pressure that is associated in the West with getting older. These populations generally tend to have a high level of physical activity, are not overweight, have a low level of animal fat in their diet, and don't take much salt (sodium) in their food. In other words, hypertension seems to be an illness of our Western way of life. The problem is, of course, that the majority of us are stuck with it. We can't suddenly emigrate to a tropical paradise, or even change our lifestyles to a significant extent.

- As long ago as 1926, it was experimentally shown that certain dietary components could be connected to hypertension. In that year, a pioneering Californian study had shown that the blood pressure of non-meat-using people could be raised – by as much as 10 per cent – in just two weeks of eating a diet that centred around meat.[8] Subsequent experiments have confirmed this effect. One was undertaken in Australia, where two groups of people were selected, one of which regularly ate meat in their diets, while the other didn't.[9] The results were extremely significant, summarised for you in Figure 6:

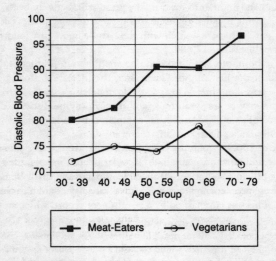

Figure 6: *Vegetarians' blood pressure compared to meat-eaters*

The top line charts the blood pressure of the meat-eaters. The bottom line shows the non-meat-eaters, and the bottom axis shows the five age groups that were surveyed. You can see that, at all ages, blood pressure is significantly lower in the non-meat-eaters. Amongst the meat-eaters, there is a steady rise in blood pressure with advancing age, but amongst the non-meat-eaters there is very little increase – and, in fact, a surprising drop in blood pressure in the oldest age group. These results were adjusted to exclude other factors, such as exercise, tea, coffee or alcohol consumption.
- Another study, carried out in Britain, again compared the blood pressure levels in people who didn't eat meat to those who did.[10] The results showed exactly the same pattern. This was true in men as well as women. Figure 7 shows the mean results that were obtained.

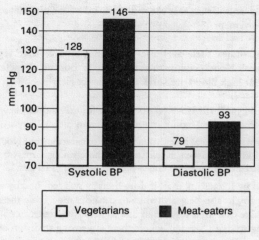

Figure 7: *Meat-eaters under pressure*

The difference between vegetarians and meat-eaters in the 'underlying' blood pressure (diastolic), which is generally thought to be a better guide to the real health of the individual, is considerable. On average, diastolic blood pressure was 15 per cent less in the non-meat-eaters compared to the meat-eaters.

• In another study, a group of 115 vegetarians was compared to a similar group of 115 meat-eaters who were closely matched to the vegetarians in all ways other than diet.[11] The results demonstrated that the systolic blood pressure of the vegetarians was 9.3 per cent lower than the meat-eaters, and diastolic pressure a massive 18.2 per cent lower.

• In America, a vegetarian diet was devised which included much fibre from whole-grain cereals, bran cereals, whole-grain breads, vegetables, beans and pulses.[12] Interestingly, the group put on this diet was allowed to use as much salt in their food as they wanted to. This group was then compared to a standard 'control' group, who carried on eating normally. The average blood pressure of the men on the plant fibre diet was 10 per cent lower than the control group.

• A recent 'crossover' trial has confirmed the results of the original 1926 study.[13] Fifty-eight patients aged between 30 and 64 with mild untreated hypertension were put onto an ovolactovegetarian diet (including dairy products and eggs). Within a few weeks, the average systolic blood pressure dropped by 5 mm. When they started to eat meat again, it rose by the same amount – very clear evidence that the meat element of the diet was responsible for the improvement.

We know, from studies such as those mentioned above, that vegetarians generally figure lower blood pressure than meat-eaters. But can a vegetarian diet also be used to treat high blood pressure? The evidence clearly shows it can.

- Scientists at the Royal Perth Hospital in Australia found that people with high blood pressure could indeed reduce it, on a vegetarian diet.[14] They wrote:

  'If the usual aim of treatment of mild hypertensives is to reduce systolic blood pressure to below 140 mmHg then 30 per cent of those eating a meat-free diet achieved this criteria compared with only 8 per cent on their usual diet.'

  They concluded by suggesting that, if drug therapy was required by a hypertensive, it might also be worthwhile to consider modifying the diet.
- Another persuasive case for a vegetarian diet to help hypertensives comes from a year-long study in Sweden, where there is a strong tradition of using dietary means to prevent or cure a number of diseases, including hypertension.[15] All the 26 subjects had a history of high blood pressure, on average for eight years. They were all receiving medication, but even so, eight of the group had excessively high readings (more than 165:95). Many of the patients complained of such symptoms as headaches, dizziness, tiredness and chest pains; symptoms which were either due to the disease or the medication they happened to be taking. They were then put on a vegan diet, from which coffee, tea, sugar, salt, chocolate and chlorinated tap water were eliminated. Their fresh fruit and vegetables had to be organic, if possible. When their diets were analysed, it was found that they were higher in vitamins and minerals than most people on a meat diet! The scientists who carried out the study reported:

  'With the exception of a few essential medicines (for example, insulin), patients were encouraged to give up medicines when they felt that these were no longer needed. Thus, analgesics were dispensed with in the absence of pain, tranquillisers when anxiety was not experienced and sleep was sound, and antihypertensive medication when the blood pressure was normal.'

The results were certainly impressive. First of all, the patients simply felt much healthier. None of them said that the treatment had left them unchanged or made them feel worse, and 15 per cent said they felt 'better'. Over 50 per cent of them said they felt 'much better', and 30 per cent said they felt 'completely recovered'. Reductions in blood pressure ranged from 7–9 mm (systolic) to 5–10 mm (diastolic). The scientist wrote:

'When the decrease in blood pressure was considered for the entire group, it was found that it occurred at the time when most of the medicines were withdrawn. Of the 26 patients, 20 had given up their medication completely after one year, while six still took some medicine, although the dose was lower, usually halved.'

Several other sorts of benefits were found, as well. The serum cholesterol levels of those studied was found to have dropped by an average of 15 per cent. And the health authorities computed that they had saved £1000 per patient over the year, by reducing the costs of drugs and hospitalisation.

Many studies have tried to identify the one, single factor which makes the vegetarian diet beneficial for blood pressure, but the evidence so far shows that neither polyunsaturated fat, saturated fat, cholesterol, potassium, magnesium, sodium, or total protein intake are independently responsible for this effect.[16] Again, we are forced to come back to the position that it is the *totality* of the vegetarian diet that is beneficial, and no single component. Hypertension is sometimes prefixed with the word 'essential', which rather confusingly means that the cause of it is not known. The studies above, and many others, give us convincing proof that a vegetarian diet can offer vital assistance in preventing, and treating, this modern silent killer.

CURBING CANCER

'Comparisons of diet and the incidence of intestinal cancer in different groups have prompted hypotheses of a possible increase in risk from the consumption of red meat or its fat content, but the evidence is inconclusive,' a junior minister for health (that's right – 'health') told the British House of Commons in 1991.[17] His words were promptly reprinted in the *Meat Trades Journal.* No doubt the junior minister will be suitably interested in the evidence that follows.

The number of cancer cases has risen dramatically in recent years until nearly one half of all cancers are suffered by just one fifth of the world's population – those who live in industrialised countries.[18] The number of new cases of most major forms of cancer in America increases by about 1 per cent every year.[19]

Growing public awareness of the epidemic nature of the problem has undoubtedly resulted in earlier diagnosis and treatment, and the importance of early treatment cannot be over-emphasised. The American Cancer Society suggests there are seven warning signs which, even if only one is present, should prompt a quick investigation. They are:

- Unusual bleeding or discharge
- Appearance of a lump or swelling
- Hoarseness of cough
- Indigestion or difficulty in swallowing
- Change in bowel or bladder habits
- A sore that does not heal
- A change in a wart or mole

What is Cancer?

It all starts as a single abnormal cell, which then multiplies out of control. This is the essential feature of cancer – a disorganised growth of cells, which multiply and may eventually destroy their host. Malignant groups of such cells form tumours and invade healthy tissue, often spreading to other parts of the body, in a process called metastasis. Benign forms do not exhibit this aggressive 'behaviour'. Because of this ability to invade and destroy healthy parts of the body, the Greek doctor Hippocrates called it *karkinos*, literally meaning 'crab', from which the modern word cancer is derived. Carcinogens are substances that promote the development of cancerous cells, and other substances known as inhibitors can keep the cells from growing. Some vitamins in plant foods are known to be inhibitors.

Cancers are broadly divided into two main types – carcinomas and sarcomas. A carcinoma is a disease affecting the tissues that cover both the external and internal body, for example breast cancer, prostate cancer, or cancer of the uterus. A sarcoma, on the other hand, is a disease that affects the body's connective tissue, such as muscles, blood vessels and bone. The prospects for survival depend, amongst other things, on the site in the body affected, the speed of diagnosis, the treatment given, and perhaps even on the attitude of the patient towards the disease.

What Causes Cancer?

It has recently been estimated that up to 60 per cent of all cancers in the Western world today may be related to environmental factors.[20] In fact, this isn't a new idea. As long ago as 1775, the eminent surgeon Sir Percival Pott suggested that there might be a link. Pott is one of the great names in the history of medicine. He was the first to notice how chimney sweeps often developed a particular form of cancer, and put forward the theory that their atrocious working conditions were responsible. So if we have reason to suspect that our environment might be a factor in the causation of cancer, shouldn't we try to do something to control it, or at least to reduce the risk? After all, we spend huge amounts of money trying to find cures or more effective treatments for cancer. Surely we should be trying to prevent the disease from appearing in the first place?

Of course we should. But that's not what happens. In 1991, a major British cancer charity spent over £42 million on research – and barely £600,000 on advising the public how they might reduce their risks. The same charity commented in its annual report: 'Although circumstantial evidence suggests that diet is linked to the cause of many human cancers, the evidence is extremely controversial'.[21] Commented the director of another major cancer charity: 'The basis for dietary effects on cancer is not understood'.[22] What dismal, discouraging words.

Well, at least one form of preventive medicine is on the agenda. 'Women worried about breast cancer should consider having healthy breasts removed before the disease has a chance to develop,' one newspaper recently reported a professor of obstetrics and gynaecology as stating.[23] It's rather like removing 'a redundant gland and pad of fat,' he said. Another proposal is to give healthy women Tamoxifen – a powerful anti-cancer drug – *before* they develop the disease.[24]

But let's get back to reality. In 1981, an epoch-making report was produced by the eminent epidemiologists Richard Doll and Richard Peto.[25] It assembled all the evidence they could find linking the occurrence of human cancers to specific identifiable factors. While the authors of the 1308-page report warn that not all causes of cancer can either be identified or avoided, it does seem from the evidence they collected that some of the causes of cancer they identify are well within our own control. The following table shows what they estimate the main risk factors to be, with their best estimate of the percentage of total cancer-caused deaths that are attributable to them.

FACTOR RESPONSIBLE FOR CANCER	PERCENTAGE OF ALL DEATHS
Diet	35
Tobacco	30
Infection	10
Reproductive and Sexual Behaviour	7
Occupation	4
Alcohol	3
Geophysical Factors	3
Pollution	2
Food Additives	1
Industrial Products	1
Medicines and Medical Products	1

You can see that 'Diet' comes right at the top of the list. 'Diet' means what we choose to eat, doesn't it? So, by informing ourselves of the evidence, and by taking steps to change our diets accordingly, we ought to be able to reduce significantly our chances of suffering from a diet-related cancer.

As researchers studied facts and figures about mortality from cancer in different countries, they were struck by an odd fact. It seemed that certain countries had a much higher mortality rate than others. What was the factor that made the United States, for example, so much worse than Japan? The researchers looked for a clue, then, they tried comparing the amount of animal protein that different nations ate and their cancer mortality.[26] What they began to see is illustrated in Figure 8 overleaf.

There is a clear relationship between the amount of animal protein in the national diet and the incidence of certain types of cancer mortality. But this wasn't the only connection. The same 'straight-line' connection seemed to exist

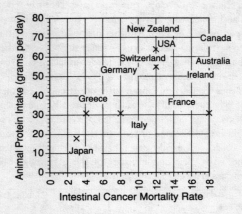

Figure 8: *More meat means more cancer*

between total fat consumption and cancer, animal fat consumption and cancer, and various other associated factors as well.

But perhaps certain nations were genetically more likely to contract cancers, no matter what they ate? To examine this possibility, studies were undertaken amongst immigrant populations. If the root cause of cancer was genetic rather than environmental, the same races should have the same occurrence of cancers, wherever they lived. The Japanese seemed to be a good subject, because they traditionally had a low incidence of most forms of cancer. So three groups of Japanese were chosen, together with a control group of Caucasians.

The first group of Japanese lived in Japan, and followed a largely traditional diet. The second group lived in the United States, but had been born abroad. The third group lived in the States and had been born there. The results (see Figure 9 opposite)[27] speak for themselves. A comparison between the extreme left and right columns shows that the Japanese living in Japan (left column) have only one quarter the risk of contracting cancer of the colon as Caucasian Americans living in the States. But even more significantly, when the Japanese move to the States, their chance of contracting colon cancer increases by three times to almost the same risk as a Caucasian. The place of birth didn't seem to matter. This was good proof that environmental, and not genetic, factors were indeed very significant.

Cancer From Meat

Now the scientific detective work really began. If the diet really was so important, then it should be possible to track down which specific factors related most strongly to increasing cancer risk, and, hopefully, to try to control them. So

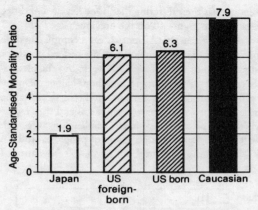

Figure 9: *Colon cancer deaths – east versus west*

the focus began to shift from international comparisons, which had pointed the way, to very specific studies amongst closely similar groups. Similar, that is, except for one or two key factors, which could be isolated, studied, and perhaps even controlled.

A group that was quickly identified as being of particular interest was the American Seventh-Day Adventist population. This group was subject to repeated studies, because the feature that distinguished them from the general

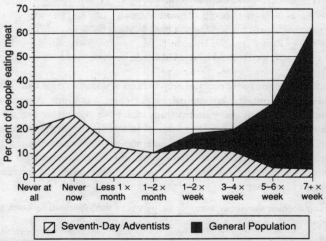

Figure 10: *Meat consumption – Seventh-Day Adventists compared to the general population*

American population was their differing diet. One key area of difference is dramatically demonstrated in Figure 10 on the previous page.[28]

The chart shows that Seventh-Day Adventists eat a completely different diet to the average American one. The vast majority of the general population use meat or poultry products seven or more times each week, but the picture is quite the reverse for the Seventh-Day Adventist group. About half of them don't consume meat or meat products at all. They do not smoke or drink (although in the survey one third of the men were previous smokers) and they tend to practise a 'healthy' lifestyle that emphasises fresh fruits, whole grains, vegetables and nuts. So now the scientists had found a good group of people to study. A seven-year scientific study tabulated the cause of death of 35,460 Adventists. This is what they found:[29]

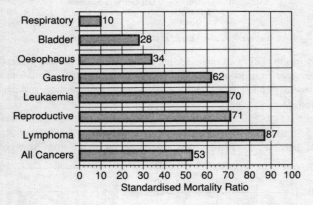

Figure 11: *Reduced cancer deaths in Seventh-Day Adventists*

The death-rate from all cancers among Adventists was amazingly half that of the general population. The bottom bar on the chart shows this, Adventists only having 53 per cent as many deaths from cancer when compared to the norm. Some of this could probably be attributed to their abstinence from smoking – cancer of the respiratory system, for example (the top bar on the chart), only being 10 per cent of the general population's. But other cancers, such as gastrointestinal and reproductive cancers, are not causatively related to smoking. The scientists concluded:

> 'It is quite clear that these results are supportive of the hypothesis that beef meat, and saturated fat or fat in general are etiologically related to colon cancer.'

Another study set out to check these remarkable findings, this time studying cancers of the large bowel, breast and prostate – the three most common ones that are unrelated to smoking.[30] Twenty thousand Seventh-Day Adventists were studied, and this time they were compared to two other population groups. Firstly, they were checked against cancer mortality figures for all US whites, and then they were compared to a special group of 113,000 people who were chosen because their lifestyle closely matched the Adventists – except, that is, for their diet. In other respects, such as place of residence, income and socio-economic status, the third group was very closely matched to the Adventists. The following graph shows how the results emerged:

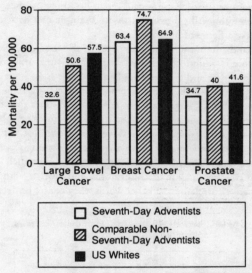

Figure 12: *Deaths from cancer in three populations*

Once again, the picture is pretty dramatic. The Adventists are shown on the graph by the white column, and the general population by the black column. The shaded column represents the special group whose lifestyle closely matched the Adventists, apart from the food they ate. You can see that for all three cancers, deaths among the Adventists were much lower than for the other groups. It is interesting that there does not appear to be a very great reduction in the risk of breast cancer among Adventists – until, that is, you compare the Adventists' results with those of the comparable group. The comparable group has a higher risk of contracting breast cancer than the national average (probably due to local environmental factors in California, where the study was undertaken). However, the Adventists have succeeded in reducing their own risk back down to below the national average – even though only half of them never consume meat.

Another Nail in Meat's Coffin

A major correlation study analysed the diets of 37 nations, and then correlated the components of the diets to mortality from cancer of the intestines.[31] Before looking at the results, let me briefly explain what a correlation study is. It's really quite simple. A correlation ends up as a number somewhere in between minus one and plus one. The higher the figure, the closer the connection between the two factors. For example, if someone is paid on an hourly basis, then the more they work, the more money they earn. This is an example of a perfect correlation, and would have a figure of plus one.

On the other hand, the more money you spend, the less you have in your bank account – this is a perfect negative correlation, since the connection between more expenditure and a decreasing bank balance is an inverse one: in this case, the correlation would be minus one. If any two factors, such as today's temperature and your bank balance, are not related at all, then the correlation would have a figure of zero. So the closer the figure gets to either plus one or minus one, the stronger the connection, positive or negative. You can see from the following chart that all the meat factors correlate very strongly with cancer. Total calories, total protein, and total fat also correlate strongly, which is not surprising, since meat is heavy in all three. But calories and protein from vegetable sources have a negative correlation, in other words, they confer protection. The study concluded:

'Animal sources of food were clearly associated with the cancer rates.'

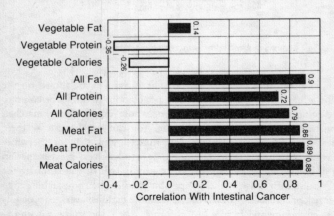

Figure 13: *How dietary components correlate to intestinal cancer*

Correlation studies like this are very important, because although we may not know precisely why and how meat in the diet contributes to various cancers (this may take many years to prove), we can see that there is a clear relationship – and this enables those of us who want to to take the necessary precautions for our own well-being.

More data, this time from an Israeli study, revealed a connection between both fats from animal sources and fats from plant sources, suggesting that saturated (and even unsaturated fats) may be connected with increased mortality.[32] The study followed the Jewish population as it grew from 1.17 million in 1949 to 3.5 million in 1975, over which period meat consumption increased by 454 per cent, and the death rate from malignant cancers doubled. You can see in the tables below how deaths from cancers rose in proportion to the amount of animal fat in the diet:

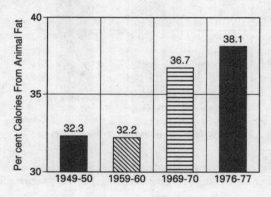

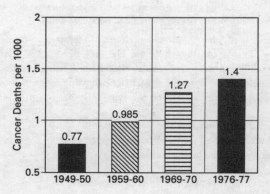

Figure 14: *Animal fat intake compared to cancer deaths*

Meat and Breast Cancer

More and more evidence was starting to accumulate. In Alberta, Canada, researchers compared the diets of women with breast cancer to a control group without the disease.[33] This study was on the lookout for a correlation between breast cancer and specific foods, in particular animal-derived foods. After analysing the data, they were able to show that the risk of contracting breast cancer increases with the amount of beef and pork in the diet:

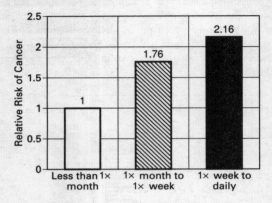

Figure 15: *Pork intake and risk of breast cancer*

For pork, an intake of anything more than once a week is associated with a doubling of the risk of contracting this form of cancer. The study concluded:

'The results suggest an association between breast cancer and the consumption of beef and pork. These findings are consistent with the higher breast cancer rates in areas of the world with higher per capita beef and pork availability.'

In Hawaii, another study showed the same pattern.[34] The study concentrated on a representative sample of the whole of Hawaii's residents – Caucasians, Japanese, Chinese, Filipinos, and, of course, Hawaiians. The wide variety of ethnic groups was useful, since it included a particularly wide range of food habits. Significant associations were established between:

• Breast cancer and all forms of fat and animal protein
• Cancer of the uterus and all forms of fat and animal protein
• Prostate cancer and fat and animal protein

The positive correlations between various forms of food and breast cancer are shown in the next chart, and the only negative correlation is between breast

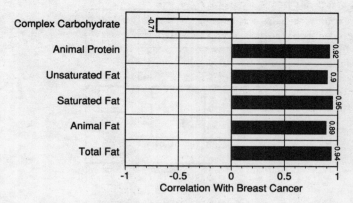

Figure 16: *How dietary components correlate to breast cancer*

cancer and complex carbohydrates – which are, of course, exclusively found in plant food. And almost exactly the same relationship emerged when the same study examined cancer of the uterus.

More Meat Equals More Risk

In 1981 yet another massive statistical world survey of 41 countries, including the US and the UK, was completed.[35] The results confirm the connection between eating meat, and the risk of contracting certain types of cancer. And yet again, they also show that plant foods seem to confer protection. Here are two charts drawn from data that the survey produced:

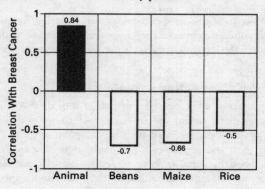

Figure 17: *How foodstuffs correlate to breast cancer*

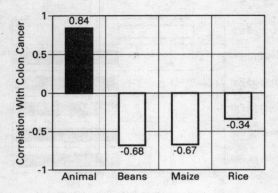

Figure 18: *How foodstuffs correlate to colon cancer*

'Less is Better'

One of the largest studies ever undertaken into the effect of meat-eating and cancer was published in 1990.[36] Over 88,000 women aged between 34 and 59 were recruited for the study (none of them had a history of cancer or bowel disease). Their health was tracked for six years, and it was found that women who ate beef, pork or lamb as a main dish every day were two and half times more likely to contract colon cancer when compared to those who ate meat less than once a month. What this study clearly demonstrated beyond reasonable doubt was this – meat in itself was a major risk factor. It wasn't that the meat-eaters were deficient in other nutrients – fibre, for example. The more meat they ate, the greater the risk.

The leader of the team of scientists commented: 'Reducing red meat consumption is likely to reduce the risk. There is no cut-off point so, really, less is better.'[37] All by itself, this study well and truly put paid to the myth that 'meat is part of a healthy diet'.

Change Your Diet – Save Your Life

So what actually happens when you start to change your diet? A clue comes from an intriguing study, carried out in Greece, which set out to measure what happened when people increased their consumption of certain types of food – including meat and vegetables.[38] The results show that an increase in consumption of spinach, beets, cabbage or lettuce actually decreases your chance of contracting colorectal cancer. But on the other hand, an increase in beef or lamb consumption increases your risk. This is what it looks like graphically:

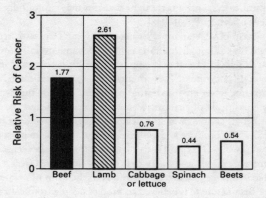

Figure 19: *Risk of colorectal cancer with doubling in food item consumption*

The study concludes as follows:

'The results of most of these studies appear to fall into two broad categories: those indicating that animal protein (mainly beef meat) and/or animal fat are conducive to the development of colorectal cancer – and those indicating that vegetables (particularly cruciferous vegetables) or, more generally, fibre-containing foods, protect against the development of this disease.'

How it Happens
We don't know every last detail of the way in which a meat-centred diet predisposes towards cancer. But we do know quite a lot. Consider:

- In just one kilogram of charcoal-grilled steak there may be as much benzopyrene (a powerful carcinogen) as in the smoke from 600 cigarettes.[39]
- Scientists at the Lawrence Livermore National Laboratory have been engaged on a five-year project cooking 'thousands of pounds of hamburgers' to see what toxic substances are produced in overcooked meat.[40] They have identified at least eight chemicals which are linked to cancer and chromosome damage. 'You don't get these structures if you cook tofu or cheese,' commented the senior investigator.
- Nitrites may be present in meat products, which can combine with other substances in the human body to form nitrosamines (extremely powerful carcinogens).[41]
- A high-meat diet lowers the age of puberty, and early puberty is associated with an increased risk of breast cancer.[42]
- Vegetarians are known to have a different composition of bile acids when

compared to meat eaters, and it is thought that this may profoundly affect the development of cancer.[43]

- The immune system of vegetarians is stronger than that of meat-eaters. One study has shown that although vegetarians have the same overall number of natural killer cells (the kind that are responsible for nipping cancer in the bud), they are *twice* as cytotoxic (i.e. potent) as those of meat-eaters.[44]

- Vegetarians consume fewer environmental pollutants than do meat-eaters. In one study, the breast milk of lactating women was analysed, and the PCB (polychlorobiphenyl) content was significantly higher in the meat-eaters.[45] Since meat, fish, dairy produce and commercial fruit are the main sources of organochlorines (PCBs, DDT and the dioxin family), it makes sense to cut out animal produce entirely and to try to buy only unsprayed fruit.[46]

- Vegetarians take in a large amount of vitamin A in the form of beta-carotene in plant foods. Beta-carotene is believed to protect people from cancers of the lungs, bladder, larynx and colon.[47]

- Vegetarians also eat diets that are rich in substances which suppress free radical formation. Molecules of oxygen are turned into free radicals inside your body by the continual process of metabolism. During this process, yet more molecules are generated which have an electron missing – called 'free radicals'. These free radical molecules immediately start to scavenge for electrons to kidnap from other molecules, and this sets in motion a continuing chain reaction which produces even more free radicals, in the process damaging cell membranes, proteins, carbohydrates and deoxyribonucleic acid (DNA). To date, some 60 diseases have been associated with free radical activity, including Alzheimer's disease, arthritis, multiple sclerosis and, of course, cancer. The vegetarian diet naturally contains substances (such as vitamin A, retinoids and protease inhibitors) which have been shown to be capable of blocking this process and halting the development of cancer.[48]

Why Don't We Do Something?

You can see from all this how difficult it is to isolate just one component of the vegetarian diet, and pin an 'anti-cancer' label on it. And yet, that is precisely what much current research tries to do. Once more, it is the totality of the healthy vegetarian diet – the whole thing – that naturally works to reduce disease.

Of course, vegetarians aren't totally immune from cancer. If you were a vegetarian who happened to live downwind from Chernobyl, then the odds would be stacked heavily against you. There are a whole host of factors which can predispose us towards this ghastly affliction, and only some of them are controllable. I personally suspect that the dangers of low-level radiation, for example, are still massively under-rated.[49] But fundamentally, avoiding cancer is all about reducing risk, and that's what evidence such as this clearly shows the vegetarian diet can do.

How long can we continue to ignore this kind of evidence? The American Heart Association, the American Cancer Society, the National Academy of Sciences, and the American Academy of Paediatrics are just a few of the professional bodies which have publicly urged a shift towards a more nearly vegetarian diet.[50] Apart from smoking, there is probably no greater personal health risk than eating meat. And yet, our politicians speak of 'inconclusive' evidence, and our medical charities fret that it is still 'extremely controversial'. And the meat trade seems to deny that studies such as those you've just read about even exist.

What it really boils down to is this: If you're not going to look after your own health, then no-one else is. It's up to you to live defensively. And let's face it, what have you got to lose by eating more healthily?

DEFEATING DIABETES

Diabetes is no longer the killer that it once was, so we tend to overlook it as a serious disease. But it *is* serious, because it is often a factor in the development of other serious illnesses, such as premature coronary disease, kidney failure, or blindness. Just because we have learnt to control its symptoms (through insulin injections), we seem to think that it's no longer a major public health problem, which it most certainly still is. Today's children are six times more likely to contract diabetes than their parents were.[51] Some authorities believe that diabetes-related complications are the third largest cause of death today.

Diabetes is a disorder in which the body is unable to control the amount of sugar in the blood, because the mechanism which converts sugar to energy is no longer functioning properly. Basically, this means that the body suffers from a deficiency or ineffective supply of insulin, which is a hormone produced in the pancreas, whose job is to lower the level of glucose (sugar) in the blood. If, over a period of time, the quantity or quality of insulin production is impaired, then diabetes will almost certainly result, and sugar will spill over into the urine. This is why the words 'Diabetes mellitus' literally mean 'sweet leaking'.

Diabetes can be divided into two types:

- Maturity onset diabetes: non-insulin-dependent. 'Overfed, overweight and underactive . . .' That is a popular summary of many, but not all, adults who develop diabetes in their middle years. Maturity onset diabetics experience the basic symptoms of thirst, fatigue, hunger and frequent urination. However, their health may improve by losing weight, increasing their level of exercise and monitoring their food intake to avoid foods high in calories, fats and sugar. In some people, diabetic symptoms can actually disappear following a strict regime of dietary control and exercise. Others must live the rest of their lives with the precautions, medications and attention to diet which have

for so long been associated with the disorder. In maturity onset diabetes, the adult need not become insulin-dependent.
• Juvenile onset diabetes: insulin-dependent. Although any age of person may develop diabetes, those who develop diabetes under the age of 40 years are most likely to suffer the more severe, insulin-dependent, form. Children who develop diabetes are almost always insulin-dependent. The insulin-dependent diabetic produces very little or no insulin and so relies on insulin injections. Without a supply of insulin, they would not survive.

Before the discovery of insulin, diabetes was considered to be invariably fatal, and most patients died within a short time of its diagnosis. Diabetes can be treated effectively today, although it does increase the risk of suffering other serious illnesses, such as cardiovascular disease, eye disorders, gangrene and other circulatory problems, nerve and muscle problems, and an increased susceptibility to ordinary infections.

An Ounce of Prevention

Some very significant research from the School of Public Health at the University of Minnesota reveals how we can reduce our risk of contracting diabetes.[52] They started a massive study of the subject in 1960, which lasted for 21 years and involved 25,698 adult Americans. They belonged to the Seventh-Day Adventist church, a group of people we've already encountered as being well known for their low meat consumption.

The results of this investigation showed that people on meat-free diets had a substantially reduced risk (45 per cent) of contracting diabetes when compared to the population as a whole.

The results also showed that people who consumed meat ran over twice the risk of dying from a diabetes-related cause. The correlation between meat consumption and diabetes was found to be particularly strong in males. The study was carefully designed to eliminate confusion arising from confounding factors, such as being over- or under-weight, other dietary habits, or amount of physical activity. The results are summarised in Figure 20 opposite. You can see that there is, of course, a striking difference between the number of people who were expected to die and the number of people who actually died.

But the study went even further than this. By analysing death certificates over the period under study, it was possible to assess the increased risk of dying from a diabetic illness that those who consumed meat ran. Figure 21 shows how it looks graphically. The graph shows that taking *any* meat in the diet increases the risk, on average, by 1.8 times. For light meat-eaters (people who only eat meat once or twice a week), the relative risk compared to a non-meat-user is 1.4 times. But for heavy meat-eaters – six or more times a week – the risk rises steeply to 3.8 times.

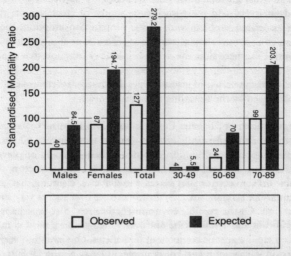

Figure 20: *Comparison of observed and expected deaths from diabetes*

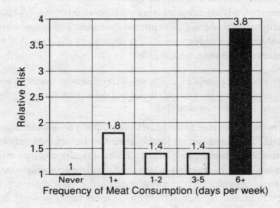

Figure 21: *Frequency of meat consumption and risk of diabetes mortality*

Meat is not just a problem because diabetics are particularly vulnerable to high blood fat levels, and meat is a prime source of saturated fat; there may also be an associated problem with excess protein consumption, too. Several clinical studies have now shown that a low-protein diet along with good blood glucose control can help slow the decline in kidney function that can occur in diabetics.[53]

While the ultimate cause of insulin-dependent diabetes remains elusive, some light has been shed by the following reports:

- A report submitted by Diabetes Epidemiology Research International (DERI) to the *British Medical Journal* suggested that between 60 per cent and 95 per cent of cases of insulin-dependent diabetes could be prevented.[54] The DERI group of scientists proposed that environmental factors are largely responsible for the recent major increase in diabetes, claiming that genetic factors could not account for such great increases over such a very short period of time. Of the possible environmental causes, diet is perhaps the most significant and certainly the one over which we have most control.
- Although there is a genetic component in diabetes, obesity is also very strongly implicated in its development. For example, in Japan almost all Sumo wrestlers become diabetic before they are 35 years old, and it is strongly suspected that this is induced by the amazingly high-fat diet they are given.
- While it seems that babies who are breastfed early in life may be less prone to developing diabetes, the converse also applies – children given cow's milk appear to be more likely to suffer from diabetes in later life.[55] A recent study examined international milk consumption patterns, and found a strong correlation with the incidence of insulin-dependent diabetes. A report by The Associated Press stated that:

'The study raises the possibility that when diabetes runs in families, parents may be able to protect their children by eliminating dairy products during the formative first nine months or so after birth.'

'If true, we should be able to do something to prevent diabetes altogether,' commented Dr. Hans-Michael Dosch, senior author of the study at the Hospital for Sick Children in Toronto. This study suggests that milk proteins cause an auto-immune reaction in which the body mistakenly attacks its own insulin-producing cells.

We also know that existing diabetics can benefit from a high-fibre vegetarian diet. A study carried out at the Veterans' Administration Medical Centre in Lexington, USA, compared two diets for the treatment of non-obese diabetic men, all of whom required insulin therapy.[56] The control diet provided 20 grams per day of plant fibre – an average Western meat-centred diet. The other diet included over three times as much fibre – 65 grams a day. The researchers found that the men on high-fibre, high-carbohydrate diets needed 73 per cent less insulin therapy than those on ordinary diets – quite a remarkable reduction.

It seems that those who suffer from diabetes might well investigate the beneficial possibilities of a largely meat-free diet. For those who wish to reduce their susceptibility to the disease, evidence suggests that a meat-free diet could help considerably.

F IS FOR FIBRE

Everyone's heard about fibre – in fact, you can't escape from it. Everywhere you look – television, posters, magazines – they're all trying to sell you fibre. You can buy specially formulated 'high-fibre' pills from the chemist, and even the most sugary breakfast cereal is now sold as 'high in fibre'. Fibre has become Big Business.

But although most people have heard about fibre, not many actually understand what it is, how they get it, how much they need, and what it really does to them. So here is a briefing for you:

• Fibre is, essentially, the cell-wall material of plants – it's what makes plant food chewy. When we cook, mill or process our food in some other way, we're breaking these walls down, and unlocking some of the nutrients in the food. And this is one of the objections to some of the processes of our modern food industry – that it destroys the fibre to such an extent that the micronutrients it encloses are exposed and destroyed.

• Meat contains absolutely no fibre. It is therefore useless from the point of view of supplying our essential dietary requirement. But there is also a further problem with a meat-based diet. Because meat is a very heavy, dense food, it tends to fill you up quickly; it is a very concentrated form of calories. So when you've eaten meat, you no longer feel like eating bulkier (but lighter) plant food. Meat therefore acts to minimise the intake of fibre in two ways. So if you cut out or cut down on your meat consumption, your natural fibre consumption will automatically tend to increase.

But why should we eat more fibre, in any case? A fascinating experiment, comparing data gathered from Western and African countries, provides the key.[57] Dr. Denis Burkitt, a famous advocate of dietary fibre, collected information from various populations concerning the size of their stools, the average time it took food to pass all the way through their bodies, and the type of diet they ate. You can see some of his results in Figure 22 overleaf.

Dr. Burkitt's findings were very exciting indeed. From left to right on the chart, the first group, with the shortest 'stool transit time', were schoolchildren living in rural Africa, eating an unrefined diet. Their food positively shot through their insides, taking on average less than a day and a half from one end to the other.

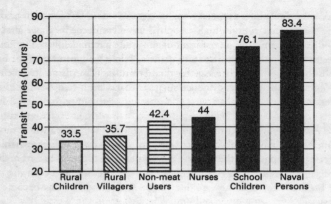

Figure 22: *Stool transit times*

Next came another group of Africans, this time adults living in villages in Uganda. Once again, their food hardly touched the sides on the way down.

But it is the third group that is so interesting from our point of view. This consisted not of Africans, eating a natural diet, but of ordinary vegetarians living in the United Kingdom. Despite enormous differences in environment and food availability, the similarity between the UK vegetarians' diet and the natural African one is very striking.

The next group on the chart consisted of nurses living and working in south India. Once again, their diet would tend towards the meat-free, and their transit times were only slightly longer than the UK vegetarians.

But then the really big jump comes. The next group has nearly twice as long a transit time as any of the preceding ones. This group was drawn from children at a boarding school in the UK, eating a refined diet typical of institutionalised catering – greasy, meat-dominated, and low in natural fibre. And the next group is even worse – naval ratings and their wives, all shore-based in the UK. This group had a mean transit time of 83.4 hours, and the longest time was 144 hours! That's six whole days for the food to hang around someone's intestines! No wonder four people out of every 10 in the United Kingdom are constipated at the moment.[58] In 20 per cent of the population, constipation is so severe that laxatives are used regularly, and over three quarters of the population only pass five to seven stools per week. No wonder the laxative manufacturers are making record profits!

Big, Fast and Regular!

You might suppose that small stools would whizz through the system quickly, but you'd be wrong. Burkitt found that the larger the stool, the faster it was processed. So, for example, the mean weight of stools passed by naval ratings was a mere 104 grams (not even 4 ounces). On the other hand, the mean weight for rural Ugandan villagers was a mind-boggling 470 grams. Over a pound! Somewhere in the middle came the UK vegetarians, with a mean weight of 225 grams (8 ounces), who compare very favourably with South African schoolchildren (275 grams, 9 ounces) and Indian nurses (155 grams, 5 ounces).

This information is crucial to our understanding of the importance of a diet high in natural fibre. Further evidence has shown that, without exception, countries which have a refined diet in which meat is predominant face a whole range of diseases that less 'advanced' countries rarely see. Some of these diseases, which can be directly associated to the Western, high-meat and high-fat, low-fibre diet include:

• Appendicitis. The commonest abdominal emergency in the West. Over 300,000 appendixes are removed every year in the United States alone. It has now been shown that a low-fibre diet makes the risk of suffering appendicitis much greater.
• Diverticular disease. Thirty per cent of all people over 45 years have symptoms of this.
• Cancer of the large bowel. After lung cancer, the most common cause of death from cancer in the West.

All these diseases were all comparatively rare in the West until the beginning of the twentieth century. Then the amount of animal fat in the diet began steadily to increase, and the amount of natural fibre began to decrease. Figure 23 overleaf shows how the American diet has changed in less than 100 years – the picture in Britain is much the same.

A hundred years ago, meat, fat and sugar between them only contributed 15 per cent of the total amount of calories in the diet. Today, the figure is nearer 60 per cent. Perhaps the biggest change in the diet has been the tremendous fall in the quantity of cereal fibre, which has dropped by 90 per cent. Most scientists now accept that there is a definite connection between the increase in modern diseases and the radical change in our fibre consumption. I asked Dr. Burkitt to explain:

'There are basically two types of fibre – insoluble fibre and water-soluble fibre. The classic insoluble fibre is wheat fibre, with bran and all the bran products. That is highly effective for combating constipation, increasing stool weight, and preventing things like haemorrhoids and diverticular diseases. It's very good for the guts. But it does almost nothing for what we call the "metabolic diseases" associated with lack of fibre, particularly diabetes, and coronary heart disease. Now soluble fibre, on the other hand, does have an effect on combating constipation, but it also has an effect on lowering raised serum lipid [i.e. fats in your blood] levels, and also

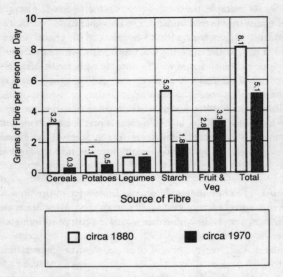

Figure 23: *The decline of fibre in our diet*

on glucose tolerance, so that it has a profoundly beneficial effect on diabetes. Now, as to how this fibre works in lowering the blood lipids, there are many suggestions. It has effects on bile acids and so on, but the main way in which soluble fibre is beneficial for diabetes is that it enormously slows down the absorption of energy from the gut. So instead of all the energy being absorbed, as it would be in sugar products, if you have a high-fibre product it makes the intestinal content into a sort of gel, so that the energy is only absorbed into the circulation very slowly, and so you don't have great and sudden demands on insulin, and so on.'

A good vegetarian diet provides enough fibre of all types to ensure a healthy intake, thus avoiding the need to worry about buying special 'added fibre' foods, and dramatically reducing the risk of suffering from many modern killer diseases. This is just one more aspect of the wholistic way in which a vegetarian diet naturally works to safeguard health.

GRAPPLING WITH GALLSTONES

Gallstones are a very common complaint, and are frequently excruciatingly painful. One sufferer compared it to 'being kicked in the guts by a horse – all the time'. But if any animals are involved in this complaint, it's more likely to be those we eat, because there is now evidence to associate the occurrence of gallstones with a low-fibre, high-meat diet.

Gallstones are mainly composed of solidified cholesterol. They can be formed in the gall bladder, where they can stay quite happily for years. However, they can also lead to infection, resulting in inflammation of the gall bladder, colic, peritonitis, gangrene of the gall bladder, and jaundice. They are more common amongst women than men – 25 per cent of all women and 10 per cent of all men will develop gallstones before they are 60 years old. The obese and diabetic are also more at risk.

It is likely that there is a strong metabolic connection in the development of gallstones. The liver secretes bile, a substance that is high in cholesterol (which literally means 'solid bile' in Greek), and is stored in the gall bladder. Lecithin (found in soybeans and corn) and bile salts together help to keep the cholesterol dissolved in the bile. However, if the level of cholesterol becomes so high that no more can be dissolved, then it begins to precipitate, and gallstones are the result.

Almost half of all those people with gallstones feel no symptoms. If a stone obstructs a passage, however, pain will be felt in the abdomen, with nausea and vomiting, particularly after eating fatty food. Sometimes cholesterol will be pre-cipitated so heavily that it is deposited around the body, especially in the eyelids. It is known that overweight people have a greater risk of suffering from gall-stones. Orientals and rural Africans, who traditionally consume a low-fat, low-cholesterol, high-fibre diet, suffer very little from them. In addition, only humans and domesticated animals have gallstones; wild animals do not. This also tends to suggest that the problem is connected with our modern, Western lifestyle.

In an experiment carried out in Oxford, two groups of women were com-pared to see if their diets could have any influence on the occurrence of gall-stones.[59] The first group, consisting of 632 women, were selected at random, and ate meat. The second group consisted of 130 women who did not eat meat, and ate a diet naturally higher in fibre. All the women were then given a thor-ough inspection, using ultrasound detection techniques, looking specifically for gallstones. The experimenters found that the meat-eaters were *two and a half times* more likely to develop gallstones than the non-meat-eaters. The scientists concluded that the low-fat, high-fibre diet of the non-meat-eating women gave them protection.

HEALING HEART DISEASE

You've already seen evidence in the first chapter of this book which clearly incriminates meat consumption in the development of heart disease. Believe me, there are many, many more reports which tell the same sad story. But rather than going over the same ground again, I'd like to present you with some truly wonderful good news (it's time we had some!).

But first, a word about the vegetarian diet in relation to coronary heart disease. Because a good vegetarian diet (such as a vegan one) is far lower in saturated animal fat than a meat-eater's diet, and because saturated fat is clearly connected with the development of heart disease, many people have assumed that this is the only factor of any importance. The meat industry, in particular, has seized upon this aspect to sell the concept of 'lean meat'.

The proposition seems to be that if you consume 'lean meat', then you'll be getting the benefits of vegetarianism, without actually *going* vegetarian. There are two points to be made here: First, the meat trade knows very well that consumers don't like the 'eating quality' of very lean meat. Consumers find meat from these carcasses 'less juicy than average, with a tendency towards toughness and poorer flavour'. And butchers themselves find it 'floppy, difficult to cut because the tissues separate, unattractive, and lacking in succulence and flavour'.[60] Despite this, the leanness of meat is constantly featured in meat trade advertising. So, as I see it, there's a danger here of consumers actually *buying* fatty meat and meat products, but *believing* that the meat on their plate is actually far lower in fat than it really is. And that, of course, could be a fatal error.

The other point to make (once more) is this: There are many factors that work together to make the vegetarian diet intrinsically healthy. If you simply focus on lowering saturated fat, and try to emulate it in a meat-centred diet, you're actually missing out on all the other co-factors. Such as vitamin E, for example. Now, it has been established that vegetarians have more vitamin E relative to their blood cholesterol than do meat-eaters.[61] And we also know that vitamin E is a 'biological antioxidant', which can probably protect us against atherosclerosis (clogged-up arteries). This is a more subtle, but nonetheless important, way in which the vegetarian diet may safeguard us against heart disease. Simply eating leaner meat is not going to have any effect on this.

All right, now for the good news:

• In the last few years, irrefutable evidence has emerged that a scrupulously followed vegetarian diet can actually heal the damage inflicted on our clogged-up arteries.

It sounds too good to be true. But here is the proof. It's worth reading closely, because the science is so neat and tidy.

• In 1985, a highly significant paper was published in the *New England Journal of Medicine*.[62] A team of scientists from the University of Leiden in the Netherlands studied a group of 39 patients, all of whom suffered from angina, and all of whom had at least one blood vessel with 50 per cent blockage, as revealed by coronary arteriography. Then they put the patients on a vegetarian diet.

After two years, the scientists took further measurements. In 21 of the 39 patients, the blockages had got worse. However, in 18 patients, things hadn't deteriorated. What was more, it was clear that the coronary lesion growth correlated with total/HDL cholesterol in the blood – the higher the total/HDL cholesterol ratio, the more the disease had progressed. By contrast, in those patients where the ratio was low, there was no progression. This evidence opened up a whole new line of tackling heart disease. First, Dr. David Blankenhorn of the University of Southern California and Dr. Greg Brown at the University of Washington both performed scientific trials which showed that the build-up of arterial plaque could be reversed, in some people, by a combination of drugs and a low-fat diet.[63] Then, in 1990, a landmark paper was published in *The Lancet*.

- For the first time, scientists irrefutably proved that a vegetarian diet – without the assistance of medication or drugs – could be used to regress coronary heart disease.[64] The science was impeccable. The study was both randomised and controlled (meaning that patients were randomly assigned to either the experimental group, or to a control group which was used for comparison). Both groups had their coronary artery lesions carefully measured at the start of the study, and after one year.

Those in the experimental group were asked to eat a low-fat vegetarian diet, consisting of fruits, vegetables, grains, legumes and soya bean products, and, remarkably, they were allowed to eat as much as they wanted to – no calorie-counting was required. Now, that's not even 'dieting' by most people's standards!

No animal products were allowed except for egg white and a maximum of one cup per day of low-fat milk or yoghurt. The diet contained 10 per cent of its calories as fat, 15–20 per cent as protein, and 70–75 per cent as complex carbohydrate. No caffeine, very little alcohol. Relaxation was encouraged, and patients were asked to exercise for a total of three hours a week, even though, at the beginning of the study, many participants suffered from such severe chest pain that they could barely walk across a room without resting.

Now for the results. After one year, blockages in the arteries of two thirds of the control group (the group that hadn't followed the vegetarian diet) had worsened. But for 18 of the 22 in the experimental group, the blockages had reduced in size, resulting in an increased blood flow to the heart. And the more severe blockages showed the most improvement.

So was this regression entirely due to a lowering of cholesterol? Dr. Dean Ornish, leader of the team, doesn't think so:

'If lowering cholesterol were the primary factor in causing reversal of heart disease, most of the patients in the studies by Dr. Blankenhorn and Dr. Brown who were taking cholesterol-lowering drugs should have shown reversal, since almost all of these patients had substantial decreases in blood-cholesterol levels. Yet only a minority showed reversal.'[65]

Once again, the suggestion is that more than one factor is at work. 'Nutritional factors other than fat and cholesterol play a role in heart disease,' believes Dr. Ornish. And one such may be beta-carotene (vitamin A). 'People who consume a low-fat vegetarian diet naturally consume not only beta-carotene,' explains Dr. Ornish, 'but other antioxidants that may play a role in preventing and reversing heart disease.'[66]

You know, in this context, I can't help thinking of the words of William Clifford Roberts, the distinguished editor in chief of *The American Journal of Cardiology*. In 1990, he wrote:

> 'Although human beings eat meat we are not natural carnivores. We were intended to eat plants, fruits and starches! No matter how much fat carnivores eat, they do not develop atherosclerosis. It's virtually impossible, for example, to produce atherosclerosis in the dog even when 100 grams of cholesterol and 120 grams of butter fat are added to its meat ration. (This amount of cholesterol is approximately 200 times the average amount that human beings in the USA eat each day!) In contrast, herbivores rapidly develop atherosclerosis if they are fed foods, namely fat and cholesterol, intended for natural carnivores. . . .
>
> 'Thus, although we think we are one and we act as if we are one, human beings are not natural carnivores. When we kill animals to eat them, they end up killing us because their flesh, which contains cholesterol and saturated fat, was never intended for human beings, who are natural herbivores.'[67]

Those words should be hung above the desk of every doctor in the country.

LEARNING ABOUT LEUKAEMIA

Leukaemia is the name used to describe a number of cancerous diseases of the blood-forming organs. The acute and chronic leukaemias, together with the other types of tumours of the blood, bone cells and lymph tissue, cause about 10 per cent of all cancer deaths and about 50 per cent of all cancer deaths in children and adults less than 30 years old.[68]

Is leukaemia infectious? Such a notion is commonly dismissed as being absurd. When a cancer charity recently released a report about public myths and misconceptions surrounding cancer, they cited a survey showing that 'one in 10 teenagers believes that cancer is infectious', presumably with the intention of proving how bizarre our beliefs about cancer can be. What it really revealed, however, was how out of touch that particular charity was itself. Because there is incontrovertible scientific evidence to show that cancers are, indeed, transmissible – both within species and across them.

It has taken a long time for much of the scientific community to accept that cancers could be caused and transmitted by a virus. In experiments conducted as far back as 1911, it was demonstrated that tumours taken from one chicken and implanted into another would infect the second chicken with a cancerous growth. In 1936, it was demonstrated that breast cancer could be transmitted between mice via a virus present in the milk of lactating mice. More recently, scientists at the University of Glasgow discovered feline leukaemia virus in cats. Today, cancer-causing viruses (oncoviruses) are now scientifically categorised as a part of the retrovirus family. Despite this, the belief still partially persists that cancer somehow 'ought not' to be capable of being virally induced. One pathologist commented:

'Indefinite statements are often expressed concerning identifying a certain cancer virus in humans by the antibodies produced in an animal. We show no such insecurity with other viruses: why should we do so with cancer viruses? If we find antibodies to smallpox virus or measles virus in an animal we confidently say the animal has had the infection of smallpox or measles. But when a cancer virus stimulates an antibody response in an animal we do not confidently state that the animal was infected by that particular virus. It is as if we are afraid to say that the virus that caused cancer in the cow or dog is the same virus that produces an identical antibody in humans.'[69]

Just like human beings, the animals that we eat suffer from various forms of cancer, sometimes caused by a virus. For example:

• Bovine leukaemia virus (BLV) causes cancer of the lymph tissue in cows.
• The avian leucosis viruses (ALV) cause leukaemia in chickens.
• Marek's disease virus (MDV) causes a cancer of the lymph and nervous systems in chickens.

One American report found that:

'Virtually all commercial chickens are heavily infected with leucosis virus. Since the tumours induced are not grossly apparent until about 20 weeks of age, this virus is not economically as important as is the Marek's disease virus, which induces tumours by 6-8 weeks of age. Bovine leukaemia virus is widespread in commercial dairy herds; more than 20 per cent of dairy cows and 60 per cent of herds surveyed in the USA are infected.'[70]

Now the key question is this: Can eating meat, or being exposed to food animals or their produce, result in a greater likelihood of contracting leukaemia or other cancer? To investigate whether these viruses can cross the species barrier and infect humans, a study was established, paid for by the US National Cancer Institute.[71] The scientists conducting this study reported:

'The viruses are widely distributed naturally in their respective hosts, and are present not only in diseased but also in healthy cattle and chickens destined for human consumption.'

Therefore, it seemed logical to examine the health of those people who would have maximum exposure to these animals – slaughtermen.

Accordingly, the health of 13,844 members of a meatcutters' unions was checked for the period 1949 to 1980. After statistical analysis, it was found that abattoir workers were nearly three times more likely to die from Hodgkin's disease (a cancer of the lymphatic system) than the general population. The scientists concluded:

> 'The excess risk was observed only in abattoir workers and seems to be associated with the slaughtering of cattle, pigs and sheep . . . Thus, the excess risk seems to be in keeping with a postulate of an infectious origin for these cases, as no other occupational exposure could adequately explain this occurrence.'

By itself, this report is very significant. But now consider the following additional evidence:

- It has been shown in laboratory experiments that bovine leukaemia virus can survive and replicate itself when placed in a human cell culture.[72]
- Scientists have found a close similarity between bovine leukaemia virus and HTLV-1 – the first human retrovirus ever shown to cause cancer.[73]
- A study recently conducted in France has concluded that children of fathers who work in the meat trade are at greater risk of developing childhood cancers.[74] The study examined over 200 cases of leukaemia diagnosed in the Lyons area, and found that a significantly larger number of fathers of children with leukaemia than expected worked as butchers or in slaughterhouses. The scientists suggest that bovine leukaemia virus could be to blame.
- In another experiment, chimpanzees were fed from birth on milk taken from cows known to be infected with bovine leukaemia virus, with the result that two out of six of them died from leukaemia.[75] (The idea is often advanced, to defend unpleasant experiments such as these, that they are necessary in order to improve human health. One is forced to wonder, however, whether 'inconvenient' results such as this are acted upon or simply swept aside.)
- Statistical analyses of human deaths from leukaemia and other cancers have shown that those people who have close contact with food animals (vets, farmers, butchers) run a significantly higher risk of dying from certain types of cancer than would be expected. For example, in a Nebraskan study, it was shown that men having regular contact with cattle were twice as likely to die from leukaemia.[76]
- In a study from Poland it has been shown that farmers, butchers, and tanners are more likely to develop leukaemia than other people.[77] And a further Polish study concluded:

> 'It should be inferred that cattle affected with leukaemia may, in favouring circumstances, be a factor disposing man to neoplasms [cancer] especially to the proliferation of the lymphatic system, either through longer contact with a sick animal or the longer ingestion of milk and milk products from cows with leukaemia. The

fact that with a rise in the incidence of leukaemia in cattle there also appears an increase in proliferating diseases of the lymphatic system is particularly worthy of attention.'[78]

- A study conducted in Minnesota amongst leukaemia sufferers showed that a higher than expected number of them were farmers having regular contact with animals.[79] And a similar study conducted in Iowa found a relationship between leukaemia in humans, cattle density, and the presence of bovine leukaemia virus in cows.[80]

- A study of mortality from leukaemia and Hodgkin's disease amongst vets has shown that they run a significantly higher risk of dying from lymphoid cancer than the norm. The vets were in clinical practice, in close contact with food-producing animals, and the authors of the report suggested that a viral cause may be responsible.[81]

- A study conducted in France and Switzerland in 1990 reveals that male sufferers from breast cancer (generally rare in men) were most likely to work as butchers.[82]

- Like the French study previously mentioned, an Italian study conducted by scientists at the University of Turin has confirmed that the children of butchers are more likely to contract cancer.[83]

All this evidence should be considered very seriously, because it has extraordinarily profound implications. Dr. Virgil Hulse, a physician who spent 15 years as a milk inspector for the state of California, writes:

'The Food and Drug Administration states that many unanswered questions remain about BLV, such as transmission, infectiousness, and whether it's a threat to humans. Some of the questions fuelling the controversy are whether pasteurisation, which inhibits infection, destroys the aspect of the virus capable of producing cancer. Also, how great is the risk of pasteurised milk being accidentally contaminated with raw milk? If we wipe out BLV, will we see a reduction of those cancers related to fat consumption? Might it be the viruses and not the fat that are linked to some human cancers?'[84]

But how could an animal cancer virus induce the disease in humans? Unfortunately, there are several possible ways. One theory suggests that a 'helper virus' can form an association with another relatively harmless one, and in the process produce a virus that can induce cancer. An animal virus may not, therefore, directly precipitate the disease in humans, but it may be able to convert otherwise harmless human viruses into killers.

It will certainly be many years before every feature of the complex process of zoonotic carcinogenesis has been resolved. And there will, no doubt, be many people who will not wish to see these rather dark and disquieting fringes of medical and veterinary knowledge examined too closely. But that, of course, is no reason not to ask questions, nor to take prudent defensive measures.

OUTWITTING OSTEOPOROSIS

When a disease turns into a marketing opportunity, truth flies out of the window. That's what's happened to osteoporosis.

Osteoporosis first hit the headlines in 1984, when the US National Institute of Health issued an advisory paper stating that women should increase their intake of calcium to prevent osteoporosis. Since one in every four women will have suffered an osteoporosis-related fracture by the time they reach 60, the demand for calcium supplements suddenly hit the roof.[85] Sensing a new promotional opportunity, the food manufacturers started to cash in (sadly for the meat trade, meat contains virtually no calcium, so they missed out on this one). The dairy industry, however, wasn't slow to appreciate the potential for increased sales. Since milk contains plenty of calcium (just the right amount for a fast-growing calf, not necessarily so right for humans) they clearly had a hit on their hands. And so, with the help of the vast promotional resources of the dairy industry, the public quickly learnt that:

- Osteoporosis is caused by a lack of calcium in the diet.
- Milk contains oodles of calcium.
- Therefore, to avoid osteoporosis you should gulp gallons of milk.

On the surface, it all sounds very plausible. Osteoporosis is caused by a slow loss of bone mass. By the age of 35 or so, your bones will be as strong as they're ever likely to be, after which age they slowly lose density. Therefore, a heavy dose of calcium should put things right again, shouldn't it?

The answer is no. Although the theory 'has an intuitive appeal', a recent article in the *British Medical Journal* stated, 'the logic is similar to that which might lead doctors to give ground-up brains for dementia.'[86]

Nevertheless, many people now seem to believe that a large intake of dairy produce will safeguard them against this crippling condition. And the myths abound, as plentifully as ever. Take this piece of advice which comes from a recent article in a health food magazine:

'Vegetarians who do not use dairy products or take supplements are especially at risk of developing osteoporosis, either because they do not get sufficient amounts of nutrients from their diet or do not absorb the nutrients properly.'[87]

Meanwhile, the recommended intakes for calcium continue to skyrocket – up to 3000 milligrams (3 grams) a day in some cases.[88] Now to get this amount of calcium from dairy produce, you'd have to drink 10 glasses of milk a day, which, even assuming you drank low-fat milk, would also give you a very unhealthy 180 milligrams of cholesterol and 30 grams of saturated fat.[89] Alternatively, you could munch your way through one pound of Cheddar cheese, which would also give you a whacking 150 grams of fat, most of it saturated![90]

Clearly, we're only being told part of the story. Do you remember the results of the China study mentioned in the first chapter? It found that 'most Chinese consume no dairy products and instead get all their calcium from vegetables. While the Chinese consume only half the calcium Westerners do, osteoporosis is uncommon in China'.

At the other end of the scale, the Eskimo population is known to have the highest dietary calcium intake in the world (over 2000 milligrams a day, mainly from fish bones), yet they also have one of the highest rates of osteoporosis in the world.[91]

So what exactly is going on? A recent study reveals all.[92] Scientists from Andrews University, Michigan, used a sophisticated technique called direct photon absorptiometry to compare the bone mass of vegetarians to meat-eaters. They studied a group of 1600 women, and found:

- By the time they reached 80, women who had eaten a vegetarian diet for at least 20 years had only lost an average of 18 per cent of their bone mineral.
- On the other hand, women who did not eat a vegetarian diet had lost an average of 35 per cent of their bone mineral.

Interestingly, there was no statistical difference in the nutrient intakes between the two groups. In other words, the vegetarians' advantage wasn't due to increased calcium intake. So what was the reason, then? Nutritionist Nathan Pritikin explains:

'African Bantu women take in only 350 milligrams of calcium per day. They bear nine children during their lifetime and breastfeed them for two years. They never have calcium deficiency, seldom break a bone, rarely lose a tooth. Their children grow up nice and strong. How can they do that on 350 milligrams of calcium a day when the recommendation is 1200 milligrams? It's very simple. They're on a low-protein diet that doesn't kick the calcium out of the body . . . In our country, those who can afford it are eating 20 per cent of their total calories in protein, which guarantees negative mineral balance, not only of calcium, but of magnesium, zinc and iron. It's all directly related to the amount of protein you eat.'[93]

At last we're getting to the truth of the matter. In fact, the difference in bone loss between vegetarians and meat-eaters can be explained by several factors:

- First, as Pritikin says, the more protein you consume, the more it 'kicks the calcium out of the body'. Since the 1920s scientists have known that diets that are high in protein cause calcium to be lost through the urine.[94] In one typical study, young men were fed experimental diets with protein contents ranging from 48 grams a day right up to 141 grams a day. It was found that the higher level of protein consumption doubled the urinary excretion of calcium.[95] And a diet that is high in animal proteins in particular increases this effect.[96] Scientists postulate that flesh foods cause an acid load in the body, which must be neutralised by a release of calcium stored in the bones.[97]

• It has recently been found that boron (a trace mineral) helps to prevent calcium loss and subsequent loss of bone mass. It is also thought to help in the manufacture of vitamin D in the body. The first study to look at the nutritional effects of boron in humans took place in 1987.[98] Twelve post-menopausal women were fed a diet very low in boron for 17 weeks, after which they were given a daily supplement of 3 milligrams for a further seven weeks. The addition of boron had a dramatic effect – the women lost 40 per cent less calcium and 30 per cent less magnesium through their urine. The study therefore concluded that boron can reduce bodily losses of elements necessary to maintain bone integrity and to prevent osteoporosis. Nutritionist Forrest Nielsen, director of the US Department of Agriculture's Human Nutrition Research Center, called it 'a remarkable effect'.[99]

What was even more extraordinary was the discovery that boron could double the most active form of oestrogen (oestradiol 17B) in the blood of the women – their oestradiol levels actually equalled those of women on oestrogen replacement therapy. Hormones in our bodies are responsible for continuously balancing the growth of new bone with the reabsorption of old bone. When levels of these hormones fall significantly, as happens in menopause, this balance is lost and bones can become very brittle and break easily. This is why Hormone Replacement Therapy (HRT) is used to combat osteoporosis.

Curtiss Hunt, of the US Human Nutrition Research Center, said he 'suspects the body needs boron to synthesise oestrogen, vitamin D and other steroid hormones. And it may protect these hormones against rapid break-down.' He also suggested that boron could be important in treating many other diseases of 'unknown cause, including some forms of arthritis'. And where can you get boron from? Why, by eating apples, pears, grapes, nuts, leafy vegetables and legumes – in other words, a healthy vegetarian diet (one medium-sized apple contains approximately 1 milligram of boron).[100]

So how can you reduce the risk of suffering from osteoporosis?

• Cut out meat, and don't go overboard on the dairy produce. There are many good sources of calcium (which is important while bones are developing) that don't involve the consumption of dairy foods. Some of them are:

> Leafy green vegetables, such as kale, Swiss chard, broccoli, spinach and sea vegetables
> Molasses
> Mineral waters
> Beans and bean products

Tofu coagulated with calcium
Sesame seeds and paste (tahini)
Pumpkin and sunflower seeds
Carob
Dried figs, currants and apricots
Almond and brazil nuts
Porridge

- Get enough exercise. Especially important are forms of exercise that are weight-bearing, such as walking, dancing, running and many sports. Swimming and chess playing, for instance, are *not* weight-bearing exercises. Proper exercise exerts the muscles around your bones, stimulating them to maintain bone density. Leading a sedentary life will increase the likelihood of osteoporosis developing later in your life.
- Avoid smoking, caffeine and excess alcohol – all these can increase your risk of suffering from osteoporosis.[101] A study of women aged 36–45 found that those who drank two cups of coffee a day suffered a net calcium loss of 22 milligrams daily. The authors concluded that a negative calcium balance of 40 milligrams a day (i.e. about four cups of coffee) was enough to explain the 1–1.5 per cent loss in skeletal mass in post-menopausal women each year.[102]
- Get regular doses of sunlight. Sunlight reacts with a substance in your skin – dehydrocholesterol – to produce vitamin D. This vitamin is essential to the proper absorption of calcium and a deficiency will cause you to lose bone mass. Most people get enough vitamin D just by being outside for part of the day with their face, hands and arms exposed.[103]
- Several commonly used drugs can induce significant calcium loss, particularly aluminium-containing antacids. If these antacids are used for prolonged periods of time, they may produce bone abnormalities by interfering with calcium and phosphorous metabolism, and so contribute to the development of osteoporosis.[104] If you must use an antacid, choose one which does not include aluminium in its ingredients.
- If you still feel you want to take a calcium supplement, consider calcium carbonate, because of its high content of calcium (40 per cent) and low price.

SIDE-STEPPING SALMONELLA

According to information which Britain's Meat and Livestock Commission give to schoolchildren:

> 'Meat itself doesn't cause disease or ill-health – it is only unprofessional or un-hygienic handling and preparation which can bring about a problem.'[105]

This is a scandalous statement, because it neatly shifts the blame for food poisoning from the supplier onto the 'handler' – usually the poor beleaguered cook.

Now if meat is smeared with animal faeces because intestines are commonly ruptured during slaughter; if knives used for cutting meat in slaughterhouses are inadequately sterilised; and if condemned meat, lesions and tumours literally spill out of waste bins – now, who's fault is that?[106] When I interviewed a slaughterhouse vet for the original *Why You Don't Need Meat*, he told me:

> 'We hear a lot about food poisoning cases these days, and in just about every case there's meat or poultry as the root cause. Now who gets the blame when patients in hospital die from it? It's almost always the cook, who's blamed for not cooking the beef long enough or for leaving it out in the open. But that's only partly true, because if the meat wasn't grossly infected with salmonella organisms to start with, there'd be no problem.'[107]

More than 1500 types of salmonella have now been identified, usually isolated from the intestinal tract of humans and animals. Salmonella organisms are responsible for a variety of human diseases, ranging from typhoid fever to food poisoning, but it is salmonella typhimurium which has accounted for most cases of food infection. Recently, a new strain – salmonella enteritidis phage type 4 – has emerged, and is apparently more virulent than others and can cause a systemic infection in chickens, invading the ovary and oviducts, not just the gut. Consequently, the bacteria can end up deep inside the egg.[108]

It is difficult to estimate how many animal carcasses are contaminated with salmonella, although, in the case of chicken, estimates have ranged from 25 per cent to 80 per cent.[109] When you learn that cases of salmonella poisoning rose by 25 per cent in 1992 compared to 1991, it is clear that we are living in the midst of an epidemic.[110] In America, there are approximately 2.5 million cases of salmonella poisoning a year, 500,000 hospitalisations, and 9000 deaths.[111]

Rather than force the industry to clean up its act, the US Agriculture Department has now announced that poultry processors will be allowed to zap chickens, turkeys and game hens with gamma rays, in an attempt to destroy disease-causing bacteria.[112]

The headline stories have become so commonplace that many of us seem to have become thoroughly inured to them. A cow with anthrax is slaughtered and left in the abattoir for 24 hours before being discovered – and even then, the slaughterhouse is allowed to continue killing and processing animals for human consumption.[113] Complacent officials allow pigs from a farm hit by the largest outbreak of anthrax for half a century to continue to go for human consumption.[114] Beef officially stamped as fit for human consumption is later found to be riddled with arthritis and septicaemia.[115] Things have got so bad that schoolchildren visiting farms should now be given an official health warning.[116] And, incredibly, officials recommend that animal 'rejects' from failed genetic engineering experiments should be sold for human consumption, to allow the experimenters to recoup some of their costs (the failure rate is colossal – for every 1000 animals experimented on, 999 are rejected).[117]

But these are merely the public scandals. It is only when you get a glimpse of the ordinarily unseen aspects of the meat business that you really begin to appreciate the enormity of the situation. A few years ago (the meat trade will undoubtedly say the situation has now improved, in which case the onus is on them to prove it), I received an anonymous bundle of photocopied documents in the post. No doubt I committed some heinous crime against the state by merely opening the envelope; if so, I hope you will send me some food in prison. It contained copies of correspondence between the Ministry of Agriculture and a major operator of slaughterhouses. As I read, my jaw literally dropped when I found that:

- Sheep were being killed in full sight of each other.
- Cattle were being shot in the head two or three times before they were stunned, many were shot in the wrong place.
- Pigs were damaged in transit or dead on arrival at the slaughterhouse.
- Electric tongs were used to goad pigs.
- Men's workwear and gloves were encrusted with fat, rarely cleaned.
- Knife sterilisers were contaminated with foul blood, and fat-encrusted.
- No soap, nail brushes or paper towels were in the washroom.
- Kosher carcasses were allowed to drag along the floor.
- The Kosher butcher was not cleaning his hands and arms before inserting them into the animal's chest.
- Pithing rods (stuck into the animal's brain) were unsterilised.
- Green algae was growing on the walls.
- 'Unsatisfactory procedures and a complete lack of regard for sanitation of the product and equipment'.

All these points, and many more besides, were from the official reports of inspection. And in an internal memorandum, one of the employees had written what was the most telling comment of all:

> 'Frankly, I am amazed that we are not already under heavy pressure to change things.'

If you think about it, there is a kind of logic to all this. An industry which treats its raw material – sentient animals – with such contempt and cruelty while they are alive is hardly likely to treat them any better when they're dead. And that means, of course, that consumers suffer too; which gives another meaning to the phrase 'meat is murder'. Apart from not eating the foodstuffs most likely to give you food poisoning, here are some further measures to consider:

- If you live in a household where some people still eat meat, insist that it is kept scrupulously away from vegetarian food. A plate of chicken or beef, for example, on an upper shelf in a refrigerator can easily splash its nasty secretions onto vegetables further down.

* Store raw and cooked foods separately. Never leave leftover canned food in its tin. Buy salads and vegetables that are unprepared and unprocessed – nature's own packaging is usually the best.
* Wash food before you eat it – even if you've grown it yourself. Vegetables and fruit can occasionally harbour bacteria from the soil.
* Never reheat food more than once. Make sure it's not underheated.
* Don't take chances. If your food smells off, throw it away.
* Be certain that frozen food is thoroughly defrosted before cooking.
* Be sure all kitchen towels, sponges, surfaces, food equipment and cutting boards are kept clean. When you're preparing a meal, it's also prudent to wash utensils and worktops between stages – don't allow the same knife or chopping board that is used for meat to be used for cooked food and fresh vegetables.
* Put all rubbish and scraps of food straight into the waste bin – and always keep the lid securely down, so that flies can't get in and germs can't get out.

Some people, of course, will never change their ways, and the vet I spoke to had some interesting advice for them:

'Do you know how easy it is to cross-infect other food in the kitchen? Well, say the housewife slices into an infected chicken, and then uses the same knife to slice some bread. She's just spread the infection. Or she puts the chicken down on a worksurface, and then later places some other food on the same spot where the chicken stood. Again, she's transferred the infection. If she pretended to herself that she was handling a lump of raw sewage in a food environment, and took all the necessary precautions, then that would be just about adequate.'

A nice little image for Sunday lunch!

X, Y, Z . . . FOR THE UNKNOWN

When AIDS was discovered in cows, I admit I had a hard time believing it. There are, after all, just about as many theories about the origin of the human immunodeficiency virus (HIV) as there are about the assassination of President Kennedy. To add one more seemed initially to me to be way past credibility.

Then I thought again. If someone had told me, just a decade ago, that a disease as bizarre and mystifying as BSE would reach epidemic proportions in the cattle population, I would have disbelieved that, too. The plain fact is, there *are* lots of new diseases out there. But they're not *staying* out there.

The evidence shows that many new viral diseases are emerging at this point in our history – over a dozen previously undescribed viral diseases in humans and other animals have been discovered in the last 10 years.[118] These can also be seen in the chart below. There are many reasons why this should be so. The exploding populations of cities in newly industrialised nations create unique breeding conditions for new diseases. Air travel around the globe provides an incredibly rapid and effective vector. And humanity's constant erosion of the world's last remaining wilderness areas may expose viruses which have been undisturbed for millennia – a kind of Gaia's revenge.

'Suddenly, it seems we are besieged with new diseases . . .' writes Edwin D. Kilbourne in a thoughtful article published in the *Journal of the American Medical Association*, 'the acquired immunodeficiency syndrome (AIDS), legionnaires' disease, Lyme disease, and others alien to these pages only a decade ago.'[119] Kilbourne believes that the most frequent cause of 'new' viral infections is 'old' viruses transmitted to us from other species. Viruses whose genetic material is contained in RNA, rather than DNA, are particularly worrying, because they mutate faster and can insert themselves directly into human genes.

'I certainly do think we should set aside resources and recognise that viral evolution is proceeding more rapidly,' Nobel laureate Joshua Lederberg told a meeting of the National Institutes of Health in Washington. 'I'm just saying there is a faint possibility that the world will fall apart tomorrow.'[120]

VIRUSES ON THE MOVE[121]		
VECTOR OR VIRUS	**TARGET**	**CARRIER**
New Viruses		
Canine Parvovirus	Dogs	Fecal material
Rocio Encephalitis	Humans	Mosquitoes
Necrotic Hepatitis	Rabbits	Rabbit blood
Rev-T	Birds	Bird droppings
Marburg Disease	Humans	Human blood,monkeys
Seal Plague	Seal	Unknown
HIV-1 (AIDS)	Humans	Blood and semen
HIV-2 (AIDS)	Humans	Blood and semen
HTLV-1	Humans	Blood and semen
HTLV-2	Humans	Blood and semen
Ebola Virus	Humans	Blood
O'Nyong Nyong	Humans	Mosquitoes
Delta Virus	Humans	Blood and semen
Old Viruses Found in a New Locale or Using a New Vector		
Lassa Fever	Humans	Mice, human blood
Hemorrhagic Fever	Humans	Mice
Seoul Hantaan	Humans	Rats
Dengue Fever	Humans	Mosquitoes
Rift Valley Fever	Humans	Insects
Human Monkey Pox	Humans	Monkeys
Kyasanur Forest Disease	Humans	Mosquitoes
Oropuche Fever	Humans	Mosquitoes

So as soon as I heard the 'Cow AIDS' story, I started to do some checking. And what I found made me feel very uneasy.

There are certainly ominous similarities with the BSE story. A disease called visna was first diagnosed in sheep and goats in Iceland in 1938–9.[122] Visna is Icelandic for 'shrinkage' or 'wasting', which pretty much describes the symptoms. The disease seemed to take two rather different forms – visna would be characterised by an infection of the brain and spinal cord, but the term maedi ('difficult breathing') would be used if the infected animal was also suffering from pneumonia-type symptoms. Thus, the lentivirus which caused this disease was sometimes known as visna-maedi, or maedi/visna. The symptoms are actually not unlike scrapie, without the persistent scratching.[123]

The discovery of visna virus in cows was only of passing, academic interest to most scientists.[124] However, with the emergence of AIDS as a serious health threat in the 1980s, scientists began to look around for other similar animal viruses, and visna came under greater scrutiny. In 1987, the name 'bovine immunodeficiency-like virus' (BIV) was proposed by researchers, 'to reflect its genetic relationship and biological similarity to HIV'.[125] A year later, a disturbing letter appeared in the pages of the *Journal of the Royal Society of Medicine*.[126] It made several key connections:

• Considerable doubt had already been cast on the 'African Green Monkey' theory for the origin of HIV.
• HIV demonstrates great genetic similarity to visna found in sheep.
• Now that visna virus had been found in cattle (BIV), the worrying possibility emerged that humans could also have been contaminated with BIV.

One way this might have happened is as follows: the manufacture of human vaccines requires the growth of virus in cell cultures using foetal calf serum. Although this serum is screened for 'contaminants', it is entirely possible that not all such infections – such as BIV – would be detected.[127] Consequently, humans might have been injected with BIV. The letter concluded:

'It seems absolutely vital that all vaccines are screened for HIV prior to use and that BVV [BIV] is further investigated as to its relationship to HIV and its possible causal role in progression towards AIDS.'

Subsequently, the *Journal* published a reply to these points, from two government scientists, who concluded:

'As far as we are aware there has been no report of the isolation of bovine visna virus [BIV] from foetal calf serum.'[128]

But this is hardly reassuring in view of the fact that BIV has been demonstrated to be capable of infecting certain types of human cells under laboratory conditions.[129] So how do things stand?

- We know that BIV (bovine immunodeficiency-like virus) is far more widespread than was first thought. Originally, it was presumed to be restricted to just one cow in Louisiana. Now, the US Department of Agriculture has admitted that 10 per cent of all cattle are infected with BIV – 'Cow AIDS'.[130] In a recent study in Mississippi, 50 per cent of cattle examined were found to be infected.[131]
- BIV is infectious. So far, in addition to the human cell culture mentioned above, it has been shown to be capable of infecting goats, sheep and rabbits.[132]
- Officials claim that 'the potential for human infection from BIV is zero'.[133] However, it has been shown that both BIV and HIV are about 35–37 per cent similar genetically (the African green monkey virus and HIV are approximately 40–42 per cent similar).[134] Both viruses have eight genes, and, magnified by an electron microscope, they look remarkably alike.[135]

Obviously, there are many more questions here than answers. What is the relationship between HIV and BIV? Did one give rise to the other, or do they both share a common genetic ancestor, such as visna in sheep? What is the true extent of BIV infection in cattle world-wide? And is it possible that BIV may have contaminated smallpox vaccines administered throughout Africa and parts of the Caribbean between 1967 and 1980, thus giving rise to AIDS?

I have no confidence that we will receive early answers to these questions, or indeed that we will receive answers at all. Today, the censorship of science is a real issue, and there is clear evidence that studies which might lead to public alarm are either not funded or never see the light of day. Recently, the magazine *Newsweek* showed how studies which came to 'unpopular' conclusions were prevented from being published.[136] A study linking childhood leukaemia to fluorescent light was banned, on the grounds that it would cause 'a possible general panic' (even though the science itself was praised as being 'an extraordinary piece of deductive reasoning'). Another study linking chlorinated water to bladder cancer was rejected by three journals because they 'were uneasy about informing people about this problem'. The same fate befell another study linking fluoride in water to bone cancer. Given the already high level of anxiety about AIDS – and the limitless desire of our politicians to appear omniscient and omnipotent – I think it is highly unlikely that the whole truth will ever be allowed to emerge.

Under the circumstances, we should be cautious, and humble. Nature is more than capable of taking our species down a peg or two, and although we may delude ourselves into believing that we now control the very secrets of life itself, a nasty surprise could be just around the next corner. What, for example, would happen if the human immunodeficiency virus suddenly learnt the tricks of airborne transmission? Recent studies suggest that this concern is not without foundation.[137] The outcome might be much the same as the devastation which myxomatosis wrought amongst rabbits.

What it comes down to is this: As long as we choose to ingest animals, we must also be prepared to ingest their diseases, and to cope with the consequences. To pretend otherwise is to indulge in pure self-deception, of the most dangerous kind.

THE TRANSFORMATION

Of course, there is another way, a better way.

As the years go by, the consumption of animal flesh will increasingly be perceived as a barbaric relic from our distant past, much in the same way as public lynchings and slavery are viewed today. Instead of trying to win a losing battle with the consequences of a meat-based diet, science and technology will provide us with healthier, more logical foodstuffs (some are already on sale today).

The revolution is already under way, in fact – but it is not really a revolution, because there are no ringleaders; no demagogues; no politicians; and no bloodshed. It is a transformation of the heart – the triumph of love over indifference; light over darkness; life over death. Mindless revolution replaced by mindful evolution.

Will you oppose it, or be part of it?

NOTES

1 *Meat Trades Journal*, 15 May 1986.

2 *Meat in the News*, Meat and Livestock Commission, Feb 1992.

3 Kjeldsen-Kragh, J. et al, 'Controlled trial of fasting and one-year vegetarian diet in rheumatoid arthritis', *The Lancet*, 12 Oct 1991, 338 (8772), pages 899–902.

4 Skoldstam, L., 'Fasting and vegan diet in rheumatoid arthritis', *Scandinavian Journal of Rheumatology*, 1986, 15, pages 219–23.

5 The *Independent*, 7 Jan 1992.

6 Mortality Statistics: Cause (1978), HMSO, 1980.

7 National Advisory Committee on Nutrition Education, 'Proposals for nutritional guidelines for health education in Britain', The Health Education Council, Sep 1983.

8 Donaldson, A.N., 'The relation of protein foods to hypertension', *Californian and Western Medicine*, 1926, 24, page 328.

9 Armstrong, B. et al, 'Blood pressure in Seventh-day Adventist vegetarians', *American Journal of Epidemiology*, 1977, 105 (5), pages 444–9.

10 Haines, A.P. et al, 'Haemostatic variables in vegetarians and non-vegetarians', *Thrombosis Research*, 1980, 19, pages 139–48.

11 Sacks, F.M. et al, 'Plasma lipids and lipoproteins in vegetarians and controls', *New England Journal of Medicine*, 1975, 292, pages 1148–51.

12 Anderson, J.W., 'Plant fiber and blood pressure', *Annals of Internal Medicine*, 1983, 98, pages 842–6.

13 Margetts, B.M. et al, 'Vegetarian diet in mild hypertension: a randomised controlled trial', *British Medical Journal* (Clinical Research Edition), 6 Dec 1986.

14 Margetts, B.M. et al, 'A randomized control trial of a vegetarian diet in the treatment of mild hypertension', *Clinical and Experimental Pharmacology and Physiology*, 1985, 12, pages 263–6.

15 Lindahl, O. et al, 'A vegan regime with reduced medication in the treatment of hypertension', *British Journal of Nutrition*, 1984, 52, pages 11–20

16 Beilin, L.J., 'Vegetarian approach to hypertension', *Canadian Journal of Physiology and Pharmacology*, Jun 1986, 64 (6), pages 852–5.

17 *Meat Trades Journal*, 13 Jun 1991.

18 *The Lancet*, 24 Aug 1990.

19 *Facts on File*, 10 Jun 1988.

20 Higson, J and Muir, C.S., 'Environmental Carcinogenesis', *Journal of the National Cancer Institute*, 1979, 63, pages 1291–8.

21 Tyler, A., 'Political Cripples', *New Statesman and Society*, 12 Jun 1992.

22 Letter, *New Statesman and Society*, 26 Jun 1992.

23 The *Independent*, 17 Apr 1992.

24 Personal communication, Mar 1992.

25 Doll, R. and Peto, R., *The Causes of Cancer*, Oxford Medical Publications, 1981.

26 Gregor, O. et al, 'Gastrointestinal cancer and nutrition', *Gut*, 10 (12), pages 1031–4.

27 Wynder, E.L. and Takao Shigematsu, 'Environmental factors of cancer of the colon and rectum',*Cancer*, 1967, 20 (9), page 1528.

28 Phillips, R.L., 'Role of Life-style and dietary habits in risk of cancer among Seventh-day Adventists', *Cancer Research*, 1975, 35, pages 3513–22.

29 Ibid.

30 Phillips, R.L. and Snowdon, D.A., 'Association of meat and coffee use with cancers of the large bowel, breast and prostate among Seventh-day Adventists: preliminary results', *Cancer Research*, 1983, 43, pages 2403–8.

31 Howell, M.A., 'Diet as an etiological factor in the development of cancers of the colon and rectum', *Journal of Chronic Diseases*, 1975, 28, pages 67–80.

32 Palgi, A., 'Association between dietary changes and mortality rates: Israel 1949–1977; a trend-free regression model', *American Journal of Clinical Nutrition*, 1981, 34, pages 1569–83.

33 Lubin, J.H. et al, 'Dietary factors and breast cancer risk', *International Journal of Cancer*, 1981, 28, pages 685–9.

34 Kolonel, L.N. et al, 'Nutrient intakes in relation to cancer incidence in Hawaii', *British Journal of Cancer*, 1981, 44, page 332.

35 Correa P., 'Epidemiological correlations between diet and cancer frequency', *Cancer Research*, 41, pages 3685–90.

36 Willett, W.C. et al, 'Relation of meat, fat, and fiber intake to the risk of colon cancer in a prospective study among women', *New England Journal of Medicine*, 13 Dec 1990, 323 (24), pages 1664–72.

37 The *Independent*, 14 Dec 1990.

38 Manousos, O. et al, 'Diet and colo-rectal cancer: A case-control study in Greece', *International Journal of Cancer*, 1983, 32, pages 1–5.

39 Lijinsky, W. and Shubik, P., 'Benzo(a)pyrene and other polynuclear hydrocarbons in charcoal-broiled meat', *Science*, 145, pages 53–5.

40 UPI, 10 May 1986.

41 *Accumulation of Nitrate*, Committee on Nitrate Accumulation, National Academy of Sciences, 1972.

42 PCRM, Mar 1992.

43 Korpela, J.T. et al, 'Fecal free and conjugated bile acids and neutral sterols in vegetarians, omnivores, and patients with colorectal cancer', *Scandinavian Journal of Gastroenterology*, Apr 1988, 23 (3), pages 277–83.

44 Malter, M. et al, 'Natural killer cells, vitamins, and other blood components of vegetarian and omnivorous men', *Nutrition and Cancer*, 1989, 12 (3), pages 271–8.

45 van Kaam, A.H. et al, 'Polychloorbifenylen (PCBs) in moedermelk, vetweefsel, plasma en navelstrengbloed; gehalten en correlaties', *Nederlands Tijdschrift Voor Geneeskunde*, 3 Aug 1991, 135 (31), pages 1399–403.

46 Hall, R.H., 'A new threat to public health: organochlorines and food', *Nutrition and Health*, 1992, 8 (1), pages 33–43.

47 'Can vitamins help prevent cancer?', *Consumer Reports*, May 1983, 48 (5), pages 243–5.

48 Troll, W., 'Prevention of cancer by agents that suppress oxygen radical formation', *Free Radical Research Communications*, 1991, 12–13, part 2, pages 751–7.

49 See the excellent *No Immediate Danger* by Dr. Rosalie Bertell, The Women's Press, 1985.

50 'Beyond Beef' Campaign, 1992.

51 The *Guardian*, 10 Sep 1985.

52 Snowdon, D.A. and Phillips, R.L., 'Does a vegetarian diet reduce the occurrence of diabetes?', *American Journal of Public Health*, 1985, 75, pages 507–12.

53 Krieb, R. and Beebe, C., 'Food choices can affect your risks for diabetes complications', *Diabetes in the News*, Sept–Oct, 1989, 6 (5), page 12 (4).

54 Diabetes Epidemiology Research International, 'Preventing insulin dependent diabetes mellitus: the environmental challenge', British Medical Journal (Clinical Research Edition), 22 Aug 1987, 295 (6596), pages 479–81.

55 Scott, F.W., 'Cow milk and insulin-dependent diabetes mellitus: is there a relationship?', *American Journal of Clinical Nutrition*, Mar 1990, 51 (3), page 489(3).

56 Anderson, J.W., 'Plant fiber and blood pressure', *Annals of Internal Medicine*, 1983, 98, pages 842–6.

57 Burkitt, D.P. et al, 'Effect of dietary fibre on stools and transit times, and its role in the causation of disease', *The Lancet*, 30 Dec 1972.

58 National Advisory Committee on Nutrition Education, 'Proposals for nutritional guidelines for health education in Britain', The Health Education Council, Sep 1983.

59 Pixley, F. et al, 'Effect of vegetarianism on development of gallstones in women', *British Medical Journal*, 6 July 1985, 291, pages 11–12.

60 *Meat Trades Journal*, 9 Jan 1987.

61 Pronczuk, A. et al, 'Vegetarians have higher plasma alpha-tocopherol relative to cholesterol than do non-vegetarians', *Journal of the American College of Nutrition*, Feb 1992, 11 (1), pages 50–5.

62 Arntzenius, A.C. et al, 'Diet, lipoproteins, and the progression of coronary atherosclerosis. The Leiden Intervention Trial', *New England Journal of Medicine*, 28 Mar 1985, 312 (13), pages 805–11.

63 *Time Magazine*, 29 Oct 1990.

64 Ornish, D. et al, 'Can lifestyle changes reverse coronary heart disease? The Lifestyle Heart Trial', *The Lancet*, 21 Jul 1990, 336 (8708), pages 129–33.

65 *Time Magazine*, 29 Oct 1990.

66 Associated Press, 13 Nov 1990.

67 Roberts, W.C., 'We think we are one, we act as if we are one, but we are not one', *American Journal of Cardiology*, 1 Oct 1990, 66 (10), page 896.

68 *Academic American Encyclopedia*, 1992.

69 Thrash, A.M. and Thrash, C.L., *The Animal Connection*, Yuchi Pines Institute, 1983.

70 Gardner, M.B., 'Viruses as environmental carcinogens: an agricultural perspective', *Basic Life Science*, 1982, 21, pages 171–88.

71 Johnson, E.S. et al, 'Cancer mortality among white males in the meat industry', *Journal of Occupational Medicine*, Jan 1986, 28 (1), pages 23–32.

72 Diglio, C.A. and Ferrer, J.F., 'Induction of syncytia by the bovine C-type leukaemia virus', *Cancer Research*, 36, pages 1056–67. Cited in *Veterinary Research Communications*, Dec 1981, 5 (2), pages 117–26.

73 Rifkin, J., *Beyond Beef*, Dutton, 1992.

74 *New Scientist*, 7 Jan 1989.

75 McClure, H.M. et al, 'Erythroleukaemia in two infant chimpanzees fed milk from cows naturally infected with the bovine C-type virus', *Cancer Research*, 34, pages 2745–57.

76 Lemon, H.M., 'Food-born viruses and malignant hemopoietic diseases', *Bacteriological Review*, 28, pages 490–2.

77 Aleksandrowicz, J., 'Leukaemia in humans and animals in the light of epidemiological studies with reference to problems of its prevention', *Acta Medica Polona*, 9, pages 217–30.

78 Muszynski, B., 'Wplyw biaLaczek bydLa na powstawanie chorob nowotworowych u ludzi na podstawie materiaLow zebranych w powiecie zlotowskim', *Przegl Lek*, 25 (9), pages 660–1.

79 Linos, A. et al, 'Leukaemia in Olmsted County, Minnesota, 1965–1974', *Mayo Clinic Proceedings*, 53, pages 714–18.

80 Donham, K.J. et al, 'Epidemiologic relationships of the bovine population and human leukaemia in Iowa', *American Journal of Epidemiology*, 1980, 112, pages 80–92.

81 Blair, A. and Hayes, H.M., 'Cancer and other causes of death among US Veterinarians 1966–1977', *International Journal of Cancer*, 25, pages 181–5.

82 Lenfant-Pejovic, M.H. et al, 'Risk factors for male breast cancer: a Franco-Swiss case-control study', *International Journal of Cancer*, 15 Apr 1990, 45 (4), pages 661–5.

83 Magnani, C. et al, 'Risk factors for soft tissue sarcomas in childhood: a case-control study', *Tumori*, 31 Aug 1989, 75 (4), pages 396–400.

84 *Vibrant Life*, May-Jun 1992, 8 (3), page 20 (3).

85 Newsletter of the National Osteoporosis Society, No. 1.

86 Quoted in *Bestways*, Feb 1990, 18 (2), page 26 (7).

87 'Osteoporosis: bone up on the facts', *Better Nutrition*, 1990.

88 Ibid.

89 Data for 2 per cent fat milk, from US Department of Agriculture Handbook No.8.

90 Data for Cheddar cheese from US Department of Agriculture Handbook No.8.

91 Mazess, R.B. and Mather, W., 'Bone mineral content of North Alaskan Eskimos', *American Journal of Clinical Nutrition*, Sep 1974, 27 (9), pages 916–25.

92 Marsh, A.G. et al, 'Vegetarian lifestyle and bone mineral density', *American Journal of Clinical Nutrition*, Sep 1988, 48 (3 Suppl), pages 837–41.

93 *Vegetarian Times*, 43, page 22.

94 Hegsted, M. and Schuette, S.A., 'Urinary calcium and calcium balance in young men as affected by level of protein and phosphorus intake', *Journal of Nutrition*, 1981, 111, pages 553–62.

95 Johnson, N.E. et al, 'Effect of level of protein intake on urinary and fecal calcium and calcium retention of young adult males', *Journal of Nutrition*, 1970, 100, page 1425.

96 Breslau, N.A. et al, 'Relationship of animal protein-rich diet to kidney stone formation and calcium metabolism', *Journal of Clinical Endocrinology and Metabolism*, Jan 1988, 66 (1), pages 140–6.

97 Scharffenberg, J.A., *Problems with Meat*, Woodbridge Press Publishing Company, 1979.

98 Nielsen, F.H. et al, 'Effect of dietary boron on mineral, estrogen, and testosterone metabolism in post-menopausal women', *FASEB Journal*, Nov 1987, 1 (5), pages 394–7.

99 UPI, 4 Nov 1987.

100 *Bestways*, Mar 1990, 18 (3), page 14 (3).

101 Lufkin, E.G. et al, 'Estrogen replacement therapy for the prevention of osteoporosis', *American Family Physician*, Sep 1989, 40 (3), page 205 (7).

102 Heaney, R.P. and Recker, R.R., 'Effects of nitrogen phosphorus and caffeine on calcium balance in women', *Journal of Laboratory and Clinical Medicine*, Jan 1982, 99 (1), pages 46–55.

103 'What everyone needs to know about osteoporosis', The National Osteoporosis Society.

104 White, J.E., 'Osteoporosis: strategies for prevention', *Nurse Practitioner*, 1986, 11 (9), pages 36–46, 50.

105 *Meat Messenger*, 12, Meat and Livestock Commission, 1992.

106 The *Guardian*, 8 Mar 1989.

107 Cox, P., *Why You Don't Need Meat*, Thorsons, 1986.

108 Mason, D. and Vines, G., 'Eggs and the fragile food chain', *New Scientist*, 17 Dec 1988.

109 *New Scientist*, 17 Sep 1987; Erlichman, J., *Gluttons for Punishment*, Penguin Books, 1986.

110 The *Independent*, 31 Jul 1992.

111 *The Top 10 Most Censored Stories of 1989*, Project Censored, Sonoma State University.

112 Associated Press, 18 Sep 1992.

113 *Meat Trades Journal*, 13 Feb 1986.

114 The *Observer*, 23 Jul 1989.

115 The *Sunday Correspondent*, 4 Nov 1990.

116 The *Sunday Telegraph*, 16 Jul 1989.

117 The *Independent*, 12 Jun 1991.

118 Miller, J.A., 'Diseases for our future: global ecology and emerging issues', *BioScience*, Sep 1989, 39 (8), page 509 (9).

119 Kilbourne, E.D., 'New viral diseases: a real and potential problem without boundaries', *Journal of the American Medical Association*, 4 Jul 1990, 264 (1), page 68 (3).

120 *Newsday*, 30 May 1989.

121 Ibid.

122 *Black's Veterinary Dictionary*, A & C Black, 1988.

123 *Veterinary Medicine*, 7th Edition, Baillière Tindall, 1989.

124 *Wall Street Journal*, 31 May 1991.

125 Gonda, M.A. et al, 'Characterization and molecular cloning of a bovine lentivirus related to human immunodeficiency virus', *Nature*, 26 Nov–2 Dec 1987, 330 (6146), pages 388–91.

126 Grote, J., 'Bovine visna virus and the origin of HIV', *Journal of the Royal Society of Medicine*, 1989, 82 (6), page 380.

127 Benz, E.W. and Moses, H.L., 'Small virus-like particles detected in bovine sera by electron microscopy', *Journal of the National Cancer Institute*, 1974, 52, page 1931.

128 Lucas, M.H. and Roberts, D.H., 'Bovine visna virus and the origin of HIV', *Journal of the Royal Society of Medicine*, May 1989, 82, page 317.

129 Georgiades, J.A., 'Infection of human cell cultures with bovine visna virus', *Journal of General Virology*, 1978, 38, pages 375–81.

130 *Foundation on Economic Trends*, 3 Jun 1991.

131 Thomas, S. et al, 'BIV and BLV infection of Mississippi dairy cattle: a seroepidemiological survey', *AIDS Weekly*, 22 Jun 1992, page 15 (1).

132 van der Maaten, M. J. and Whetstone, C. A., 'Infection of rabbits with bovine immunodeficiency-like virus', *Veterinary Microbiology*, 1992, 30 (2–3), pages 125–35.

133 *New York Times*, 1 Jun 1991.
134 UPI, 19 Sep 1987.
135 *San Francisco Chronicle*, 1 Jun 1991.
136 *Newsweek*, 14 Sep 1992.
137 Kilbourne, E.D., op. cit; Lederberg J., 'Pandemic as a natural evolutionary phenomenon', *Social Research*, 1988, 55, pages 344–59.

CHAPTER FIVE
GREENER CUISINE!

'What can I eat if I don't eat meat?' That's the most daunting question of all for new or would-be vegetarians, and I'm going to answer it for you in this chapter. We'll look at the way you should structure a healthy vegetarian diet, and I'll give you some of my personal favourite recipes to try. But first, let's spend a moment examining how our food preferences are actually created.

One of life's biggest deceptions is to make us believe that we choose the food we eat for ourselves. Nothing could be further from the truth. You may think that your taste preferences reflect your own likes and dislikes, but in all probability they owe more to your parents than they really do to you. As human infants, we are more dependent upon our parents, and more vulnerable over a far greater period of time, than any other species on the face of the Earth. It takes us years to achieve the same degree of control over even the most basic of our activities that the young of other species manage to achieve in a matter of months or even weeks.

Year after year, we rely upon adults to make most of our simple everyday decisions for us. Now, biologically speaking, this works out very well because we have a lot more growing to do than most other species do, and while we're doing all this growing we need the protection that parents can give us.

But parents give you more than just protection. They pass on to you their own values and beliefs, to the extent that by the time you're old enough to make informed decisions for yourself, you've acquired a whole set of inherited likes, dislikes and habits that, by sheer force of repetition, you've grown to regard as your own. Habits such as what you eat, the way you eat and, indeed, what you think about what you eat. All these habits have been largely predetermined for you. Other people have made these decisions, because you weren't able to at the time. But you are able to now.

MEAT HOOKED?

The difficult thing with meat is that, just like tobacco and some other drugs, although you may not enjoy it at the beginning, your taste buds get hooked on its fatty, salty flavour quite quickly. Humans are not the only animals to respond like this. Gorillas are naturally gentle vegetarians, but when captive ones in zoos have been forcibly fed a meat diet they, too, will develop carnivorous appetites – the more meat they eat, the more they must have. These behavioural changes are also accompanied by physical changes in their digestive system whereby the ciliate protozoa (useful micro-organisms we all need) in their intestines, which would normally help to digest the fibre in their natural diet, disappear, so returning to plant food isn't very easy for them.

It isn't so strange, then, that when young humans are fed animal flesh, they also become accustomed to the taste of it, and grow up believing that large quantities of flesh are an indispensable part of their diet. However, what has really happened is that we have been 'taught' to eat meat, taught to regard its taste as palatable, and taught to consider it (if, indeed, we think about it at all) as a perfectly normal part of our diet. Many young children instinctively resist eating meat – I did, and perhaps you did, too – but by the time you were old enough to think objectively about the issue, you may already have been hooked.

The only way to break into this cycle is to do precisely what you're doing now – to examine the evidence, and to take an informed decision based on your own personal feelings. I seriously believe that this may be the most important decision that you've ever taken. It may be the first time that you've ever had the chance consciously and rationally to reclaim control of a crucial area of your daily activity, that has, until now, been pre-programmed by a pattern of behaviour that someone else decided upon decades ago. It's a great opportunity to get things right!

MAKING THE BREAK

So what happens? Do you come home at six o'clock one Friday evening and have a nut cutlet instead of a lamb cutlet? Do you have to sign a pledge that meat will never pass your lips again? Or do you just do it in private, with consenting adults? Here are some ideas for making the break that I know have worked very well for other people. But do remember that fundamentally it's *your* decision – you're trying to find what genuinely suits you. So you should take everything that follows as suggestions, not as firm rules.

Method One : The Cut-Back

This may be the most useful method to try if you want to kick the meat habit, but are worried that you wouldn't know where to start. Basically, you should aim to reduce your meat intake by about 50 per cent per week. For example, if you used to eat meat at two meals a day, cut this back during the first week to just one meal a day. Then, for the second week, cut it back again to one meal every other day. The third week you probably won't eat more than a couple of meat meals in total. And the fourth week you'll be free. This gradual process of cutting back gives you the chance to spread the transition over several weeks, and so allows you to experiment with lots of new meat-free recipes. It also creates some thinking space in your life – something which is in very short supply for most of us these days. When you're eating meat, just think about what you're *really* eating – where it came from, what it really tastes like. When I was writing *Linda McCartney's Home Cooking*, I spent much time discussing these things with Paul and Linda, and something Paul said stuck in my mind:

'If you really want to know what meat actually tastes like, just try biting the inside of your cheek. That's all it is, really. The thing that gives meat its taste is the way it's cooked, and the sauce that's used with it. There's no reason why you can't cook vegetarian food just the same way, and use the same sauces.'

Many Eastern religions and cultures are fundamentally different to us in the way they encourage people to develop an awareness of the food they eat. In the West, we barely have time to bite the food as we shovel it down, before we bulldoze the next mouthful in (our eating implements are particularly well designed to ensure maximum intake in the minimum possible time). Once you start to become aware of what sort of food you're really putting inside your body, you may be rather shocked. For example, try chewing your food rather longer, leaving it in your mouth a few more seconds, to experience the full range of tastes it offers. You'll find that a mouthful of dead flesh is really rather repugnant to hold there for more than a second or so. And you'll find that much of the processed food now sold to us tastes truly awful after a few seconds. Chemicals in the food give us an initial 'hit' (producing maximum salivation, thus deceiving our senses into thinking that the food is appetising), and then quickly dissolve into a bitter, synthetic disintegration of unwholesome tastes.

When you allow your mind and body to guide you in your food choices, then you will, quite naturally, stop eating flesh.

Method Two: The Green-Out

This method creates one tightly defined point in your life when you take charge of your food intake; it represents a kind of personal watershed.

What you do is to eat a completely raw diet for 7–10 days. Nothing – absolutely nothing – that has been cooked, processed or preserved is allowed down! Meat, of course, is automatically excluded. Similarly, bread, biscuits, jams, butter, tea, coffee, alcohol and canned food are all absolutely out. However, the diet *must* contain lots of fresh fruit and vegetables – there is no limit to the quantity, eat as much as you can take. Aim to buy only organic vegetables. Don't overlook nuts and seeds, and try making salad dressings using only cold-pressed oils and lemon juice.

You'll find it very difficult to overeat on raw food. Although you'll be taking in lots of vitamins from all this fresh food, it might be a good idea to supplement your diet with a good multi-vitamin and mineral pill as well. During this period all sorts of things may start to happen. You may feel wonderfully elated or (very rarely) rather depressed. Your body may start to feel lighter and younger. On the other hand, you may very well have some kind of 'deferred reaction' – you may get spots and pimples, a bad headache, or diarrhoea (don't be surprised if you do, because your intestinal micro-flora is changing!). Stick with it, because it won't last very long, and you'll feel much better when you've been through it.

After 7–10 days, you can start to add some cooked food. You'll also find that your tastes have started to change. You will almost certainly have developed an appreciation for fresh food, and a healthy desire to eat something fresh at least once a day. And then you won't look back! You will have made the break, given your body a thorough detoxification, and started to set the pattern for a better, healthier life.

Method Three: The Switcheroo

What you do here is to is substitute a non-meat product for meat, on every occasion a recipe calls for meat. It's really easy.

The range of meat replacement products is getting better all the time, and finding an acceptable substitute shouldn't be a problem. You should be able to purchase textured vegetable protein (TVP) from any health food shop, in a wide variety of flavourings and textures. There are also many commercial mixtures, such as Sosmix, VegeBurger, Quorn, etc., that are all 100 per cent non-meat. Most health food stores are full of them, although some may be rather expensive. The cheapest substitute is probably to purchase loose TVP granules (unflavoured), and to experiment with flavourings that please you. Some to try are yeast extract, tamari (a natural kind of soy sauce), or even an ordinary commercial gravy mixture (surprisingly, many of them contain no meat!).

This method can be very useful in situations where the cook is totally unfamiliar with anything other than meat-based cooking. If there's just one vegetarian in the family (a younger son or daughter, for example), then this may be the way to go. What frequently happens is that the rest of the family come to the conclusion that the vegetarian option is actually much better than meat (and cheaper), and very often they'll go vegetarian too. *Linda McCartney's Home Cooking* has plenty of recipes which will be useful here.

THE WELL-NOURISHED VEGETARIAN

The table opposite shows what you should eat if you want to be a very well nourished vegetarian – I hope your mouth waters just looking at it! Devised by Dr. Michael Klaper, one of America's foremost experts on achieving optimum health through pure vegetarian nutrition, this table shows you how easy it is to eat well on a vegetarian diet.[1] Actually, if you look closely, you'll see that it includes *no animal produce whatever* – it's a 'pure vegetarian' or vegan diet. Personally, I prefer a vegan diet, but it took me some years to realise that. If you want to consume some dairy produce in addition to the below, that, of course, is your decision – although please avoid eating high-fat dairy foods, such as hard cheeses.

GROUP	PROVIDES	EXAMPLES	QUANTITY
Whole grains and potatoes	Energy, protein, oils, vitamins, fibre	Brown rice, corn, millet, barley, bulghur, buck-wheat, oats, muesli, bread, pasta, flour, etc.	2–4 servings daily
Legumes	Protein, oils	Green peas, lentils, chickpeas, kidney beans, baked beans, soy products (milk, tofu, tempeh, Textured Vegetable Protein, etc.)	1–2 servings daily
Green and yellow vegetables	Vitamins, minerals, protein	Broccoli, Brussel sprouts, spinach, cabbage, carrots, marrow, sweet potatoes, pumpkins, parsnips, etc.	1–3 servings daily
Nuts and seeds	Protein, oils, calcium, trace minerals	Almonds, pumpkin seeds, walnuts, peanuts, sesame seeds, nut butters, tahini, sunflower seeds, etc.	1–3 servings daily
Fruit	Energy, vitamins, minerals	All kinds	3–6 pieces daily
Vitamin and mineral foods	Trace minerals and vitamin B_{12}	(a) Sea vegetables (b) B_{12}-fortified foods, such as soy milk, TVP, breakfast cereals, soy 'meat' products	1 serving of (a) and (b) 3 times a week

In the next chapter, I'll try to answer some of the most frequent dietary questions that people sometimes have about their new way of living. But now, let's look at some of the social questions.

BREAKING THE NEWS

Sometimes, people worry about how they're going to tell other people that they don't indulge in quasi-cannibalism anymore. Let's look at a few situations.

PARENTS

Depending on your relationship with them (and this depends not only on your real age, but also on how old they *think* you are), you may or may not have problems. As humans get older, they generally become less concerned about freedom, justice and morality and more concerned with mortgages, pensions and prosthodontics. They also tend to lose things more often, forget what they were saying in mid-sentence, and, er, where was I? Oh, yes. What this all means is: they're worried about the practicalities, the details. They've spent a lot of time and money to make you the sort of person you are today (yes, they really do think like that) and they don't want to see their investment starve itself to death. Basically, there seem to be two sorts of problems. One is the 'Oh-My-God-How-Are-You-Going-To-Survive?' reaction, and the other is the 'Oh-My-God-What-Am-I-Going-To-Cook?' reaction. Both can verge on the hysterical, so try to disarm them early on.

If we take the first reaction, which mainly comes from family or friends, as basically being a sign of well-meaning concern, then it shouldn't upset you too much. Maybe they can't imagine what you're going to live on, and they're obviously expecting you to shrivel up and die at any moment. In a word, they're ignorant. So enlighten them! Lend them this book, talk to them about it *ad nauseam*, try to get them interested and involved. Tell them about famous people such as Leonardo Da Vinci, Voltaire, George Bernard Shaw, and all the many others who did pretty well for themselves without eating flesh. If there's a good wholefood restaurant nearby, suggest that you all go out and have a meal together. Don't try to sell them the idea if they're not ready for it, but do try to reassure them, which is all that's really needed.

The second reaction is usually found among mothers who find they've suddenly got to cope with a meat-free menu. Not surprisingly, they feel as if they've been dropped in at the deep end. The best advice here is to discuss the situation with them as early as you can. Tell them that more and more people are finding a better, healthier way of eating, and for a variety of reasons you'd like to try it too. You'll probably have to lead the way by obtaining a few recipe books, and also by doing your fair share of kitchen work. Providing you take it slowly and don't panic them, you'll probably find that they are extremely interested in what you're doing, and may try it as well.

THE SPOUSE

If you're in a relationship, and one of you is going to go vegetarian, the most important thing is to talk it through, together. Don't underestimate the impact this change will have on your lifestyles. Eating is one of the most fundamental of all human activities, and any major change is bound to have considerable repercussions. By talking it through together, and planning it together, you'll ensure that all the consequences of your choice are good ones.

The couples who seem to have most problems at this time are the ones who don't normally share other things together – where the wife always does the cooking, for example, and the husband always does the eating. Conversely, the couples who share the food preparation are invariably the ones who get the most pleasure out of it, and there's at least as much pleasure in making food as in eating it. It's easy to forget that, with so much instant and fast food around these days. So try rediscovering this special sort of togetherness for yourselves. If that sounds like a tall order, have a go at some of the suggestions that follow. A useful trick to involve males in the kitchen (who may have been badly spoiled by their mothers) is to appeal to their vanity. Certain aspects of cookery are more 'technical' than others, and these can be presented as an intellectual challenge for them. Here are some ideas:

• Give him a book about making soy milk and soybean curd. This is a fascinating process, and will provide him with many happy hours! Hopefully, it may even produce some food!
• Some dishes have an arbitrary cultural association with masculinity (you know, in the same way as 'going down to the pub with his mates' is alleged to be male bonding at its most profound – by the way, did you know that men who drink excessively grow breasts?). Use this image, if you have to, to get him involved in food preparation. For example, try all the many different forms of curry (there are thousands), try barbecued foods, get him to make some bread (it's very physical), or anything with alcohol in, etc.
• Tell him that he should think about opening a restaurant. Again, this seems to cut straight to the quick of the pathetically easy-to-flatter sense of male vanity. Most men seem to have their own distinct ideas about this, and you never know, he might just end up doing it! Even if he doesn't, he'll have learnt something useful in the kitchen, like how to turn the light on.
• Try creating recipes together. You can start by asking him to suggest 'improvements' to standard recipes, and then ask him to show you what he means. It'll set you both thinking!
• When you're feeling reasonably confident, throw a small party at which he can show off his newly acquired skills.

In a way, I'm sorry to have to suggest ruses like these to you, but the fact is they've worked for many other couples. Well, to tell you the truth, there was *one* couple who I could do nothing for, at all. She phoned me up on a radio programme and told me the problem:

'I've gone vegetarian,' she said, 'but my husband's refusing to follow.'

'Have you told him that it's healthier, cheaper, kinder to the environment – that he'll feel fitter, lose weight, look 10 years younger, and will be smouldering with passion all night long?' I asked.

'Oh, yes,' she said, 'he knows all that. The trouble is, he owns a butcher's shop.'

'Then, my dear,' I said, 'you'll just have to leave him.'

'I'm going to,' she said, and I believe she did.

KNOW WHY YOU'RE DOING IT

When they know that you've 'gone green', people will immediately ask you 'Why?' You may get fed up with this, but please think very carefully about this situation – which I guarantee will be repeated hundreds or even thousands of times throughout your life. Vegetarianism is a very attractive, even fashionable, way of living, and most people are keenly interested.

I want to put a little idea into your head, which goes like this:

'Merely to content oneself with personal abstention is to become part of the problem, rather than part of the solution.'[2]

When I read those words a few years ago, they immediately stuck in my mind – I hope they stick in yours, too. What they're really telling us is that time is too short not to seize every chance you get to tell people about this saner way of living.

I can explain this by telling you about one of my many encounters with meat trade spokespeople. This one happened during a debate in a radio studio a few years ago. Sometimes I only half-listen to them, because I know their well-worn arguments so well. And this one was running well and truly to form.

'There is no evidence,' he was saying, '*absolutely* no evidence that meat is in any way bad for you.'

Before I could make a reply, he cut me off.

'Now look here,' he said, affecting an avuncular manner. 'I don't mind you people at all. Not in the slightest. Live and let live, that's what I say. But for heaven's sake, let people eat what they want to. Why don't you live and let live like I do?'

Momentarily, I was stunned, then the irony of the situation struck me, and I burst out laughing. A butcher just told me to live and let live.

And that's precisely what's wrong with those words – 'live and let live'. It's all too easy to use them as an excuse for inaction. And if you're not actively part of the solution, then you're actually part of the problem.

So when someone next asks you why you've chosen to go vegetarian, please tell them honestly. Don't argue, don't be provoked, and don't hesitate to seize the opportunity to spread the word a little bit more.

EATING WITH FRIENDS

If you're invited for a meal at some friends' house, it's probably best to tell them in advance that your food preferences have changed, and so avoid any problems. Usually a phone call something like this is all that it takes: 'I thought I'd give you a call just to let you know in advance that I've (we've) given up eating meat. I hope it won't be a problem for you?' Usually, your friends will thank you for being so thoughtful and letting them know. Just occasionally, they will be stumped for an answer, in which case you have various possibilities. You could offer to drop by beforehand for a chat, with a few recipe books, which could be another enjoyable social occasion in its own right. Or, if you're feeling brave, you could offer to cook something yourself, and bring it for everyone to try (be warned – cook enough, or you won't get any yourself!). Whatever you and your friends decide, it's likely to be the conversational and culinary centrepiece of the evening, and will almost certainly make you the party's expert, who everyone will want to talk to!

EATING OUT

Most restaurants offer a selection of meat-free meals, and many offer vegan meals, too. Most restaurateurs know a good thing when they see one, and a meat-less meal is actually more profitable for them to prepare and serve than all the fuss and wastage involved in cooking meat. So more and more restaurants are quickly realising that what's good for their customers is also good for their bank account. Most Indian, Chinese, Italian, Mexican, Greek, Jewish, Middle Eastern and, of course, health food restaurants will prove particularly easy to eat in. If you don't see anything you fancy on the menu, speak to the owner or the head chef, who in my experience will be only too delighted to try and expand his culinary repertoire. You may even get to try some unique ethnic dish that is usually reserved for 'the regulars' (that's how I first got to try tofu in a Chinese restaurant).

Speaking of food, would you like to try some?

WELCOME TO COX'S KITCHEN

I thought long and hard about what sort of recipes you'd like to see here, and came to the conclusion that the best thing would be to show you how my own family eats most of the time. All these recipes are very 'user-friendly', and will tolerate a wide margin of error. Although most recipe books still seem to think that we all sit down to three full meals every day, very few of us actually live like that any more. Most of us 'graze' – we feed when we can, not necessarily always eating together. These recipes are ideal for grazing – they'll keep for a day or more, and some of them actually improve their flavour if kept longer (the marinades, for example). And, of course, they're all terribly healthy. Most of them are low in fat, yet still provide important amounts of protein and other nutrients.[3] So – enjoy!

175

EASY FIVE-BEAN SALAD

Most of us still think of salads as anaemic little concoctions, a soggy slice of tomato and a limp lettuce leaf. Well, meet Mega-Salad – it's back, and this time it's personal. You don't want to get on the wrong side of this little beauty, I'm telling you. Although it's pretty energy-dense, as you can see from the per-portion analysis, it's remarkably low in fat – under 30 per cent of its calories come from fat (if you're counting calories, just eat a smaller portion). It keeps well in the fridge and is an excellent snack, lunch or main meal accompaniment, and a great favourite with children. Serve it with dry toast or toasted pitta breads.

Serves	8
Calories per serving	530
Protein (g)	16
Total fat (g)	16
Saturated fat (g)	2
Per cent calories from fat	26

INGREDIENTS

1 × 440 g (1 lb) can red kidney beans, partly drained
1 × 440 g (1 lb) can green beans, drained
1 × 440 g (1 lb) can chickpeas, partly drained
1 × 440 g (1 lb) can butter beans, partly drained
1 × 440 g (1 lb) can borlotti or pinto beans, drained
5 spring onions, thinly sliced
1 eating apple, grated
3 garlic cloves, crushed
2 tsp fresh root ginger, grated
1 tbsp brown sugar
2 tsp freshly ground black pepper
juice of 1 lemon
285 ml (½ pint) cider vinegar
140 ml (5 fl oz) olive oil

METHOD

Measure all the ingredients into a large glass or enamel saucepan and stir well. Place the pan over a medium heat, cover and bring to a slow boil. Reduce the heat and simmer, covered, for 10 minutes.

Leave the pan covered and remove from the heat. Allow to cool, then, if desired, spoon into a separate dish and chill in the refrigerator. This salad improves as it cools and may be kept, chilled, for up to three days.

Allow 30 minutes, plus cooling and chilling time

LENTIL PÂTÉ

If you're used to thinking of pâtés as sheer heart-attack food (most of them are 80 per cent fat, or more), this is going to blow your mind. No, there's no mistake – it really is just 5 per cent calories from fat. This is the basic pâté, delicious and attractive as it stands. You can, however, adjust it to suit your own tastes: make it more spicy; add finely chopped red and green pepper; slice some olives into it or add your own blend of herbs. Over to you!

Serves	8
Calories per serving	110
Protein (g)	8
Total fat (g)	0.5
Saturated fat (g)	0.1
Per cent calories from fat	5

INGREDIENTS
570 ml (1 pint) water
pinch of salt
1 tsp ground turmeric
225 g (8 oz) dried red lentils, washed and drained
50 g (2 oz) porridge oats
50 g (2 oz) rice flakes
1 tsp freshly ground black pepper
1 tsp ground ginger

METHOD
Put the water, salt and turmeric in a large saucepan. Add the lentils and bring to a slow boil over a medium heat. Cover, reduce the heat and simmer for 30 minutes, stirring often.

Add the remaining ingredients, stir well and simmer for another 5–10 minutes. Remove from the heat, stir very well and spoon into a serving dish, pressing down as you fill the dish. Allow the pâté to cool, then chill or serve immediately with bread or crackers.

Allow 45 minutes, plus chilling time

NINE-TO-FIVE STEW

Most of us dream of eating luscious stews but think they're impossible for all but rural folk with an Aga. A hearty stew is warming, wholesome, full of flavour and with an aroma that turns a house into a home. Make this in the morning and go to work knowing you can tuck in as soon as you get home.

Serves	6
Calories per serving	322
Protein (g)	7
Total fat (g)	2
Saturated fat (g)	0.4
Per cent calories from fat	6

INGREDIENTS
455 g (1 lb) swedes, cubed
455 g (1 lb) carrots, chopped
455 g (1 lb) parsnips, cubed
455 g (1 lb) turnips, cubed
455 g (1 lb) potatoes, cubed
1 tbsp whole cloves
455 g (1 lb) very small onions, peeled
12 whole peppercorns
1 × 455 g (1 lb) can chestnut puree
2 litres (4 pints) water

METHOD

Preheat the oven to 140°C/275°F/Gas Mark 1, or a very low temperature of your choice.

Mix the first five ingredients together in a large bowl or pan. Press one whole clove into each end of each onion. Mix the vegetables, onions and peppercorns together and turn into a large ovenproof stew pot or flameproof casserole. Blend the chestnut purée and water together in a jug and pour over the vegetable mixture.

Cover the pot tightly and cook in the oven for 6–8 hours or until the vegetables are tender. Don't lift the lid! At the end of this time, serve the stew hot with lots of fresh bread.

Allow one whole day

PEA AND CORIANDER SOUP

For those who thought pea soup had to have a bone. This soup makes the most of the pea – nutritious, colourful and a perfect companion to the coriander.

Serves	4
Calories per serving	220
Protein (g)	16
Total fat (g)	4
Saturated fat (g)	0.6
Per cent calories from fat	15

INGREDIENTS
1 tbsp olive oil
3 garlic cloves, finely chopped
1 medium onion, finely chopped
225 g (8 oz) dried split green peas
1 litre (2 pints) vegetable stock or water
2 tsp yeast extract
1 bunch fresh coriander, finely chopped
2 tsp freshly ground black pepper

METHOD

Heat the oil in a deep saucepan over a medium heat. Add the garlic and onion and sauté for about 5 minutes or until tender, stirring frequently.

Wash and drain the peas and add them to the sauté with the stock and yeast extract. Stir the mixture well. Cover the pan, increase the heat and bring the soup to the boil.

When the soup begins to foam, reduce the heat and simmer, covered, for about 45 minutes or until the peas are very tender. Stir occasionally and add a little extra water, if necessary.

When the peas are very tender, add the coriander and the black pepper to the soup. Stir well and simmer for a further 5 minutes. Serve hot with fresh bread or croûtons.

Allow 1 hour

POMMES ANNABELLE

This dish has the richness, heady flavours and aroma of its buttery sister but without the animal products or the fat content. Don't skimp on the garlic, though!

Serves	4
Calories per serving	125
Protein (g)	5
Total fat (g)	4
Saturated fat (g)	0.5
Per cent calories from fat	25

INGREDIENTS
900 g (2 lb) potatoes, peeled and thinly sliced
1 whole garlic bulb, finely chopped
2 tsp freshly ground black pepper

FOR THE SAUCE
1 tbsp olive oil
1 tbsp cornflour
350 ml (12 fl oz) vegetable stock
2 tsp yeast extract

METHOD
Preheat the oven to 180°C/350°F/Gas Mark 4. Layer the potatoes, garlic and ground pepper in a casserole dish.

To make the sauce, heat the oil in a saucepan over a medium heat and sprinkle the cornflour over it. Stir well to make a thick paste (*roux*). Gradually add the vegetable stock, stirring well after each addition. Bring slowly to the boil, stirring constantly, until smooth and thickened. Add the yeast extract, stir well and remove from the heat.

Pour sauce over the layered potatoes, cover the casserole and bake in the oven for 45 minutes. Serve hot with steamed green and yellow vegetables.

Allow 1 hour

ROASTED VEGETARIAN LOAF

Apart from being easy, this loaf is versatile: you may use it hot, cold or reheated to meet any meal occasion. It is light, as loaves go, without any of the nuts or seeds that can sometimes make vegetable roasts hard going.

Serves	4
Calories per serving	200
Protein (g)	15
Total fat (g)	5
Saturated fat (g)	0.1
Per cent calories from fat	23

INGREDIENTS
55 g (2 oz) dried red lentils, washed and drained
2 packets No-Salt VegeBurger dry mix (or similar)
55 g (2 oz) porridge oats
1 tsp chilli powder
1 tsp freshly ground black pepper
2 tsp dried parsley
400 ml (14 fl oz) water

FOR THE SAUCE
30 ml (2 tbsp) tomato purée
140 ml (5 fl oz) water
2 tsp mixed dried sweet herbs

METHOD
Lightly oil a 900 g (2 lb) loaf tin and preheat the oven to 170°C/325°F/Gas Mark 3.

Stir the first six ingredients together in a mixing bowl. Add the water, stir well and leave the mixture to sit for 10 minutes.

Meanwhile, to make the sauce, mix the tomato purée, water and herbs together in a small bowl.

Stir the loaf mixture once again, then press firmly into the loaf tin. Bake for 10–15 minutes, then remove from the oven and pour the sauce over the loaf. Cover the loaf with kitchen foil and return to the oven. Bake for a further 40 minutes.

Leave the loaf in the tin on a cooling rack, uncovered, for 5 minutes, then turn it out onto a serving platter. Serve in slices – hot with vegetables, or cold in sandwiches. Or slice the cold loaf, lightly sauté each slice and serve with baked beans and grilled tomatoes.

Allow 1 ¼ hours

SIMPLE VEGETABLE KEBABS

These kebabs are quick and easy to make, exceptionally pretty and can be served to accompany any number of other dishes. They are perfect for a party, picnic or as a late-night supper.

Serves	4
Calories per serving	96
Protein (g)	2.5
Total fat (g)	3.5
Saturated fat (g)	0.5
Per cent calories from fat	28

INGREDIENTS
2 medium onions, quartered
1 medium red pepper, de-seeded and coarsely chopped
1 medium green pepper, de-seeded and coarsely chopped
115 g (4 oz) button mushrooms, cleaned (large ones halved)
2 medium courgettes, quartered lengthways and cut into chunks
115 g (4 oz) pitted olives, drained
1 large orange, peeled, divided into segments and halved

METHOD
Push a selection of the prepared vegetables and fruit onto each of eight long kebab skewers (metal or bamboo). Support the kebabs (e.g. across a baking tray) and place under a hot grill.

Grill the kebabs for 2–3 minutes, then turn or rotate them and grill them for another 2 minutes. Serve immediately (two kebabs per person) with rice or salad and a sauce of your choice (e.g. tomato or peanut sauce).

Allow 25 minutes

VEGETARIAN SHEPHERD'S PIE

Tradition without trauma. A delicious example of how textured vegetable protein (TVP) can be used to make traditional dishes vegetarian. This meal gets finished in one sitting, even when you thought you weren't very hungry.

Serves	4
Calories per serving	288
Protein (g)	16
Total fat (g)	4.8
Saturated fat (g)	1
Per cent calories from fat	15

INGREDIENTS
55 g (2 oz) dried red lentils, rinsed and drained
55 g (2 oz) soya mince (TVP)
2 tsp freshly ground black pepper
1 tbsp mixed dried sweet herbs
1 tbsp yeast extract
710 ml (1¼ pints) water
455 g (1 lb) potatoes, peeled and quartered
1 tbsp olive oil
3-5 garlic cloves, finely chopped
2 medium onions, finely chopped
225 g (8 oz) carrots, sliced
225 g (8 oz) Brussels sprouts, halved
170 ml (6 fl oz) soya milk
1 tbsp margarine

METHOD
Pre-heat the oven to 180°C/350°F/Gas Mark 4 and lightly oil a deep casserole dish.

Stir the lentils, soya mince, pepper and herbs together in a large mixing bowl. Dissolve the yeast extract in the water and pour this over the lentil and soya mixture. Stir well and leave to one side.

Steam the potatoes until tender. Meanwhile, heat the oil in a frying pan and sauté the garlic and onion until tender. Add the carrots and Brussels sprouts and cook for a further 5-7 minutes, stirring frequently. Now stir this sauté into the lentil mixture and turn into a deep casserole dish.

Mash the cooked potatoes with the milk and margarine. Spread the mash over the mixture in the casserole and draw a fork across the top to give it texture.

Cover the dish and bake in the oven for 30 minutes, then remove the cover and bake for a further 10 minutes to brown the top. Serve hot with steamed vegetables.

Allow 1¼ hours

CHILLI CON BEANY

This dish is so tasty you'll wonder what all the *con carne* fuss was about. It is quick enough to seem a spur-of-the-moment winter supper but also benefits from time; it improves in flavour (and spiciness) when it's transported to a picnic or party, for instance.

Serves	4
Calories per serving	225
Protein (g)	15
Total fat (g)	4.5
Saturated fat (g)	0.6
Per cent calories from fat	17

INGREDIENTS
1 tbsp olive oil
5 garlic cloves, finely chopped
1 medium onion, finely chopped
55 g (2 oz) soya mince (TVP)
570 ml (1 pint) water
140 g (5 oz) tomato purée
1 × 400 g (14 oz) can chopped tomatoes
1 tbsp soy sauce
1 tsp chilli powder
1 × 455 g (1 lb) can kidney beans
30 ml (2 tbsp) cider vinegar

METHOD
Heat the oil in a deep saucepan over a high flame. Add the garlic and onion and sauté for 3–5 minutes or until tender. Add the soya mince and stir for a further 2–3 minutes.

Mix the water, tomato purée and chopped tomatoes and stir into the sauté. Reduce the heat and add the soy sauce, chilli powder and beans. Stir well, then cover the pan, reduce the heat and leave to simmer for 20 minutes. Five minutes before serving, add the vinegar and stir once again. Serve in bowls with a plate of corn chips and a side salad.

Allow 40 minutes

GREEK-STYLE SPINACH AND CHICKPEA SAUTÉ

I don't think I could ever love anyone who didn't love chickpeas. This is a variation of a classic Greek dish which, in its simplicity, captures the full flavours and nutritional benefits of its ingredients. For a quicker dish, canned chickpeas and frozen spinach may be used, though you may then need to adjust the seasoning.

Serves	4
Calories per serving	181
Protein (g)	14
Total fat (g)	2.7
Saturated fat (g)	0.3
Per cent calories from fat	12

INGREDIENTS
115 g (4 oz) dried chickpeas
900 g (2 lb) fresh spinach, washed and trimmed
1 tbsp olive oil
5 garlic cloves, finely chopped
2 medium onions, finely chopped
2 tsp freshly ground black pepper
1 tsp ground cumin
3 tsp soy sauce (or Rechard or tandoori paste)
140 ml (5 fl oz) water

METHOD

Wash the chickpeas and soak them in cold water all day or overnight. Drain the chickpeas and put them in a saucepan. Cover with fresh water and cook them over a medium heat until tender, or pressure cook according to your cooker's instructions. Drain well. Coarsely chop the spinach and leave to drain.

Heat the oil in a deep saucepan over a medium heat. Add the garlic and onion and sauté until tender, stirring frequently. Add the pepper and cumin and sauté for a further 3 minutes. Add the cooked chickpeas and cook for a further 5 minutes, stirring often. Mix the soy sauce with the water and add to the sauté.

Place the spinach on top of the chickpea mixture, cover the pan and leave over a medium-low heat for 15 minutes. Do not remove the cover.

At the end of this time, stir the spinach into the chickpeas, cover and continue cooking for a further 5 minutes. Serve immediately by itself, or over rice or noodles.

Allow 35 minutes, plus soaking and cooking chickpeas

HUNGARIAN GOULASH

Another traditionally 'meaty' dish well served by the use of TVP chunks. Don't hurry this dish; it should be very thick and the soya chunks very tender.

Serves	4 – 6
Calories per serving	380
Protein (g)	16
Total fat (g)	5
Saturated fat (g)	1
Per cent calories from fat	12

INGREDIENTS
1 tbsp olive oil
5-7 garlic cloves, finely chopped
1 large onion, finely chopped
125 g (4½ oz) soya chunks
5 medium potatoes, peeled and cubed
2 large carrots, sliced
2 tbsp paprika
1 tsp cayenne
1 × 455 g (1 lb) can chestnut purée
1 litre (2 pints) water
3 tsp soy sauce or yeast extract
1 large green pepper, de-seeded and coarsely chopped
plain, vegan yoghurt (optional)

METHOD

Heat the oil in a large saucepan over a medium heat. Add the garlic and sauté for 2–3 minutes or until it begins to turn golden. Add the onion and sauté for a further 2–3 minutes. Add the soya chunks and stir often as they absorb the oil. Add the potatoes, carrots, paprika, cayenne and chestnut purée. Add the water and soy sauce or yeast extract and stir well. Simmer, covered, over a medium heat for 30–40 minutes, stirring frequently.

Add the green pepper to the goulash and simmer for a further 5 minutes. Serve hot, with a dollop of yoghurt, if desired, over rice, noodles or in a bowl on its own, with slices of fresh bread.

Allow 1 hour

CREAM OF SPINACH SOUP

Look, no cow's milk! An utterly irresistible soup.

Serves	4
Calories per serving	274
Protein (g)	16
Total fat (g)	9
Saturated fat (g)	1.3
Per cent calories from fat	27

INGREDIENTS

455 g (1 lb) fresh spinach
2 tsp yeast extract
60ml (2 fl oz) water
3 garlic cloves, finely chopped
2 medium onions, finely chopped
2 tsp freshly ground black pepper
570 ml (1 pint) vegetable stock or water
1 tbsp olive oil
2 tsp caraway seeds
1 tbsp cornflour
570 ml (1 pint) soya milk
1 tsp ground coriander

METHOD

Wash the spinach and slice it into coarse strips. Leave to drain in a colander until needed.

Dissolve the yeast extract in the water and bring to a rapid simmer in a deep saucepan over a medium heat. Add the garlic and onion and cook for about 2 minutes or until the onions are just tender. Add the black pepper and stock and bring to a slow boil. Add the drained spinach, reduce the heat, cover the pan and leave to simmer gently, stirring occasionally.

Meanwhile, heat the oil in a small saucepan and lightly sauté the caraway seeds over a low heat. Sprinkle over the cornflour and stir thoroughly to make a smooth paste (*roux*). Add the soya milk, a little at a time, stirring well after each addition. Add the coriander and stir until a smooth consistency is reached.

Add the white sauce to the soup and stir together very well. Cook the soup for another 5-10 minutes, stirring often. Serve hot with fresh wholewheat bread rolls.

Allow 45 minutes

GREEN BEANS AND CARROTS IN SPICY TOMATO SAUCE

This is a very colourful and spicy dish which looks and tastes like it takes hours to prepare. Just perfect to serve unexpected guests.

Serves	4
Calories per serving	164
Protein (g)	5
Total fat (g)	5
Saturated fat (g)	1
Per cent calories from fat	26

INGREDIENTS
4 tsp olive oil
1 whole garlic bulb, finely chopped
1 large onion, finely chopped
455 g (1 lb) carrots, peeled and thinly sliced
1 × 400 g (14 oz) can peeled tomatoes
1 tsp dried oregano
1 tsp dried basil
1– 3 tsp freshly ground black pepper, to taste
2 × 400 g (14 oz) cans green beans, drained

METHOD

Divide the oil equally between two saucepans and place both over a medium heat. Add half the chopped garlic to each saucepan and sauté for 2–3 minutes, until the garlic begins to turn slightly golden. Add half the onions to each saucepan and continue to sauté for a further 2–3 minutes, stirring both sautés frequently.

Add the carrots to one saucepan, stir well, cover and reduce the heat. Cook the carrot sauté for about 15 minutes or until the carrots are just tender.

Add the tomatoes with their juice, herbs and pepper to the other saucepan, stir well, cover and reduce the heat. Simmer for 15 minutes, then remove from the heat.

Add the tomato sauce and the beans to the carrot sauté. Stir, cover and simmer gently for a final 5 minutes, stirring occasionally.

Serve hot over rice, pasta, toast or baked potato.

Allow 40 minutes

SCRAMBLED TOFU

This quick dish is wonderfully pretty and makes the best Sunday morning breakfast you could dream of. It can be served as a main course with a selection of vegetables, or on toast as a snack, and is a healthy alternative to omelettes and scrambled eggs.

Serves	2
Calories per serving	300
Protein (g)	17
Total fat (g)	11
Saturated fat (g)	1.2
Per cent calories from fat	32

INGREDIENTS

2 tsp olive oil
1 small onion, finely chopped
1 tsp ground turmeric
1 medium carrot, thinly sliced
1 medium courgette, sliced
55 g (2 oz) mushrooms, quartered
1 × 297 g (10½ oz) carton tofu, drained
1 tsp freshly ground black pepper
1 tbsp finely chopped fresh parsley
4 slices wholewheat toast, dry

METHOD

Heat the oil in a frying pan and sauté the onion until tender. Add the turmeric and stir well to distribute the colour evenly. Add the carrot, courgette and mushrooms and continue the sauté for a further 10 minutes.

Crumble the tofu into the sauté and add the black pepper. Stir over a medium heat for about 5 minutes. Add the chopped parsley and stir for 1 minute. Place the toast slices on two warmed serving plates and spoon over the tofu mixture. Serve immediately with a dash of soy sauce and accompanied by grilled tomatoes. Try this dish over rice or baked potatoes instead of toast, too.

Allow 20 minutes

TEMPEH MARINADE

Tempeh (pronounced 'tem-pay') is a fermented soya bean cake. Like cheese, yoghurt and ginger beer, it is made with a cultured 'starter'. It is highly digestible, smells like fresh mushrooms and tastes remarkably similar to chicken. You can pan-fry it and have it with sauce, dice and deep-fry it like potato chips, add it to a stir-fry, and it is very good roasted. You can buy it from the freezer cabinets of almost all health food shops. It's also a significant source of vitamin B_{12}. This unusual dish builds on the subtle, slightly nutty flavour of tempeh to create a rich, fragrant tempeh morsel. Use it as a protein-rich cornerstone to your meal.

Serves	4
Calories per serving	519
Protein (g)	20
Total fat (g)	17
Saturated fat (g)	2
Per cent calories from fat	29

INGREDIENTS
225 g (8 oz) tempeh, defrosted
170 ml (6 fl oz) cider vinegar
60 ml (2 fl oz) olive oil
60 ml (2 fl oz) soy sauce
1 tsp mustard seed, slightly crushed
12 whole cloves
12 whole peppercorns, partly crushed
6 garlic cloves, finely chopped
2 small onions, finely chopped

TO SERVE
225 g (8 oz) long-grain rice, rinsed and drained
455 g (1 lb) broccoli, trimmed and cut into florets
455 g (1 lb) carrots, sliced
2 medium onions, peeled and quartered
soy sauce (optional)

METHOD
Cut the tempeh into 2.5 cm (1 inch) cubes and place in a casserole dish.

Mix the remaining ingredients together in a jug. Stir well and pour over the tempeh pieces. Cover the casserole and leave the tempeh to marinate for 4–18 hours.

Bake the casserole, covered, in the oven at 170°C/325°F/Gas Mark 3 for 30 minutes. Remove the cover and bake for a further 10 minutes. If a drier, crisper surface texture is desired, remove the cover earlier in the baking time.

Meanwhile, put the rice in a saucepan and cover it with twice its volume of water. Cook over a medium heat until tender. Steam the broccoli and carrots until tender. Place the pieces of onion in a shallow baking tray and grill them under a hot grill for 5–7 minutes, turning once or twice in that time. Spoon the rice onto hot plates. Top with the tempeh and its marinade, the roasted onions and steamed vegetables. Add a dash of soy sauce if desired.

Allow 1 hour, plus marinating time

TOFU MARINADE

Tofu is very high in protein, low in saturated fats and entirely cholesterol-free. It's used in East Asia in the same ways you would use eggs or meat. Tofu plays a tasty supporting role to almost anything in pies, dips, fritters and sauces, complementing the main ingredients. Or you can use tofu as your star – in a tofu burger, tofu salad, tofu fried rice, chilli con tofu. It is also extremely good barbecued. This simple dish is a gourmet delight that turns plain tofu into an exquisite appetiser or side dish.

Serves	2–4
Calories per serving	123
Protein (g)	9
Total fat (g)	4
Saturated fat (g)	1
Per cent calories from fat	23

INGREDIENTS
1 × 297 g (10½ oz) carton tofu, drained and cubed
285 ml (½ pint) cider vinegar
1 tbsp soy sauce
1 tsp whole allspice, roughly crushed
1 tsp peppercorns, roughly crushed
6 whole cloves
1 cardamom pod
1 garlic clove, thinly sliced (optional)
1 thin slice fresh root ginger (optional)

FOR THE CRUDITÉS
1 medium carrot, cut into sticks
2 celery stalks, cut into sticks
½ cucumber, sliced
1 apple, cored and sliced
8 lettuce leaves, washed and drained

METHOD
Arrange the tofu cubes in a deep bowl or dish (a single layer is preferable). Mix the remaining ingredients together in a jug and pour over the tofu. Ideally, the cubes should be covered by the marinade. Cover the dish and leave the tofu to marinate for 6–24 hours, preferably in the fridge. Agitate the dish occasionally during this time if possible.

Carefully remove the tofu cubes from the marinade and arrange on a large plate with the crudités.

You may reuse the marinade once again, immediately, if you wish. (Do not use more than twice.)

Allow 20 minutes, plus marinating time

VEGETABLE MARINADE
Make this on Sunday evening and snack on it until Thursday lunch! This dish is improved by keeping it in the fridge and, provided you stir it each day and serve it with a clean spoon, it will keep in the fridge for 3–4 days, *if* you can let it alone for that long. It's very pretty and easy to transport to school or work.

Serves	7
Calories per serving	215
Protein (g)	9
Total fat (g)	8
Saturated fat (g)	1.2
Per cent calories from fat	31

INGREDIENTS
1 medium cauliflower, cut into florets
455 g (1 lb) broccoli, coarsely chopped
1 × 440 g (1 lb) can red kidney beans, drained
1 × 395 g can sweetcorn, drained
1 bunch spring onions, thinly sliced
5 garlic cloves, finely chopped

FOR THE MARINADE
430 ml (¾ pint) cider vinegar
60 ml (2 fl oz) oil
2 tsp coarsely ground black pepper
1 tsp coarsely ground mustard seed
1 tsp coarsely ground caraway or fennel seed
juice and grated zest of 1 lemon

METHOD

Measure the marinade ingredients into a large enamel or glass saucepan and place over a medium heat. Bring to a gentle simmer. Add the vegetables, stir well and cover the pan. Cook covered, over a medium heat, for 15 minutes, stirring twice in that time.

Remove the pan from the heat and allow the vegetables to cool without removing the cover. Place in the fridge once the marinade has cooled to blood temperature. Serve as a starter, light lunch or accompaniment to other dishes.

Allow 45 minutes, plus cooling and chilling

CHOCOLATE AND ALMOND CAKE

Now for a couple of desserts, if you've got room, that is. This one is a hit wherever it goes. It is light and slightly crumbly and very pretty.

Serves	8
Calories per serving	300
Protein (g)	7
Total fat (g)	12
Saturated fat (g)	1.3
Per cent calories from fat	34

INGREDIENTS
350 g (12 oz) whole wheat flour
55 g (2 oz) oat flakes
115 g (4 oz) dark brown (molasses) sugar
2 tsp baking powder
100 g (3½ oz) flaked almonds
100 g (3½ oz) carob or plain chocolate drops
2 tbsp olive oil
430 ml (¾ pint) water
1 tsp almond essence

METHOD

Preheat the oven to 180°C/350°F/Gas Mark 4 and lightly oil a 15 cm (6 in) round cake tin.

Mix the first four dry ingredients together in a large bowl. Add the almonds and carob or chocolate drops and stir well.

Stir the oil, water and almond essence together in a jug and pour into the centre of the dry mixture. Stir well, then tip the batter immediately into the cake tin.

Bake in the oven for 35–40 minutes, then turn out onto a wire rack and leave to cool.

Allow 50 minutes

FAMILY FRUIT AND NUT CAKE

This is a tasty, substantial cake with lots of texture. It is simple to make; perfect for the kids to throw together. Serve it with soya pudding, soya ice cream or on its own.

Serves	8
Calories per serving	215
Protein (g)	5
Total fat (g)	8
Saturated fat (g)	1
Per cent calories from fat	30

INGREDIENTS
115 g (4 oz) raisins or currants
55 g (2 oz) brown sugar
2 tbsp olive oil
285 ml (½ pint) apple juice
225 g (8 oz) whole wheat flour
55 g (2 oz) porridge oats
2 tsp baking powder
1 tsp ground ginger
55 g (2 oz) citrus peel
55 g (2 oz) flaked almonds

METHOD

Preheat the oven to 190°C/375°F/Gas Mark 5 and lightly oil a 23 cm (9 inch) cake tin.

Put the first four ingredients together in a mixing bowl and stir well. Leave to one side to soak.

Mix the remaining ingredients together in a large mixing bowl and stir well. Stir the raisin mixture into the dry mixture to make a thick batter. Turn the batter into the cake tin and bake for 40 minutes. Let the cake cool in the tin for 10 minutes before turning it out onto a wire rack to cool completely before serving.

Allow 1 hour

NOTES

1 Klaper M., *Pregnancy, children and the vegan diet*, Gentle World, 1987.
2 Regan T., *The Case for Animal Rights*, University of California Press, 1983.
3 Recipes analysed by Nutri-Calc Plus, using USDA data and manufacturers' analyses.

CHAPTER SIX
EVERYTHING YOU REALLY NEED TO KNOW

Whenever I take a workshop or do a phone-in, I can guarantee that some of the following questions will keep on cropping up. So here they are – The Questions That Will Not Die – together with the answers. I hope this selection will help you, your family or your friends to experience an easy transition to the new way of living!

'Does it take longer to prepare vegetarian meals?'

Only if you want it to! The example most often given to illustrate the convenience of a meat-based diet goes something like this: 'I can come home from a long, tiring day, take a chop out of the fridge and slap it under the grill. It's ready to eat by the time the potatoes and peas have cooked – *and* I've slipped into something more comfortable.'

Well, for those who wish to keep this particular schedule, you can prepare exactly the same meal using a vegetable burger or grill instead, and nothing else will change. However, you'd be missing a lot of the fun and flavour of eating without meat. In the same period of time, about 30 minutes, you can make:

- Pasta topped with tomato, vegetable or meat-free bolognaise sauce accompanied by a green salad.
- Stir-fried vegetables over basmati rice with a peanut sauce.
- A robust Chef's salad followed by onion soup.
- Tempeh or vegetable burgers in a sesame bun with the full compliment of garnishes and a side salad.
- Scrambled tofu on toast with grilled tomatoes and mushrooms, followed by fresh fruit salad topped with yoghurt.

The list could go on and on. And if you are willing to spend just 15 minutes longer over your meal you can anticipate luscious home-made pizzas, a variety of vegetable quiches, curries, casseroles and even tantalising Mexican meals, such as tacos or tostadas with bean paste filling and spicy tomato sauce. Of course, most of these can be prepared even quicker using those handy appliances of the modern kitchen, the pressure cooker, microwave and food processor. And quicker still if some of the meals make partial use of canned or frozen foods; the choice is yours. Certainly fresh is better if you can manage it, but speed and convenience are often paramount – thank goodness for handy foods at those times.

Just a few words about beans. Many people suffer from fear of beans on three accounts. First, that a vegetarian diet relies almost entirely on them. Second, that they have drastic and dire effects on the digestive system. And third, that they take hours and hours to cook. On all three counts, 'not true', if you follow a few basic tips.

First, beans are a marvellous source of protein, fibre, iron, the B vitamins and, when sprouted, vitamin C. With such a healthy nutritional profile, who wouldn't be tempted to include them in every meal? It is important to remember, however, that beans can either look like beans or they can be transformed into one of the many delicious and nutritious bean products which supply the food value but not the same beany experience. Among these are soya milk, tofu, TVP (textured vegetable protein), tempeh, soya yoghurt and ice cream, marinated tofu and soya cheeses of every description.

Second, it is true, beans can be a musical vegetable if you don't follow the three golden rules of cooking them:

• Let them soak overnight. Do it last thing before turning in, and you'll be able to use them any time in the next day or so. Isn't this a terrible hassle? Not really, it's the very small price we pay for what is practically the ultimate in convenience foods. What other foodstuff would be perfectly happy to wait around your kitchen for months on end, and still be in great nutritional shape when you eventually decide to use it? Certainly not meat!
• Don't cook beans in the soaking water, and never cook them without rinsing them.
• People often under-cook beans (help-yourself salad bars are particularly bad). Here's the secret: do the tongue test. Put one bean in your mouth, and try to squash it against the roof of your mouth using normal pressure from your tongue. If you can't squash it, it isn't well cooked. Anything less may create gas and digestive discomfort.

Third, cooking beans in an open pot can indeed take hours, which is why I'd suggest you use a pressure cooker. You can cook most beans in a pressure cooker in less than 30 minutes. Buy a stainless steel one.

'My six-year-old won't eat meat. What should I do?'

Celebrate, quickly learn a few new meatless recipes (see below for some cookbooks) and then join him. Most children are naturally vegetarian, especially so if they have realised that the meat on their plates is actually dead animal. Because they empathise so strongly with animals (and we encourage them to do so with our gifts of teddy bears and tales of cute animals with human characteristics), they cannot easily bear the transition from 'friend' to 'food'. By supporting your child in his or her choice you are enabling them to live with one less hypocrisy in their life. You are also giving them a chance to avoid the horrific catalogue of diseases that accompanies a meat-based diet. Children come to us to be with us and learn from us but also to change us. Accept your child's decision gladly for the changes and improvements it will inevitably make to your diet, your health and your whole life.

'What are the best vegetarian cookery books?'

The Farm Vegetarian Cookbook, Louise Hagler (ed.), The Book Publishing Co. This excellent cookbook is full of vegan recipes which use the soya bean and its various products in a wholesome and imaginative way. It describes in simple terms how to make soya milk, tofu, soya yoghurt, ice cream and tempeh, and gives you recipes that are so tasty and aromatic you'll have the whole street knocking on your door.

The Vegan Cookbook, Alan Wakeman and Gordon Baskerville, Faber and Faber. Over 200 recipes that will ensure you never feel you've missed out on anything by adopting the vegan diet. This creative pair have done magnificent things with salads, breads and desserts and have even come up with scrumptious vegan double cream. A must for your shelves.

The Caring Cook, Janet Hunt, The Vegan Society. A tidy little cookbook for the complete beginner with lots of tips to help you set up and run a meat-free kitchen. These recipes are so reliable, you'll probably use it for years.

365 Plus One Vegetarian Main Meals, Janet Hunt, Thorsons. An easy-to-use cookbook with one recipe for each day of the year, arranged by seasons and therefore making the most of in-season fruits and vegetables. Most of the recipes are either vegan already or may be easily converted to vegan by substituting soya products for dairy products. A good book to browse through. Any cookbook by Janet Hunt is certain to be a good one.

Linda McCartney's Home Cooking, Linda McCartney and Peter Cox, Bloomsbury. Over 200 recipes that give classic dishes a new incarnation. Creative use of meat substitutes will enable you to prepare traditional-style meals, and this book has been widely and successfully used by families with one or more members who are not vegetarian. This beautifully illustrated book has a lengthy section dealing with cooking techniques, nutrition and the new non-meat foods, as well as tips on choosing and storing grains, fruits, vegetables, nuts and seeds. A marvellous book which has helped many people make the transition and saved many a dinner party. A beautiful gift. Please buy one for every room in your house.

The Rose Elliot cookbooks, all published by HarperCollins, are thoroughly reliable collections of recipes created by a woman who obviously loves cooking and knows how to prepare food that people love to eat. There are several of her cookbooks to choose from, here are just a few:

Mother and Baby Book gives you practical, tried and tested advice and recipes on how to rear healthy vegetarian kids.

The New Simply Delicious is a good starting point for new vegetarians or new cooks, including a wide selection of recipes for all occasions and all levels of confidence.

The Bean Book introduces about two dozen different beans and then shows you how to make tempting and often very un-beany dishes from them. Ours is very well thumbed.

Gourmet Vegetarian Cooking takes the vegetarian cook a stage further, providing recipes that are just that little bit more unusual, time-consuming or challenging to make. I can't read this without getting hungry.

Rose Elliot's Vegetarian Cookery is a beautifully illustrated and comprehensive collection that will make a wonderful gift for the new or established vegetarian cook. Again, Rose Elliot's name on a cookbook means it's a winner.

'My doctor's told me I must eat meat. What should I do?'

Change your doctor.

'Isn't vegetarian food more expensive?'

No, it's inevitably cheaper. In order to understand why this should be so, you have to grasp the underlying economics of the meat industry. A meat animal is treated as nothing more or less than a machine – a machine that the industry uses to convert vegetable protein into animal protein. As a machine, it is deplorably inefficient. For every kilo of meat protein that is produced as a steak, twenty kilos of vegetable protein have to be put into it (twenty kilos that *could* have gone to feed human mouths). It is a disgraceful, obscene waste of food. Figure 24 (overleaf) shows you just what inefficient 'machines' food animals are.[1] As you can see, beef animals are extraordinarily wasteful converters of vegetable protein, only managing to convert a miserly 6 per cent of it into meat protein. This is the reason behind the desperate use of chemicals (and genetic engineering techniques) in animal rearing – it's an attempt by the farmer to improve on a process that is notoriously inefficient.

So when someone buys a steak, they're actually paying not just for the meat, but also for a vast amount of wasted vegetable food that the cow has consumed and excreted. Which means that consumers are actually paying to create all the

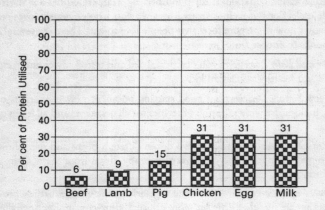

Figure 24: *How much vegetable protein animals convert to meat protein*

pollution problems associated with that excreta, too. For example, here's just one problem that most people don't know about: the global cattle population emits 100 million tonnes of methane gas each year.[2] Concentrations of methane in the air have been rising at 1 per cent per year since 1950, four times the rate of increase of carbon dioxide. Scientists fear that soon methane may be the prime greenhouse gas, responsible for global warming.

But let's return to the faulty economics of meat production, because this is where the story gets very intriguing. Although the meat industry constantly stresses the 'naturalness' of meat, they go to great pains to conceal the fact that they themselves are violating one of the most fundamental laws of nature – the law that explains why large, fierce animals are rare, and why smaller, vegetable-feeding ones are much more numerous.

In the wild, food chains exist whereby one level of the chain consumes something on a lower level of the chain. Close to the bottom of the chain, there are lots and lots of animals feeding on a profusion of plant foods (for example, rabbits feeding on grass). At the top of the chain, there are just a few carnivorous animals who feed on the lower levels (foxes feeding on rabbits). Thus, the foxes are indirectly eating the lowest level of the chain – grass. If things get out of balance, and the fox population suddenly expands, there won't be enough rabbits to go round. So the foxes starve until things get back into balance again.

Therefore, if humans were designed to be predominantly carnivores, there would be relatively few of us, and we would live at a considerable distance from each other. That way, there would be enough flesh to go round. Now, does that sound like the way most of us live today?

Of course it doesn't. But that's just where the meat industry tries to buck nature. The more meat they sell us, the more cattle they have to breed, feed and bleed. There are now 1.3 billion cattle on the face of our planet, consuming its food resources at a truly incredible rate.[3] As biologist Dr. David Hamilton Wright of Emory University observes, 'An alien ecologist observing Earth might conclude that cattle is the dominant animal species in our biosphere.' As long as we continue to eat at such a perilously high position in the food chain, there will be more and more cattle, creating bigger and bigger problems for our ravaged planet. Jeremy Rifkin, President of the Greenhouse Crisis Foundation, shares this view:

'In all of the ongoing public debates around the global environmental crisis a curious silence surrounds the issue of cattle, one of the most destructive environmental threats of the modern era. Cattle grazing is a primary cause of the spreading desertification process that is now enveloping whole continents. Cattle ranching is responsible for the destruction of much of the earth's remaining tropical rain forests. Cattle raising is indirectly responsible for the rapid depletion of fresh water on the planet, with some reservoirs and aquifers now at their lowest levels since the end of the last Ice Age. Cattle are a chief source of organic pollution; cow dung is poisoning the freshwater lakes, rivers, and the streams of the world. Growing herds of cattle are exerting unprecedented pressure on the carrying capacity of natural ecosystems, edging entire species of wildlife to the brink of extinction. Cattle are a growing source of global warming, and their increasing numbers now threaten the very chemical dynamics of the biosphere. Most Americans and Europeans are simply unaware of the devastation wrought by the world's cattle. Now numbering over a billion, these ancient ungulates roam the countryside, trampling the soil, stripping the vegetation bare, laying waste to large tracts of the earth's biomass.'[4]

Unfortunately, most of us don't realise that the meat industry is playing this grotesque game with nature. But it is, and consumers are paying for it. I hope you can now see that the original question raises some extremely profound yet little-known issues.

And now, to answer that question directly: The cheapest way to eat vegetarian food is to buy in-season fruits and vegetables and use them, with a few grains and pulses, to design your menu. This approach can allow you to provide truly hearty meals for two to four persons every day of the week for something like £1 per meal. Of course, they're nothing fancy nor are they five courses, but such meals are wholesome, tasty, nutritious and, of course, cheap.

Next cheapest is to buy the same fruit and veg but to add a few special things from the supermarket (tofu, soya milk, pastas, nuts, VegeBurger mix, etc.) and the 'ethnic' grocers (Asian, Chinese, Jamaican) to liven up your cooking repertoire. A revolution has been taking place over the past few years that has meant foods such as herbs, spices, soya and meat-replacement products are stocked in most of the major supermarket chains, and at reasonable prices.

'If everyone was vegetarian, what would happen to all the farm animals?'

That's a silly question.

'No, I'm serious! What would happen to them?'

I pull my hair out when people ask me this in all seriousness. It reveals such an astounding depth of ignorance about the wretched lives of today's food animals. And implicit in it lies the logic of the lunatic asylum. It suggests:

- We have a duty to eat animals, because if we didn't eat them they wouldn't be born.
- If they weren't born, they wouldn't get a chance to experience life, even though that existence is wretched and painful.
- It is a greater evil to deny the possibility of existence to a potential animal than it is to inflict real suffering on a live animal.

The simple answer is this: If everyone was vegetarian, fewer and fewer food animals would be bred. Eventually, we would have just a few cows, a few pigs, and so on – all valued as beautiful creatures in their own right, rather than as lumps of flesh to be consumed.

'Which kitchen utensils should I buy?'

These are the ones I find most useful:

- Pressure cooker (a must for beans and pulses).
- Cast-iron pans (or enamel, but not aluminium or copper).
- Wok (for really fast stir-frying of vegetables using minimal oil, or with care can even be used without any oil if you watch over it and sprinkle water on when necessary).
- Garlic press (releases the flavour better than chopping).
- A food processor (allows you to make delicious raw salads including tasty vegetables such as beetroot, turnip, swede, carrot, etc., in a flash).
- Steamer (never boil the nutrients in vegetables away again!).
- Mortar and pestle (for grinding spices).

'I've heard that plants scream when you dig them up. Isn't that just as cruel as eating meat?'

No. Unlike animals, plants don't have a central nervous system, and they can't feel pain. This crackpot idea is trotted out by meat-eaters with guilty consciences. They figure that, if they can only convince themselves that chickpeas suffer just as much as chickens, there's really no difference between meat-eating and vegetarianism, so they might just as well carry on eating flesh. Nevertheless, it's a little piece of lunacy that regularly comes up, so I tried to find out where it originated. As far as I can discover, it first arose when that expert self-publicist L. Ron Hubbard (the founder of Scientology) came to England. The following is an extract from Russell Miller's biography of Hubbard.

'It was not long before television and Fleet Street reporters were beating a path to Saint Hill Manor [his chosen home near East Grinstead, Sussex] demanding to interview Hubbard about his novel theories. Always pleased to help the gentle-men of the press, he was memorably photographed looking compassionately at a toma-to jabbed by probes attached to an E-meter – a picture that eventually found its way into *Newsweek* magazine, causing a good deal of harmless merriment at his expense. Alan Whicker, a well-known British television interviewer, did his best to make Hubbard look like a crank, but Hubbard contrived to come across as a rather likeable and confident personality. When Whicker moved in for the kill, sarcasti-cally inquiring if rose pruning should be stopped lest it cause pain and anxiety, Hubbard neatly side-stepped the question and drew a parallel with an essential life-preserving medical operation on a human being. He might have wacky ideas, Whicker discovered, but he was certainly no fool.'⁵

So it seems as if that particularly deranged notion is all part of L. Ron's rich legacy to humankind.

'How can I lose weight on a vegetarian diet?'

In my experience, people who change to the vegetarian way of living automati-cally start to 'normalise' their body weight. And it's true that you see very few overweight vegetarians. So don't bother about trying to lose weight if you've recently gone meat-free, you may find that your body will naturally stabilise at an optimum weight, and it's certainly more difficult to overeat without all those calories from saturated fat in meat.

Meanwhile, give a little thought to the whole question of *why* we overeat. Did you know that humans are the only species on the face of the planet to have a chronic weight problem? No other animal has as much excess body fat as we do, and no other animal so regularly commits suicide by overeating. One very successful evolutionary strategy our bodies have developed is the ability to store food energy in the form of fat all over our bodies. Humans actually have more fat cells (known as adipocytes) in proportion to their body mass than almost *any other creature* – only hedgehogs and whales have a greater proportion of fat in their bodies! Even animals which we traditionally think of as 'fatties', such as pigs, seals, bears and camels, all have less fat than we do!

This evolutionary adaptation is an outstanding success story, which has allowed us to cope with the uncertainty of a highly variable food supply. Today, of course, some of us have access to far more food than we know what to do with. However, our old survival instincts tell us to go on eating, just in case . . .

People who sell us foodstuffs know very well how to exploit these deep-rooted, unconscious instincts of ours. Take, for instance, the routine use of the word 'NEW' on food products of all types. Have you ever wondered why this word is repeated so incessantly? The answer again lies in an old and once-useful behaviour pattern which we and other successful omnivores have evolved. Omnivores are animals which have adapted to eat a highly varied diet. Their successful survival strategy is to continually search out *new* types of food, so that if one staple in the diet fails for any reason, there is always another food-source ready to replace it. A great idea! But, in nature, the urge to experiment with new food is precisely counterbalanced with the desire to be cautious. You can see both of these clever survival mechanisms demonstrated in the behaviour patterns of that other hugely successful omnivore feeder – the common rat.

Rats, just like humans, are always ready to experiment with anything new. When a rat finds something unusual to eat (and, like us, they are on the look-out all the time), he will carefully nibble a small amount, then leave it strictly alone for a day or so. If the animal suffers no ill effects in the meantime, he will return to the new food and start to include it in his diet. However, if the rat feels sick or ill after eating that small sample, he will *never* return to that food again. This is a highly successful feeding strategy, which is still present deep inside all of us. Here's the proof: Think back to a time when you felt sick shortly after eating a particularly distinctive type of food. It doesn't really matter whether your sick feeling was actually *caused* by the food you remember eating – what matters is that you now *associate* that food with the unpleasant feelings afterwards, and in all probability, you will never eat that particular food again. It's in ways like this that our old survival mechanisms try to keep us away from food which may be dangerous for us. However, it's the other side of this mechanism – the strong desire which we all have to try out anything marked 'NEW' – that gets us into so much trouble these days, simply because there's so *much* temptation all around us. It is this desire to constantly look for – and experiment with – new foodstuffs which makes us so vulnerable to all those advertising come-ons.

When you go vegetarian, you're reclaiming control over what you eat – it's an act of personal liberation. For probably the first time in your life, you are consciously deciding what feeding strategy you will adopt. Now this, all by itself, gives you power, and takes away power from those food industries which seek to limit your food choices, and exploit our instinctive feeding strategies for their own purposes. Again, this makes it far less likely that you will continue to be overweight.

Now, here are two ridiculously simple but highly effective ways of losing weight. First, try a mainly raw food diet. It's almost impossible to overeat on raw food. Secondly, don't eat after sundown (you should drink, of course, but not alcohol). This, also, allows your body to normalise itself. Obviously, if you have diabetes or another metabolic disease, or are pregnant, don't diet without first discussing it with an advisor.

'Is it OK to wear leather?'

Only if you think it's OK to eat meat. Bear in mind that 25–50 per cent of slaughterhouse profits come from the leather industry.[6] Leather alternatives are increasingly available in the shops, and the more you ask for them, the more there will be.

'Hitler was a vegetarian, wasn't he?'

I wouldn't call anyone who had regular injections of bull's testicles a vegetarian.

'Should I buy vegetarian cheeses?'

Many cheeses contain rennet, an enzyme taken from the stomachs of slaughtered baby calves, a small amount of which helps to coagulate milk into cheese. Not all cheese includes animal rennet but if you want to be sure, buy a specially marked cheese from a health food shop, or a growing number of supermarkets. Vegetable rennet is sometimes called rennin to distinguish it from the animal-derived type. In my view, the rennet controversy tends to be exaggerated by some vegetarians. If you're buying dairy produce, you are economically contributing to the meat industry in any case. There's a growing range of completely animal-free cheeses on the market (try your local health food shop) which I would urge you to investigate. Personally, I ate a vegetarian diet for many years before I went vegan, and I found that dairy produce simply became less and less important in my diet, until one day I realised that I'd made the transition.

'How can I get enough vitamin D?'

Although vegetarians who consume dairy produce get some in milk, most people – vegetarian, vegan or meat-eating – get all the vitamin D they need from the action of sunlight on their skin. In fact, food sources of vitamin D are relatively unimportant. Comments expert on vegan nutrition Dr. Gill Langley:

'Bright sunshine is not necessary: even the 'skyshine' on a cloudy day will stimulate the formation of some vitamin D_2 in the skin, while a short summer holiday in the open air will increase serum levels of vitamin D two- or three-fold.'[7]

In recent years, there have been moves to reclassify this substance as a steroid hormone rather than as a vitamin.[8] And there have been scientific suggestions that, when taken by non-deficient people, low level ingestion of calciferol (the name for this group of substances) may accelerate the ageing of arteries, kidneys and bones.[9]

'How can I get enough vitamin B₁₂?'

Vitamin B_{12} is chemically the most complex of all the vitamins. Almost all organisms need this vitamin, but only in very small amounts. The human daily requirement is currently set at one and a half millionths of a gram.[10] Infrequent cases of B_{12} deficiency are usually due to the lack of an 'intrinsic factor' – a substance manufactured by the stomach and necessary to bond with B_{12} before it can be absorbed by the body. Meat-eaters or vegans are equally likely (or unlikely) to suffer from this. The human liver stores enough B_{12} to sustain the body for six years or so.[11]

Vitamin B_{12} has become something of a *cause célèbre* in vegan nutrition (but not for vegetarians, who consume B_{12}-containing dairy produce) because there are few B_{12} sources in the plant kingdom. The meat industry has not been slow to capitalise on this and to distort the facts to suit themselves. But don't be misled. In nature, B_{12} is synthesised by micro-organisms, so anything which has been fermented may well contain it (it is also possible that the human body can manufacture its own B_{12} – research is continuing). If you're eating a vegan diet, the easiest thing to do is to buy one or two products which you know contain B_{12} – such as fortified soya milk, yeast extract, margarine and textured vegetable protein. Just read the label!

'How can I get enough iron?'

Lack of iron is the one of the most common nutritional deficiencies in Western countries, although you should note that, contrary to much-repeated propaganda, vegans and vegetarians are no more at risk of iron deficiency than meat-eaters.[12 13] Once again, the meat industry has reprehensibly spread confusion here. The major cause of iron deficiency is an inadequate diet, with an over-reliance on highly refined foods. In extreme form, iron deficiency will result in anaemia. Nevertheless, the body manages its iron supply remarkably well. In normal circumstances, only about 10 per cent of the iron in the diet will be absorbed, but in circumstances of deficiency the body will compensate by absorbing more. Also, the body will recycle its own supply of iron, generally only losing small amounts. Since over half of the body's supply of iron is normally found as haemoglobin in the blood, any loss of blood will deplete the body's store of iron. Vitamins C and E help the body absorb iron. Interestingly, the use of iron cooking utensils will provide a dramatic increase in the amount of iron found in foods cooked in them. It has been calculated that if spaghetti sauce is cooked for three hours in an iron pan, it will contain 29 times as much iron as it would have contained if it had been cooked in glassware. However, since the dietary requirement for iron is basically a 'topping-up' function, it should be noted that it is possible to suffer from an overdose of iron. This may be caused by either a personal metabolic problem, or from excessive and prolonged dietary supplementation. As always, the best answer for normal individuals is to ensure that their diet naturally provides an optimum quantity of the element. The main sources of iron in the diet are bread, flour and other cereal products, potatoes and green leafy vegetables, and:

• prune juice
• rolled oats
• brewer's yeast powder
• dried apricots
• raisins
• broccoli
• plain chocolate (I knew there must be a good reason to eat it!)

Until very recently, it was assumed by almost everyone – including doctors and nutritionists – that the only problem associated with iron was not getting enough – in other words, anaemia. This is precisely what they used to think about protein, too. Now, however, disturbing new evidence is emerging that 'excess iron in the bloodstream may be more deadly than cholesterol or high blood pressure as a cause of heart disease,' to quote one of the first reports.[14] A study of 1931 men was conducted at the University of Kuopio in eastern Finland between 1984 and 1989. At the beginning of the study, the men were aged 42, 48, 54 or 60 and were free of heart disease. At the end of the study, men with high amounts of iron in their bodies had twice as much risk of having suffered heart attacks as men with lower amounts of iron – even after known heart disease risk factors, such as cholesterol levels and smoking, were taken into account.[15] Dr. Jerome Sullivan, director of clinical laboratories at the Veterans Affairs Medical Center in Charleston, USA, believes this may explain why most women are safe from heart attacks until after the menopause. 'When women stop menstruating,' he says, 'iron stores start accumulating quite rapidly. They convert from a protected state to one in which their heart attack rate is much more similar to men's.'[16]

Common sense surely tells us that we can have too much of a good thing. Plant foods contain all the iron we need as humans, so why consume the concentrated amounts found in certain flesh foods? And if, in a few years' time, it becomes generally accepted that excess iron consumption is just as deadly as saturated fat or cholesterol, how humorous present-day meat advertisements will seem to us, then. Ironic, even . . .

'What's the difference between saturated and unsaturated fat?'

The term 'fat' includes all oils and fats, whether or not they're solid or liquid. All fats consist of long chains of carbon and hydrogen atoms. When all the available sites on the carbon atoms are filled with hydrogen atoms, the fat is saturated. If there are unfilled spaces, the fat is unsaturated. The more empty spaces, the more unsaturated the fat is.

Saturated fat is known to raise the level of cholesterol in your blood. The more you eat, the higher your cholesterol level, and the greater your chances of suffering a stroke or heart attack. As a guide, saturated fat is usually solid at room temperature. Animal fat (lard, meat, butter and so on) contains lots of saturated fat. A few plant fats also contain significant amounts – principally coconut and palm oil.

Monounsaturated fat is much healthier that saturated fat. A major dietary source is olive oil. Experiments on humans show that switching to mono-unsaturated fat from the saturated kind can not only decrease the risk of heart disease, but it may also be able to lower your blood pressure. It is also less prone to go rancid than other types of fat, and rancidity may promote cancer.

Polyunsaturated fat is also better for you than saturated fat. However, polyun-saturated oils should only be used raw, because they can be damaged by exposure to heat, air and light, and may then form free radicals, which have been con-nected to a growing range of human diseases. Good sources of polyunsaturated fats include sunflower and corn oil.

The way oils are produced is important, too. There are three main methods of extraction. The first, cold pressing, is the traditional hydraulic pressing process where the temperature is kept low throughout, thus preserving heat-sensitive nutrients. The end product is expensive, mainly because there is a high per-centage of waste in the discarded pulp, but the oil is nutritious and tastes and smells good. This is by far the best type to buy.

Another method, the screw or expeller process, involves high pressure pressing, which generates high temperatures. Nutrients are destroyed during this process and, although it enables more oil to be extracted, it is dark, strong-smelling and needs further refining and deodorising.

The most commonly used process is solvent extraction, because it produces the highest yields. The grains or seeds are ground, steamed and then mixed with solvents. The solvents used are either the petroleum-based benzene, hexane or heptane. The mixture is then heated to remove the solvents and then washed with caustic soda. This has the effect of destroying its valuable lecithin content. After this it is bleached and filtered which removes precious minerals as well as any coloured substances. Finally, the oil is heated to a high temperature to deodorise it. Although manufacturers insist that any chemical residues are minimal, they are of substances which are known to be carcinogenic. One other aspect of vegetable oils produced by solvent extraction is that their vitamin E content is diminished. This vitamin helps stop the oil from going rancid, and rancid oils are dangerous because they provide the raw material for producing free radicals in our bodies. Sometimes chemical retardants are added to stop the oil from turning rancid.

Sometimes, manufacturers 'hydrogenate' fat to increase its shelf life, but it also has the effect of increasing its saturated fat content. The process of hydrogenation fills some unsaturated bonds of the fat molecule with hydrogen atoms, making it more similar to a saturated fat. For example, soya bean oil in its natural state is only 15 per cent saturated, but when it's partially hydrogenated it's closer to 25 per cent saturated, similar to vegetable shortenings. Hydrogenation produces 'transformed unsaturated fats', usually called 'trans fatty acids', which may be just as bad for your heart as saturated fats are. So avoid hydrogenated fats where you can.

'Can my cat or dog go vegetarian?'

Yes, both animals can be fed a vegetarian diet. Cats need food fortified with an amino acid called taurine, found in the muscles of animals. Synthetic taurine has been developed, and vegetarian cats should be fed it as a supplement. For information on these products, contact:

Happidog Petfoods (dog food)
Bridgend, Brownhill Lane, Preston, PR4 4SJ

Wafcol (dog food)
The Nutrition Bakery, Haigh Avenue, Stockport, SK4 1NU

The Watermill (dog food)
Little Salkeld, Penrith, CA10 1NN

Wow-Bow Distributors (cat and dog food)
309 Burr Road, East Northport, NY 11731 USA
Tel. (516) 449-8572

Natural Life Pet Products, Inc. (dog food)
Frontenac, Kansas 66762, USA

'Is organic food better?'

Yes. A third of the fruit and vegetables on sale in this country contains detectable traces of pesticide residue, according to a recent survey, including such notorious compounds as DDT. You can avoid this problem by buying (or, better still, growing) organic fruit and vegetables. Also, you won't believe how different they taste – it's a real revelation! Get information on organic gardening from the **Henry Doubleday Research Association**, Convent Lane, Bocking, Braintree, Essex CM7 6RW (Tel. Braintree 24083); or from **The Soil Association Ltd**, Walnut Tree Manor, Haughley, Stowmarket, Suffolk IP14 3RS (Tel. 0449-673235); or from the **Good Gardeners' Association**, CRG Shewell-Cooper, Arkley Manor Farm, Rowley Lane, Arkley, Barnet, Herts EN5 3HS (Tel. 081-449 7944), who also run training courses with diplomas.

'How can I start my own wholefood restaurant?'

If you find the idea of starting your own restaurant or wholefood business attractive, you should remember that both these activities involve extremely hard work, often for little or no reward, and major risk to your own capital. Someone once said that there are just three important things to consider when setting up a restaurant – location, location and location. But to secure the best location, you may have to pay dearly. Which means that you have to be sure of having a sizeable number of customers. It is essential to do some budgets early on, as you may otherwise find that you cannot possibly make any money, no matter how hard you work. The **Hotel and Catering Industry Training Board** runs introductory courses, PO Box 18, Ramsey House, Central Square, Wembley, Middlesex HA9 7AP (Tel. 081-902 8865). If you're planning a co-operative venture, contact the **National Co-Operative Development Agency**, Broadmead House, 21 Panton Street, London SW1Y 4DR (Tel. 071-839 2985), and ask at your local town hall as well. Your local **Small Firms Information Service** may be able to help as well. Dial the operator and ask for Freephone 2444. Other organisations of possible interest include: **National Association of Shopkeepers**, Lynch House, 91 Mansfield Road, Nottingham, NG1 3FN (Tel. 0602-475046); **Manpower Services Commission Enterprise Allowance Scheme** – contact your local Job Centre for current details. If you have a friendly bank manager, use him (or her) for all he (or she) is worth as a source of up-to-date information. And don't take no for an answer – from anyone!

'Which organisations should I join?'

It depends what you want. Some organisations campaign effectively on behalf of animals, and achieve some real successes, others merely seem to generate in-fighting and waste energy and resources. Some I can recommend are:

Animal Aid
7 Castle St, Tonbridge, Kent TN9 1BH

Farm Animal Welfare Network
PO Box 40, Holmfirth, Huddersfield, West Yorkshire HD7 1QY

Respect for Animals
PO Box 500, Nottingham, NG1 3AS

VEGA (Vegetarian Economy & Green Agriculture)
PO Box 39, Godalming, GU8 6BT

The Vegan Society
7 Battle Road, St Leonards On Sea, TN37 7AA

The Vegetarian Society
Parkdale, Dunham Road, Altrincham, WA14 4QG

Don't stop when you join an organisation. Real change is brought about by effective local action and that means you should join or establish a group in your local community. If you are worried that you don't know enough to organise such a group, then you're probably *exactly* the right person to do it! All you need to start with is one or two more like-minded people to yourself. To find them, try putting up a card or poster in suitable locations (wholefood shops, restaurants, etc.), saying something like:

> Giving Meat The Chop? Local group starting for people who want to spread the word about meat-free cooking and living. Needs help and ideas from people like you . . . Contact me for a chat: (your name and phone number).

Then fix a date for the first planning meeting (possibly held at your house). Make the meeting very informal. Everyone must have the chance to express themselves. Don't vote on anything – come to an agreed consensus instead. You may have to lead the discussion, but *don't dictate*! These are some suggestions for the areas you should aim to discuss:

• The name of the group (have at least three suggestions ready).
• Its objectives (suggestion: to convert more people in your area to meat-free living).
• Activities (suggestions: four large public events a year; four issues of a magazine a year; cookery demonstrations, special restaurant evenings, all-day food fairs with stands for other organisations, leafleting).
• Who is going to be responsible for certain areas (e.g. finance, chairperson, publicity, secretary).

The real key to success is publicity, and that means hard work for everyone. Draw up a 'Contact List' together, with all the people who need to be notified in advance of your events (TV, radio, press) and all the places where you can display posters (aim to get at least 50 posters out four weeks before any event, in all libraries, shops, large businesses, public buildings, etc.). Street leafleting a day or two before the event is very worthwhile, too, if you can get several thousand handbills cheaply produced. There is a lot of satisfaction in helping to run a local group, and it will certainly open new horizons for you.

'What are hot dogs made from?'

Many people are rather nauseated when they find out what goes into burgers, but hot dogs have, so far, escaped much scrutiny. That is about to change! A good friend of mine had the opportunity to tour a factory where these especially loathsome comestibles were made, and in the process somehow seemed to acquire a copy of the recipe (I put it down to chaos theory). It passed to me in due course, and I reprint it here, with no further comment, since nothing I could say could be more revolting than the recipe itself:

Formula
- Fish
- Chicken feet
- Chicken carcasses
- Chicken heads
- Lungs and trachea
- Udder
- Liver
- Fresh blood
- Wheat flour
- Salt
- PP530
- Water
- Phosphate
- Carageenan

'What is "hamburgerisation"?'

An ugly word for an ugly reality. In their quest for ever cheaper meat supplies, the international beef industry is always looking for usable tracts of grazing land – anywhere in the world. As one rancher puts it: 'Here's what it boils down to – $95 per cow per year in Montana, $25 in Costa Rica.'[17] So forests are felled, land is cleared, grass is planted, cheap beef is produced, and consumer demand is satisfied. But recently, a few consumers have begun to wonder how much their hamburgers really cost. Not in dollars and cents, but in terms of natural resources consumed or destroyed. Then someone did a simple, but very shocking, calculation. They reasoned that a hectare of rain forest – the sort of land regularly cleared for ranching in remote areas – supports about 800,000 kilograms of plants and animals. When the same hectare has been felled, torched, razed and seeded with grass for grazing, it will produce at most 200 kilograms of meat a year – enough flesh to make about 1600 hamburgers. The grazing won't last long, however, because the land is quickly leached of its nutrients and left barren by over-grazing, so that in a few years it becomes useless. By then, of course, more forest will have been destroyed, and converted to stopgap pasture land. This makes the true cost of a hamburger to be something in the region of half a tonne of rain forest for each burger – or about nine square metres of irreplaceable natural wealth, forever destroyed for the price of a quick, unhealthy snack. Enough to give you indigestion, isn't it?

'Where cattle move in, we move out – cattle mean hunger,' say the Amazonian Caboclo, people who lived in the forest before the settlers and ranchers came. The fact that forest areas do have indigenous populations – people who peacefully co-exist with their environment and who know how to utilise it – is often forgotten in the indecent haste to destroy and develop. Beef is, of course, a cash crop – it goes to foreign markets (that means people like you and me) who can afford to pay good prices for hamburger meat. The money earned is often put aside to pay off development loans. Ethiopia, recently the target of much Western media attention, has suffered a similar fate. From a forest area covering 16 per cent of the land in the 1950s, less than 4 per cent is now left. Multinationals, which were welcomed with generous tax incentives, proceeded to develop the best and most fertile areas, forcibly evicting the indigenous population who then struggled to survive on fragile upland areas. The consequent tree felling, over-grazing and land degradation has exacerbated the country's problems. Nevertheless, Ethiopian beef was still recently being exported to the West. The *Meat Trades Journal* says that the importer is 'well aware of the sensitive nature of the product. It has been shipping the corned beef to Europe on a regular basis for the past two years and on the continent the tins that go onto shop shelves often have the country of origin obscured.'[18]

Organisations such as the World Bank, the United Nations Development Fund, the African Development Bank and the Inter-American Development Bank have all provided assistance for livestock production and meat processing in developing countries. It is essential that pressure is brought to bear on these development organisations to curtail such catastrophic loans programmes. Some organisations, such as the Environmental Defense Fund, are doing just that, and having a degree of success. A newly formed group, Conservation International, pulled off a significant success when it purchased $650,000 of Bolivian debt in exchange for a promise to set aside 3.7 million acres for conservation. Fundamentally, of course, it is Western consumers, who by producing a demand for cheap beef, make the whole process economically worthwhile. In the words of James Nations and Daniel Komer, two staff members at the Center for Human Ecology, ironically based in the heart of Texas beef country: 'Consumers must be made aware that when they bite into a fast-food hamburger or feed their dogs, they may also be consuming toucans, tapirs and tropical rain forests.'[19]

Actually, meat consumption is implicated in a huge range of environmental problems, some of which never cross the minds of the majority of meat consumers. Consider the American cowboy, for example – that rugged icon of individualism and independence. In reality, he was likely to be employed by one of the multitude of British meat companies that spearheaded and bankrolled the invasion and destruction of the American West, and the resultant genocide of the native American people.

The combined developments of refrigerated transport and the westward expansion of the American railroads (which the British also helped to finance) opened up a huge source of supply for the insatiable British craving for fatty beef. 'British bankers launched a financial invasion onto the western range of America in the early 1880s,' writes Jeremy Rifkin.[20] 'They set up giant cattle companies across the plains, securing millions of acres of the best grasslands for the British market. While the west was made safe for commerce by American frontiersmen and the US military, the region was bankrolled by English lords and lawyers, financiers and businessmen who effectively extended the reach of the British beef empire deep into the short grass of the western plains.'

The consequences of this 'green goldrush' were, and still are, catastrophic. Vast tracts of public lands were illegally appropriated, enclosed by the newly invented barbed wire. The buffalo was bloodily exterminated, as was any other creature which competed for food with the bovine invasion. And the corn-fed feedlot system of cattle rearing was developed, a truly obscene waste of grain. In the words of Edward Abbey, conservationist and author:

> 'Most of the public lands in the West, and especially the Southwest, are what you might call "cow burnt". Almost anywhere and everywhere you go in the American West you find hordes of cows. They are a pest and a plague. They pollute our springs and streams and rivers. They infest our canyons, valleys, meadows, and forests. They graze off the native bluestems and grama and bunch grasses, leaving behind jungles of prickly pear. They trample down the native shrubs and cacti. Even when the cattle are not physically present, you see the dung and the flies and the mud and the dust and the general destruction. If you don't see it, you'll smell it. The whole American West stinks of cattle.'[21]

In livestock areas, cowboys set traps for coyotes, mountain lions, bobcats and bears. Traps, of course, are indiscriminate; any animal attracted to the bait will fall victim. And so is poison. During the 1950s and 1960s, poisons became the rangeland rage. Contaminated livestock carcasses were routinely left on grazed land across the West. From trucks, horses, trail bikes and aeroplanes, millions of strychnine-laced tallow pellets were scattered over the Western landscape, even where no livestock grazed, in a secret orgy of continuing slaughter. Soon, there will be little else left in the lonely rangelands – no coyotes, no wolves, no foxes, no eagles, no hawks, no owls, no badgers . . . nothing but the ever-present cow.

'I'm a Christian, and the Bible says I can eat meat.'

Well, the Bible says a great many things, and with selective quotation you can prove almost anything – nineteenth-century American slave owners frequently justified themselves with pious Biblical quotations. Many books have been written about this subject, but I'd like to make three points here. First, if we loved the animals today in the way humans loved them in the garden of Eden, we would not eat them. It is only after Adam and Eve had been expelled from the garden of Eden that they started killing for food. Other cultures have essentially similar tales of a fall from a once-perfect state of universal kinship. The Cherokee Indians have a tribal myth that says that humans once lived in perfect harmony with all their fellow creatures and plants, and all of them could speak to each other. In China, the Taoist Chuang Tsu wrote in the fourth century BC of a past 'age of virtue' when all humankind lived a common, co-operative life with the birds and the beasts. And in Greece, the philosopher Empedocles wrote of a 'golden age, an age of love' when 'no altar was wet with the shameful slaughter of bulls', and he maintained that the primal sin was man's slaughter of animals. The Old Testament tells us that the world that God creates is, initially, perfect. It is quite clear that in this perfect state, it is not intended for humans to eat the flesh of other creatures: 'Behold I have given you every herb yielding seed, which is upon the face of the earth, and every tree, in which is the fruit of a tree yielding seed; to you it shall be for meat.' Clearly, our task is to strive for that state of perfection once again, and a good starting point is to stop slaughtering our fellow creatures.

Secondly, as Rev. Dr. Andrew Linzey puts it: 'The Christian argument for vegetarianism is simple: since animals belong to God, have value to God and live for God, then their needless destruction is sinful.'[22] Andrew Brown, religious affairs correspondent for a British newspaper, rather brilliantly describes this 'needless destruction' in the following passage:

> 'They [factory-farmed animals] are martyred in the cause of mediocrity, confined and tortured to make our diets blander. Not even the weariest and most jaded epicure has ever crowned a lifetime of increasingly decadent sensuality by reaching, as his final perverted thrill, for a handful of chicken McNuggets. This is, of course, pleasingly moralistic: the fruits of sin turn to fast food in our mouths, as they should.'[23]

And thirdly, I would urge you to consider what is, for many people, the kernel of Christianity, expressed in Christ's Sermon on the Mount. I won't indulge in selective quotation from it here; read the whole of it (Matthew, chapter 5) and then consider these words written by Richard Whitehead:

> 'Let us take two images and place them side by side. The first is of a young preacher, healing the sick, washing his disciples' feet, preaching gentleness and humility, and going to the cross without so much as raising a hand against his persecutors. The second is of an industry which systematically and mercilessly slaughters millions of animals which it has compelled to spend their brief lives in cramped, dingy, smelly sheds, and stuffed full of every chemical under the sun in order to produce meat to the maximum economic efficiency, a product which it spends millions of pounds persuading the public to buy. Clearly the two images are not simply incompatible; they are diametrically opposed. Can we really envisage Jesus being anything *but* a vegetarian if he were to be born into our society?'[24]

Well, there we are. I don't imagine for a moment that I've answered all the questions that you may have, although I hope some of the most common ones have been put to rest. What's really important now is that you feel confident enough to go ahead and live the new lifestyle for yourself, finding out things along the way. You'll enjoy it!

Finally, I have a request to make of you. Now you know the truth about all this, you have a responsibility to pass it on to someone else. This is the best way, and perhaps the only way, that things will ever really change for the better.

NOTES

1 From data in Wilson, P.N., 'Biological Ceilings and Economic Efficiencies for the Production of Animal Protein AD 2000', *Chemistry & Industry*, 6 Jul 1968.

2 *New Scientist*, 1989.

3 *Foundation on Economic Trends*, Washington DC, 1992.

4 *Beyond Beef Campaign*, Washington DC, 1992.

5 Miller, R., *Bare-Faced Messiah*, Michael Joseph, 1987.

6 Howlett, L., *Cruelty-Free Shopper*, Bloomsbury, 1989.

7 Langley, G., *Vegan Nutrition*, Vegan Society, 1988.

8 Norman, A.W., *Vitamin D, the calcium homeostatic steroid hormone*, Academic Press, 1979.

9 Moon, J., 'Factors affecting arterial calcification associated with atherosclerosis', *Atherosclerosis*, 16, pages 199–226

10 Dietary reference values for food energy and nutrients for the United Kingdom, Department of Health, 1991.

11 Langley, G., op cit.

12 Levin, N. et al, 'Mineral intake and blood levels in vegetarians', *Israel Journal of Medical Sciences*, Feb 1986, 22 (2), pages105–8.

13 Mcendree, L.S., 'Iron utilization by vegetarians and omnivores', PhD Thesis 1982, The University of Nebraska, Lincoln.

14 *Toronto Daily Star*, 8 Sep 1992.

15 *Los Angeles Times*, 9 Sep 1992.

16 *Toronto Daily Star*, 8 Sep 1992.

17 Leppé, F.M. and Collins, J., *Food First*, Souvenir Press, 1980.

18 *Meat Trades Journal*, 7 Jan 1988.

19 *The Ecologist*, Jul–Nov 1987, 17 (4–5), pages 161–7.

20 Rifkin, J., *Beyond Beef*, Dutton, 1992.

21 Speech before cattlemen at the University of Montana in 1985.

22 Linzey, A., *Christianity and the Rights of Animals*, Crossroads Publishing Company, New York.

23 The *Independent*, 7 Jul 1990.

24 Personal correspondence, Jan 1988.

CHAPTER SEVEN
THE COMPOSITION OF VEGETARIAN FOODS

This chapter provides you with the nutritional composition for over 600 vegetarian food items, which can be used to analyse your own diet, or to find foods which are rich in particular nutrients.

Vegetarians generally have a more varied diet and therefore a better nutrient intake than meat-eaters, and, providing they are regularly eating from the six food groups mentioned on page 171, should feel confident that they are amongst the most well-nourished members of the population. For guidance, suggested nutrient intakes are shown in the table on page 220, based on guidelines issued by the UK Department of Health.

The foods are grouped into the following main divisions:

- Beans and Pulses
- Breads and Cakes
- Dairy Products
- Drinks
- Fruit
- Grains and Grain Products
- Spices and Herbs
- Nuts and Seeds
- Oils, Spreads and Dressings
- Soups
- Sweet Things
- Vegetables

The following abbreviations have been used:

Amt	= Amount	C	= Cup
Appx	= Approximately	Ca Phos	= Calcium Phosphate
Aprct	= Apricot	Ca	= Calcium
Art	= Artificial	Cals or Kcals	= Calories
Asc	= Ascorbic Acid	Carbo	= Carbohydrate
Ban	= Banana	Cherr	= Cherry
Bisct	= Biscuit	Choc	= Chocolate
Bkd	= Baked	Cinn	= Cinnamon
Bld	= Boiled	Ckd	= Cooked
Brld	= Broiled	Cl	= Class
Brn	= Brown	Cn	= Can
Bttr	= Batter	Cnd	= Canned
Bx	= Box	Cntnr	= Container

Cocnt	= Coconut		Marg	= Margarine
Commer	= Commercial		mcg	= microgram
Conc	= Concentrate		Med	= Medium
Crckrs	= Crackers		mg	= milligram
Crm	= Cream		Mill	= Milled
Crmbld	= Crumbled		Mon/fat	= Monounsaturated fat
Croq	= Croquette		Na	= Sodium
Cu In	= Cubic Inch		Na Cas	= Sodium Casienate
Dehy	= Dehydrated		Nfdm	= Non-Fat Dried Milk
Di Na Phos	= Di Sodium Phosphate		Nfdms	= Non-Fat Dried Milk Solids
Diam	= Diameter		Nfms	= Non-Fat Milk Solids
Dil	= Diluted		Niac	= Niacin
Drnd	= Drained		Nt Wt	= Net Weight
Enr	= Enriched		oz	= ounce
Eq	= Equal		Past	= Pasteurised
Fe	= Iron		Pc	= Piece
fl oz	= fluid ounce		Pdr	= Powder or Powdered
Flav	= Flavoured		Pk	= Pack
Flr	= Flour		Pkd	= Packed
Fol	= Folate		Pkg	= Package
Fort	= Fortified		Pkt	= Packet
Frstd	= Frosted		Pnappl	= Pineapple
Frt	= Fruit		Pol	= Polished
Frz	= Frozen		Pol/fat	= Polyunsaturated fat
Ft	= Fat		Ppd or Prep	= Prepared
g	= gram		Preswt	= Presweetened
Gal	= Gallon		Proc	= Processed
Gl	= Glass		Ptty	= Patty
Gr	= Grain		Qt	= Quart
Gran	= Granules		R-T-C	= Ready-To-Cook
Grpe	= Grape		Rd	= Rounded
Hi-Prot	= High-Protein		Rec	= Recipe
Hlvs	= Halves		Rect	= Rectangular
Hydr	= Hydrogenated		Reg	= Regular
Ic	= Icing		Rf	= Refuse
Imitn	= Imitation		Ribo	= Riboflavin
In	= Inch		Rstd	= Roasted
Instnt	= Instant		Rw	= Raw
Jnr	= Junior		Sat/fat	= Saturated fat
lb	= pound		Sect	= Sections
Lf	= Loaf		Sflwr	= Safflower
Lg or Lrg	= Large		Skn	= Skin
Liq or Lq	= Liquids		Sl or Slc	= Slice or Slices
Ln	= Lean		Slt	= Salt
Lng	= Long		Sm	= Small

Snflwr	= Sunflower	Veg-S	= Vegetable Shortening
Sol	= Solids	Veg	= Vegetables
Sq	= Square	Vit A	= Vitamin A
Stbl	= Stabilisers	Vit B_6	= Vitamin B_6
Stk	= Stick	Vit C	= Vitamin C
Str	= Strained	Vol	= Volume
Sug	= Sugar	W/	= With
Swtn	= Sweetened	Wdg	= Wedge
Sybn	= Soybean	Whl	= Whole
Sz	= Size	Wht	= White
Tblts	= Tablets	Wo/	= Without
Tbsp	= Tablespoon	Wt	= Weight
Thiam	= Thiamin	Wtr	= Water
Tsp	= Teaspoon	Yel	= Yellow
Unenr	= Unenriched	Yld	= Yield
Var	= Variety or Varieties	–	= No data available

Age	Protein (g)	Vitamin A (mcg retinol equivalent)[1]	Thiamin (mg)	Riboflavin (mg)	Niacin (mg)	Vitamin B6	Folate (mcg)	Vitamin B12 (mcg)	Vitamin C (mg)	Calcium (mg)	Iron (mg)	Zinc (mg)
MALES												
11–14 years	42.1	600	0.9	1.2	15	1.2	200	1.2	34	1000	11.3	9.0
15–18 years	55.2	700	1.1	1.3	18	1.5	200	1.5	40	1000	11.3	9.5
19–50 years	55.5	700	1.0	1.3	17	1.4	200	1.5	40	700	8.7	9.5
50+ years	53.3	700	0.9	1.3	16	1.4	200	1.5	40	700	8.7	9.5
FEMALES												
11–14 years	41.2	600	0.7	1.1	12	1.0	200	1.2	35	800	14.8	9.0
15–18 years	45	600	0.8	1.1	14	1.2	200	1.5	40	800	14.8	7.0
19–50 years	45	600	0.8	1.1	13	1.2	200	1.5	40	700	14.8	7.0
50+ years	46.5	600	0.8	1.1	12	1.2	200	1.5	40	700	8.7	7.0
Pregnancy	+6	+100	+0.1 (last trimester only)	+0.3	no increment	no increment	+100	no increment	+10	no increment	no increment	no increment
Lactation 0–4 months	+11	+350	+0.2	+0.5	+2	no increment	+60	+0.5	+30	+550	no increment	+6.0
4+ months	+8	+350	+0.2	+0.5	+2	no increment	+60	+0.5	+30	+550	no increment	+2.5

[1] Food composition values of vitamin A are given in IU (international units). One IU equals 0.3 mcg retinol, or 0.6 mcg of beta-carotene

Food Item & Measure	Cals	Protein (g)	All fat (g)	Sat/fat (g)	Pol/fat (g)	Mon/fat (g)	Carbo (g)	Vit A (IU)	Thiam (mg)	Ribo (mg)	Niac (mg)	Vit B₆ (mg)	Fol (mcg)	Vit B₁₂ (mcg)	Vit C (mg)	Ca (mg)	Iron (mg)	Zinc (mg)
BEANS AND PULSES																		
Adzuki, Boiled, W/Salt 1 C:230 g	294	17.3	0.2	0.1	—	—	57	13.8	0.26	0.15	1.65	0.22	278.5	0	0	64.4	4.6	4.07
½ C:115 g	147	8.7	0.1	0	—	—	28.5	6.9	0.13	0.07	0.82	0.11	139.3	0	0	32.2	2.3	2.04
Adzuki, Boiled, Wo/Salt 1 C:230 g	294	17.3	0.2	0.1	—	—	57	13.8	0.26	0.15	1.65	0.22	278.5	0	0	64.4	4.6	4.07
½ C:115 g	147	8.7	0.1	0	—	—	28.5	6.9	0.13	0.07	0.82	0.11	139.3	0	0	32.2	2.3	2.04
Baked, Cnd 1 C:254 g	236	12.2	1.1	0.3	0.5	0.1	52.1	434.3	0.39	0.15	1.09	0.34	60.7	0	7.9	127	0.74	3.56
½ C:127 g	118	6.1	0.6	0.2	0.3	0.1	26.1	217.2	0.19	0.08	0.54	0.17	30.4	0	3.9	63.5	0.37	1.78
Black, Boiled, W/Salt 1 C:172 g	227	15.2	0.9	0.2	0.4	0.1	40.8	10.3	0.42	0.1	0.87	0.12	255.9	0	0.	46.4	3.61	1.93
½ C:86 g	114	7.6	0.5	0.1	0.2	0	20.4	5.2	0.21	0.05	0.43	0.06	128	0	0	23.2	1.81	0.96
Black, Boiled, Wo/Salt 1 C:172 g	227	15.2	0.9	0.2	0.4	0.1	40.8	10.3	0.42	0.1	0.87	0.12	255.9	0	0	46.4	3.61	1.93
½ C:86 g	114	7.6	0.5	0.1	0.2	0	20.4	5.2	0.21	0.05	0.43	0.06	128	0	0	23.2	1.81	0.96
Broad, Boiled, Wo/Salt 1 C:170 g	187	12.9	0.7	0.1	0.3	0.1	33.4	25.5	0.16	0.15	1.21	0.12	177	0	0.5	61.2	2.55	1.72
½ C:85 g	94	6.5	0.3	0.1	0.1	0.1	16.7	12.8	0.08	0.08	0.6	0.06	88.5	0	0.3	30.6	1.28	0.86
Broad, Cnd 1 C:256 g	182	14	0.6	0.1	0.2	0.1	31.8	25.6	0.05	0.13	2.46	0.12	83.7	0	4.6	66.6	2.56	1.59
½ C:128 g	91	7	0.3	0.1	0.1	0.1	15.9	12.8	0.03	0.06	1.23	0.06	41.9	0	2.3	33.3	1.28	0.79
Chickpeas, Boiled, Wo/Salt 1 C:164 g	269	14.5	4.3	0.4	1.9	1	45	44.3	0.19	0.1	0.86	0.23	282.1	0	2.1	80.4	4.74	2.51
½ C:82 g	134	7.3	2.1	0.2	1	0.5	22.5	22.1	0.1	0.05	0.43	0.11	141	0	1.1	40.2	2.37	1.25
Chickpeas, Cnd 1 C:240 g	286	11.9	2.7	0.3	1.2	0.6	54.3	57.6	0.07	0.08	0.33	1.14	160.3	0	9.1	76.8	3.24	2.54
½ C:120 g	143	5.9	1.4	0.1	0.6	0.3	27.1	28.8	0.03	0.04	0.17	0.57	80.2	0	4.6	38.4	1.62	1.27
French, Boiled, W/Salt 1 C:177 g	228	12.5	1.4	0.2	0.8	0.1	42.5	5.3	0.23	0.11	0.97	0.19	132.2	0	2.1	111.5	1.91	1.13
½ C:86 g	111	6.1	0.7	0.1	0.4	0	20.7	2.6	0.11	0.05	0.47	0.09	64.2	0	1	54.2	0.93	0.55
French, Boiled, Wo/Salt 1 C:177 g	228	12.5	1.4	0.2	0.8	0.1	42.5	5.3	0.23	0.11	0.97	0.19	132.2	0	2.1	111.5	1.91	1.13
½ C:86 g	111	6.1	0.7	0.1	0.4	0	20.7	2.6	0.11	0.05	0.47	0.09	64.2	0	1	54.2	0.93	0.55
Green, Boiled, Wo/Salt ½ C:62 g	22	1.2	0.2	0	0.1	0	4.9	412.9	0.05	0.06	0.38	0.03	20.7	0	6	28.5	0.79	0.22
1 C:125 g	44	2.4	0.4	0.1	0.2	0	9.9	832.5	0.09	0.12	0.77	0.07	41.6	0	12.1	57.5	1.6	0.45
Green, Raw ½ C:55 g	17	1	0.1	0	0	0	3.9	367.4	0.05	0.06	0.41	0.04	20.1	0	9	20.4	0.57	0.13
1 C:110 g	34	2	0.1	0	0.1	0	7.9	734.8	0.09	0.12	0.83	0.08	40.2	0	17.9	40.7	1.14	0.26
Kidney, Boiled, Wo/Salt 1 C:177 g	225	15.4	0.9	0.1	0.5	0.1	40.4	0	0.28	0.1	1.02	0.21	229.4	0	2.1	49.6	5.2	1.89
½ C:88 g	112	7.6	0.4	0.1	0.2	0	20.1	0	0.14	0.05	0.51	0.11	114.1	0	1.1	24.6	2.59	0.94
Kidney, Cnd 1 C:256 g	207	13.3	0.8	0.1	0.4	0.1	38.1	0	0.28	0.18	1.29	0.18	126	0	3.1	69.1	3.15	1.41
½ C:128 g	104	6.7	0.4	0.1	0.2	0	19.1	0	0.14	0.09	0.64	0.09	63	0	1.5	34.6	1.57	0.7

Food item & Measure	Cals	Protein (g)	All fat (g)	Sat/fat (g)	Pol/fat (g)	Mon/fat (g)	Carbo (g)	Vit A (IU)	Thiam (mg)	Ribo (mg)	Niac (mg)	Vit B6 (mg)	Fol (mcg)	Vit B12 (mcg)	Vit C (mg)	Ca (mg)	Iron (mg)	Zinc (mg)
Beans and Pulses cont.																		
Lentils, Boiled, Wo/Salt 1 C:198 g	230	17.9	0.8	0.1	0.4	0.1	39.9	15.8	0.33	0.14	2.1	0.35	358	0	3	37.6	6.59	2.51
½ C:99 g	115	8.9	0.4	0.1	0.2	0.1	19.9	7.9	0.17	0.07	1.05	0.18	179	0	1.5	18.8	3.3	1.26
Lima, Boiled, Wo/Salt 1 C:188 g	216	14.7	0.7	0.2	0.3	0.1	39.3	0	0.3	0.1	0.79	0.3	156.2	0	0	32	4.49	1.79
½ C:94 g	108	7.3	0.4	0.1	0.2	0	19.6	0	0.15	0.05	0.4	0.15	78.1	0	0	16	2.25	0.89
Lima, Cnd 1 C:241 g	190	11.9	0.4	0.1	0.2	0	35.9	0	0.13	0.08	0.63	0.22	121.5	0	0	50.6	4.36	1.57
½ C:120 g	95	5.9	0.2	0.1	0.1	0	17.9	0	0.07	0.04	0.31	0.11	60.5	0	0	25.2	2.17	0.78
Miso 1 C:275 g	567	32.5	16.7	2.4	9.4	3.7	76.9	239.3	0.27	0.69	2.37	0.59	90.8	0	0	181.5	7.54	9.13
½ C:138 g	284	16.3	8.4	1.2	4.7	1.9	38.6	120.1	0.13	0.35	1.19	0.3	45.5	0	0	91.1	3.78	4.58
Mung, Boiled, W/Salt 1 C:202 g	212	14.2	0.8	0.2	0.3	0.1	38.7	48.5	0.33	0.12	1.17	0.14	320.8	0	2	54.5	2.83	1.7
½ C:101 g	106	7.1	0.4	0.1	0.1	0.1	19.3	24.2	0.17	0.06	0.58	0.07	160.4	0	1	27.3	1.41	0.85
Mung, Sprouted, Raw ½ C:52 g	16	1.6	0.1	0	0.1	0	3.1	10.9	0.04	0.06	0.39	0.05	31.6	0	6.9	6.8	0.47	0.21
12-oz Pkg:340 g	102	10.3	0.6	0.2	0.2	0.1	20.2	71.4	0.29	0.42	2.55	0.3	206.7	0	44.9	44.2	3.09	1.39
Mung, Sprouted, Stir-Fried ½ C:62 g	31	2.7	0.1	0	0	0	6.6	19.2	0.09	0.11	0.74	0.08	43.2	0	9.9	8.1	1.18	0.56
1 C:124 g	62	5.3	0.3	0.1	0.1	0.1	13.1	38.4	0.17	0.22	1.49	0.16	86.3	0	19.8	16.1	2.36	1.12
Natto 1 C:175 g	371	31	19.3	2.8	10.9	4.3	25.1	0	0.28	0.33	0	0.23	14	0	22.8	379.8	15.05	5.3
½ C:88 g	187	15.6	9.7	1.4	5.5	2.1	12.6	0	0.14	0.17	0	0.11	7	0	11.4	191	7.57	2.67
Navy, Boiled, Wo/Salt 1 C:182 g	258	15.8	1	0.3	0.5	0.1	47.9	3.6	0.37	0.11	0.97	0.3	254.6	0	1.6	127.4	4.51	1.93
½ C:91 g	129	7.9	0.5	0.1	0.2	0.1	23.9	1.8	0.18	0.06	0.48	0.15	127.3	0	0.8	63.7	2.26	0.96
Pinto, Boiled, Wo/Salt 1 C:171 g	234	14	0.9	0.2	0.3	0.2	43.9	3.4	0.32	0.16	0.68	0.27	294.1	0	3.6	82.1	4.46	1.85
½ C:85 g	116	7	0.4	0.1	0.2	0.1	21.8	1.7	0.16	0.08	0.34	0.13	146.2	0	1.8	40.8	2.22	0.92
Pinto, Cnd 1 C:240 g	187	10.9	0.8	0.2	0.3	0.2	34.9	2.4	0.24	0.15	0.7	0.18	144.5	0	1.7	88.8	3.86	1.66
½ C:120 g	94	5.5	0.4	0.1	0.1	0.1	17.5	1.2	0.12	0.08	0.35	0.09	72.2	0	0.8	44.4	1.93	0.83
Refried Beans 1 C:253 g	271	15.8	2.7	1	0.4	1.2	46.8	0	0.12	0.14	1.23	0.25	211.3	0	15.2	116.4	4.48	3.47
½ C:126 g	135	7.9	1.4	0.5	0.2	0.6	23.3	0	0.06	0.07	0.61	0.13	105.2	0	7.6	58	2.23	1.73
Soy Milk 1 C:240 g	79	6.6	4.6	0.5	2	0.8	4.3	76.8	0.39	0.17	0.35	0.1	3.6	0	0	9.6	1.39	0.55
½ C:120 g	40	3.3	2.3	0.3	1	0.4	2.2	38.4	0.19	0.08	0.18	0.05	1.8	0	0	4.8	0.7	0.28
Soy Sauce (Shoyu) 1 Tbsp:18 g	10	0.9	0	0	0	0	1.5	0	0.01	0.02	0.6	0.03	2.8	0	0	3.1	0.36	0.07
¼ C:58 g	31	3	0.1	0	0	0	4.9	0	0.03	0.08	1.95	0.09	9	0	0	9.9	1.17	0.21
Soy Sauce (Tamari) 1 Tbsp:15 ml	11	1.9	0	0	0	0	1	0	0.01	0.03	0.71	0.04	3.3	0	0	3.6	0.43	0.08
¼ C:58 g	35	6.1	0.1	0	0	0	3.2	0	0.03	0.09	2.29	0.12	10.6	0	0	11.6	1.38	0.25

Food Item & Measure	Cals	Protein (g)	All fat (g)	Sat/fat (g)	Pol/fat (g)	Mon/fat (g)	Carbo (g)	Vit A (IU)	Thiam (mg)	Ribo (mg)	Niac (mg)	Vit B6 (mg)	Fol (mcg)	Vit B12 (mcg)	Vit C (mg)	Ca (mg)	Iron (mg)	Zinc (mg)
Beans and Pulses cont.																		
Soy Sauce Made From Hydrolysed Vegetable Protein																		
1 Tbsp:18 g	7	0.4	0	0	0	0	1.4	0	0.01	0.02	0.51	0.03	2.3	0	0	0.9	0.27	0.06
1/4 C:58 g	24	1.4	0.1	0	0	0	4.5	0	0.02	0.06	1.64	0.08	7.5	0	0	2.9	0.86	0.18
Soybeans, Boiled, W/Salt 1 C:172 g	298	28.6	15.4	2.2	8.7	3.4	17.1	15.5	0.27	0.49	0.69	0.4	92.5	0	2.9	175.4	8.84	1.98
½ C:86 g	149	14.3	7.7	1.1	4.4	1.7	8.5	7.7	0.13	0.25	0.34	0.2	46.3	0	1.5	87.7	4.42	0.99
Soybeans, Boiled, Wo/Salt 1 C:172 g	298	28.6	15.4	2.2	8.7	3.4	17.1	15.5	0.27	0.49	0.69	0.4	92.5	0	2.9	175.4	8.84	1.98
½ C:86 g	149	14.3	7.7	1.1	4.4	1.7	8.5	7.7	0.13	0.25	0.34	0.2	46.3	0	1.5	87.7	4.42	0.99
Soybeans, Steamed, Wo/Salt 1 C:94 g	76	8	4.2	0.5	2.3	0.5	6.1	10.3	0.19	0.05	1.03	0.1	75	0	7.8	55.5	1.23	0.98
½ C:47 g	38	4	2.1	0.2	1.2	0.2	3.1	5.2	0.1	0.02	0.51	0.05	37.5	0	3.9	27.7	0.62	0.49
Split Peas, Boiled, W/Salt 1 C:196 g	231	16.4	0.8	0.1	0.3	0.1	41.4	13.7	0.37	0.11	1.74	0.09	127.2	0	0.8	27.4	2.53	1.96
½ C:98 g	116	8.2	0.4	0.1	0.2	0.1	20.7	6.9	0.19	0.05	0.87	0.05	63.6	0	0.4	13.7	1.26	0.98
Split Peas, Boiled, Wo/Salt 1 C:196 g	231	16.4	0.8	0.1	0.3	0.2	41.4	13.7	0.37	0.11	1.74	0.09	127.2	0	0.8	27.4	2.53	1.96
½ C:98 g	116	8.2	0.4	0.1	0.2	0.1	20.7	6.9	0.19	0.05	0.87	0.05	63.6	0	0.4	13.7	1.26	0.98
Tempeh 1 C:166 g	330	31.5	12.8	1.8	7.2	2.8	28.3	1138.8	0.22	0.18	7.69	0.5	96.3	1.66	0	154.4	3.75	3
½ C:83 g	165	15.7	6.4	0.9	3.6	1.4	14.1	569.4	0.11	0.09	3.84	0.25	43.2	0.83	0	77.2	1.88	1.5
Tofu; Dried-Frozen (Koyadofu), Prepared W/Calcium Sulphate 1 Piece:17 g	82	8.2	5.2	0.8	2.9	1.1	2.5	88.1	0.08	0.05	0.2	0.05	15.6	0	0.1	362.8	1.65	0.83
Tofu; Fried 1 Piece:13 g	35	2.2	2.6	0.4	1.5	0.6	1.4	0	0.02	0.01	0.01	0.01	3.5	0	0	48.4	0.63	0.26
Tofu; Fried, Prepared W/Calcium Sulphate 1 Piece:13 g	35	2.2	2.6	0.4	1.5	0.6	1.4	0	0.02	0.01	0.01	0.01	3.5	0	0	124.9	0.63	0.26
Tofu; Okara C:122 g	94	3.9	2.1	0.2	0.9	0.4	15.3	0	0.02	0.02	0.12	0.14	32.2	0	0	97.6	1.59	0.68
½ C:61 g	47	2	1.1	0.1	0.5	0.2	7.7	0	0.01	0.01	0.06	0.07	16.1	0	0	48.8	0.79	0.34
Tofu; Raw, Firm ½ C:126 g	183	19.9	11	1.6	6.2	2.4	5.4	209.2	0.2	0.13	0.48	0.12	36.9	0	0.3	258.3	13.19	1.98
¼ Blk:81 g	117	12.8	7.1	1	4	1.6	3.5	134.5	0.13	0.08	0.31	0.07	23.7	0	0.2	166.1	8.48	1.27
Tofu; Raw, Firm, Prepared W/Calcium Sulphate ½ C:126 g	183	19.9	11	1.6	6.2	2.4	5.4	209.2	0.2	0.13	0.48	0.12	36.9	0	0.3	860.6	13.19	1.98
¼ Blk:81 g	117	12.8	7.1	1	4	1.6	3.5	134.5	0.13	0.08	0.31	0.07	23.7	0	0.2	553.2	8.48	1.27
Tofu; Raw, Regular ½ C:124 g	94	10	5.9	0.9	3.4	1.3	2.3	105.4	0.1	0.06	0.24	0.06	18.6	0	0.1	130.2	6.65	0.99
¼ Blk:116 g	88	9.4	5.5	0.8	3.1	1.2	2.2	98.6	0.09	0.06	0.23	0.05	17.4	0	0.1	121.8	6.22	0.93

Food Item & Measure	Cals	Protein (g)	All fat (g)	Sat/fat (g)	Pol/fat (g)	Mon/fat (g)	Carbo (g)	Vit A (IU)	Thiam (mg)	Ribo (mg)	Niac (mg)	Vit B6 (mg)	Fol (mcg)	Vit B12 (mcg)	Vit C (mg)	Ca (mg)	Iron (mg)	Zinc (mg)
Beans and Pulses cont.																		
Tofu; Raw, Regular, Prepared W/Calcium Sulphate ½ C:124 g	94	10	5.9	0.9	3.4	1.3	2.3	105.4	0.1	0.06	0.24	0.06	18.6	0	0.1	434	6.65	0.99
½ Blk:116 g	88	9.4	5.5	0.8	3.1	1.2	2.2	98.6	0.09	0.06	0.23	0.05	17.4	0	0.1	406	6.22	0.93
Tofu; Salted And Fermented (Fuyu) 1 Blk:11 g	13	0.9	0.9	0.1	0.5	0.2	0.6	18.2	0.02	0.01	0.04	0.01	3.2	0	0	5.1	0.22	0.17
Tofu; Salted And Fermented (Fuyu), Prep W/Calcium Sulphate 1 Blk:11 g	13	0.9	0.9	0.1	0.5	0.2	0.6	18.2	0.02	0.01	0.04	0.01	3.2	0	0	135.2	0.22	0.17
BREADS AND CAKES																		
Biscuit Dough; Plain, Chilled, Bkd 4 Biscuits:48 g	238	1.9	12	3	–	–	31.2	33.6	0.1	0.08	0.86	–	–	–	0	17.3	0.86	–
Biscuits; Assorted, Commer 11-oz Pkg (Appx 36):312 g	1498	15.9	63	12.5	–	–	221.5	249.6	0.09	0.16	1.25	–	–	–	0	115.4	2.18	–
Biscuits; Butter, Thin, Rich 10 Biscuits:50 g	229	3.1	8.5	4	–	–	35.5	325	0.12	0.11	1.1	–	–	–	0	63	1.15	–
8-oz Pkg (Appx 45):227 g	1037	13.9	38.4	18.2	–	–	160.9	1475.5	0.54	0.48	4.99	–	–	–	0	286	5.22	–
Biscuits; Chocolate Chip, Commer 10 Biscuits, 1⅜ In Diam:53 g	250	2.9	11.1	3.4	–	–	36.9	63.6	0.09	0.08	0.8	–	–	–	0	20.7	1.48	–
Biscuits; Coconut Bars 10 Biscuits:90 g	445	5.6	22.1	8.4	–	–	57.5	144	0.17	0.15	1.53	–	–	–	0	64.8	2.88	–
Biscuits; Fig Bars 4 Biscuits:56 g	200	2.2	3.1	0.9	–	–	42.2	61.6	0.1	0.1	0.9	–	–	–	0	43.7	1.23	–
Biscuits; Gingersnaps 10 Biscuits:70 g	294	3.9	6.2	1.6	–	–	55.9	49	0.2	0.15	1.68	–	–	–	0	51.1	2.24	–
Biscuits; Macaroons 11-oz Pkg (Appx 16):312 g	1482	16.5	72.4	49.9	–	–	206.2	0	0.12	0.47	1.87	–	–	–	0	84.2	2.81	–
Biscuits; Oatmeal, W/Raisins 14-oz Pkg (Appx 30):397 g	1790	24.6	61.1	15.7	–	–	291.8	198.5	1.11	0.79	7.54	–	–	–	0.8	83.4	12.7	–
Biscuits; Peanut 10-oz Pkg (Appx 23):284 g	1343	28.4	54.2	11.4	–	–	190.3	568	0.94	0.68	13.35	–	–	–	0	119.3	7.95	–
Biscuits; Sandwich-Type 4 Biscuits 1⅜ In Diam:40 g	198	1.9	9	2.4	–	–	27.7	0	0.09	0.07	0.84	–	–	–	0	10.4	0.8	–
Biscuits; Shortbread 10 Biscuits:75 g	374	5.4	17.3	4.3	–	–	48.8	60	0.34	0.2	2.33	–	–	–	0	52.5	2.25	–
10½-oz Pkg (40):291 g	1449	21	67.2	16.8	–	–	189.4	232.8	1.31	0.79	9.02	–	–	–	0	203.7	8.73	–

Food item & Measure	Cals	Protein (g)	All fat (g)	Sat/fat (g)	Pol/fat (g)	Mon/fat (g)	Carbo (g)	Vit A (IU)	Thiam (mg)	Ribo (mg)	Niac (mg)	Vit B6 (mg)	Fol (mcg)	Vit B12 (mcg)	Vit C (mg)	Ca (mg)	Iron (mg)	Zinc (mg)
Breads & Cakes cont.																		
Biscuits; Vanilla Wafers																		
10 Reg, 1¾ in Diam:40 g	185	2.2	6.4	1.6	–	–	29.8	52	0.1	0.09	0.84	–	–	–	0	16.4	0.92	–
Bread Pudding; W/Raisins 1 C:265 g	496	14.8	16.2	7.7	–	–	75.3	795	0.19	0.5	1.59	–	–	–	2.7	288.9	2.92	–
Bread; French 1 Sl:35 g	102	3.2	1.1	0.2	–	–	19.4	0	0.03	0.03	0.28	–	–	–	0	15.1	0.25	–
Bread; Raisin 1 Sl/1 lb Loaf:25 g	66	1.7	0.7	0.2	–	–	13.4	0	0.1	0.06	0.6	–	–	–	0	17.8	0.73	–
Bread; Raisin, Toasted 1 Sl/1 lb Loaf:21 g	66	1.7	0.7	0.2	–	–	13.6	0	0.08	0.06	0.59	–	–	–	0	18.1	0.74	–
Bread; Rye, Pumpernickel 1 Reg Sl/1 lb Loaf:32 g	79	2.9	0.4	0.1	–	–	17	0	0.09	0.07	0.61	–	–	–	0	26.9	0.93	–
Bread; Salt-Rising 1 Reg Sl/ 1 lb Loaf:24 g	64	1.9	0.6	0.2	–	–	12.5	2.4	0.01	0.01	0.14	–	–	–	0	5.5	0.14	–
Bread; White, Enr, 3–4%Nfdms, Soft Crumb 1 Reg Sl/1 lb Loaf:25 g	68	2.2	0.8	0.2	–	–	12.6	0	0.1	0.06	0.83	–	–	–	0	21	0.7	–
Bread; White, Enr, 5–6%Nfdms, Firm Crumb 1 Reg Sl/1 lb Loaf:23 g	63	2.1	0.9	0.2	–	–	11.6	0	0.09	0.06	0.76	–	–	–	0	22.1	0.64	–
Bread; White, Enr, 3–4%Nfdms, Soft Crumb, Toasted 1 Reg Sl/1 lb Loaf:22 g	69	2.2	0.8	0.2	–	–	12.9	0	0.08	0.06	0.84	–	–	–	0	21.6	0.73	–
Bread; White, Enr, 5–6%Nfdms, Firm Crumb, Toasted 1 Reg Sl/1 lb Loaf:20 g	64	2.1	0.9	0.2	–	–	11.7	0	0.07	0.06	0.76	–	–	–	0	22.4	0.66	–
Bread; White, Unenr, 3–4%Nfdms, Toasted 1 Reg Sl/1 lb Loaf:22 g	69	2.2	0.8	0.2	–	–	12.9	0	0.01	0.02	0.28	–	–	–	0	21.6	0.18	–
Bread; White, Unenr, 5–6%Nfdms, Firm Crumb, Toasted 1 Reg Sl/lb Loaf:20 g	64	2.1	0.9	0.2	–	–	11.7	0	0.01	0.03	0.21	–	–	–	0	22.4	0.16	–
Bread; White, Unenr, 3–4%Nfdms 1 Reg Sl/1 lb Loaf:25 g	68	2.2	0.8	0.2	–	–	12.6	0	0.02	0.02	0.28	–	–	–	0	21	0.18	–
Bread; White, Unenr, 5–6%Nfdms, Firm Crumb 1 C Crumbs:45 g	124	4.1	1.7	0.4	–	–	22.6	0	0.03	0.06	0.41	–	–	–	0	43.2	0.32	–
1 Reg Sl/1 lb Loaf:23 g	63	2.1	0.9	0.2	–	–	11.6	0	0.02	0.03	0.21	–	–	–	0	22.1	0.16	–
Bread; Whole wheat, 2%Nfdms, Firm Crumb 1 Sl (Flat Top)/1 lb Loaf:23 g	56	2.4	0.7	0.1	–	–	11	0	0.06	0.03	0.64	–	–	–	0	22.8	0.69	–

226 COMPOSITION OF VEGETARIAN FOODS

Food item & Measure	Cals	Protein (g)	All fat (g)	Sat/fat (g)	Pol/fat (g)	Mon/fat (g)	Carbo (g)	Vit A (IU)	Thiam (mg)	Ribo (mg)	Niac (mg)	Vit B6 (mg)	Fol (mcg)	Vit B12 (mcg)	Vit C (mg)	Ca (mg)	Iron (mg)	Zinc (mg)
Breads and Cakes cont.																		
1 Sl (Rnd Top)/1 lb Loaf:25 g	61	2.6	0.8	0.2	–	–	11.9	0	0.06	0.03	0.7	–	–	–	0	24.8	0.75	–
Bread; Whole wheat, Made W/Water, Soft Crumb 1 Reg Sl/																		
1 lb Loaf:28 g	67	2.6	0.7	0.1	–	–	13.8	0	0.08	0.03	0.78	–	–	–	0	23.5	0.84	–
Bread; Whole wheat, 2%Ntdms, Firm Crumb, Toasted 1 Sl (Flat Top)/																		
1 lb Loaf:19 g	55	2.4	0.7	0.1	–	–	10.8	0	0.05	0.03	0.65	–	–	–	0	22.4	0.68	–
1 Sl (Rnd Top)/1 lb Loaf:21 g	61	2.6	0.8	0.2	–	–	11.9	0	0.05	0.03	0.71	–	–	–	0	24.8	0.76	–
Bread; Whole-Wheat, Made W/Water, Soft Crumb, Toasted																		
1 Reg Sl/1 lb Loaf:24 g	69	2.6	0.7	0.2	–	–	14.1	0	0.07	0.03	0.79	–	–	–	0	24	0.86	–
Breadcrumbs; Dry, Grated 1 C:100 g	392	12.6	4.6	1.1	–	–	73.4	0	0.35	0.35	4.8	–	–	–	0	122	4.1	–
Cake Icing; Caramel 1 C:340 g	1224	4.4	22.8	12.5	–	–	260.1	952	0.03	0.2	0	–	–	–	1	346.8	6.8	–
Cake Icing; Chocolate 1 C:275 g	1034	8.8	38.2	21.3	–	–	185.4	577.5	0.06	0.28	0.55	–	–	–	0.6	165	3.3	–
Cake Icing; Coconut 1 C:166 g	604	3.2	12.8	11	–	–	124.3	0	0.02	0.07	0.33	–	–	–	0	10	0.83	–
Cake; Angelfood, Home Rec																		
1 Cake, Vol 247 Cu In:716 g	1926	50.8	1.4	0	–	–	431	0	0.07	1	1.43	–	–	–	0	64.4	1.43	–
Cake; Chocolate, Uniced, Veg-S Cpcake, 2¾-In Diam:33 g	121	1.6	5.7	2	–	–	17.2	49.5	0.01	0.03	0.07	–	–	–	0.1	24.4	0.3	–
Cake; Chocolate, Choc-Icing, Unenr, Veg-S, (Icing Made W/Butter)																		
Cpcake, 2¾-In Diam:44 g	162	2	7.2	2.9	–	–	24.6	–70.4	0.01	0.04	0.09	–	–	–	0.1	30.8	0.44	–
Cake; Coffee Cake, Bkd From Mix, W/Eggs, Milk, Unenr 1 Cake, Vol 55 Cu In:430 g	1385	27.1	41.3	11.9	–	–	225.3	688	0.22	0.47	2.58	–	–	–	0.9	262.3	2.58	–
Cake; Fruit, Dark, Butter Whole, Vol 77 Cu In:1361 g	4940	65.3	187.8	64.7	–	–	813.9	4355.2	2.18	2.18	14.97	–	–	–	5.4	993.5	38.11	–
Cake; Fruit, Dark, Veg-S Whole, Vol 77 Cu In:1361 g	5158	65.3	208.2	43.6	–	–	812.5	1633.2	2.18	2.18	14.97	–	–	–	5.4	979.9	38.11	–
Cake; Fruit, Light, Butter Whole, Vol 77 Cu In:1361 g	5131	81.7	213.7	77.2	–	–	766.2	4083	1.77	1.77	13.61	–	–	–	1.4	939.1	25.86	–

Breads and Cakes cont.

Food Item & Measure	Cals	Protein (g)	All fat (g)	Sat/fat (g)	Pol/fat (g)	Mon/fat (g)	Carbo (g)	Vit A (IU)	Thiam (mg)	Ribo (mg)	Niac (mg)	Vit B$_6$ (mg)	Fol (mcg)	Vit B$_{12}$ (mcg)	Vit C (mg)	Ca (mg)	Iron (mg)	Zinc (mg)
Cake; Fruit, Light, Veg-S Whole, Vol 77 Cu In:1361 g	5294	81.7	224.6	50.4	–	–	781.2	952.7	1.77	1.77	13.61	–	–	–	1.4	925.5	25.86	–
Cake; Gingerbread, Butter Whole, Vol 162 Cu In:1055 g	3133	41.2	103.4	53.4	–	–	542.3	4220	1.69	1.48	13.72	–	–	–	0	728	28.49	–
Cake; Gingerbread, Veg-S Whole, Vol 162 Cu In:1055 g	3344	40.1	112.9	28.8	–	–	548.6	949.5	1.69	1.48	13.72	–	–	–	0	717.4	28.49	–
Cake; Plain or Cupcake, Enr, Butter, Boiled Wht Icing 1 Cake, Vol 200 Cu In:1028 g	3516	39.1	99.7	51.5	–	–	630.2	4214.8	1.23	1.44	11.31	–	–	–	1	514	11.31	–
Cake; Plain or Cupcake, Enr, Veg-S, Boiled Wht Icing 1 Cake, Vol 200 Cu In:1028 g	3619	39.1	107.9	30	–	–	635.3	1336.4	1.23	1.44	10.28	–	–	–	1	503.7	11.31	–
Cake; Plain or Cupcake, Veg-S, Wo/Icing 1 Cake, Vol 162 Cu In:777 g	2828	35	108	30	–	–	434.3	1320.9	0.16	0.7	1.55	–	–	–	1.6	497.3	3.11	–
Cake; Pound, Old Fash, Fat, Unenr Ck Fl, Veg-S 1 Loaf, Vol 89 Cu In:514 g	2431	29.3	151.6	39	–	–	241.6	1439.2	0.15	0.46	1.03	–	–	–	0	107.9	4.11	–
Cake; Sponge 1 Cake, Vol 164 Cu In:524 g	1556	39.8	29.9	9.4	–	–	283.5	2358	0.26	0.73	1.05	–	–	–	0	157.2	6.29	–
Crackers; Cheese 10 Crackers, 1-In Sq:10.8 g	52	1.2	2.3	0.9	–	–	6.5	38.9	0.04	0.04	0.4	–	–	–	0	36.3	0.38	–
Danish Pastry, 4¼-In Diam:65 g	274	4.8	15.3	4.5	–	–	29.6	201.5	0.18	0.2	1.63	–	–	–	0	32.5	1.24	–
Doughnuts; Cake-Type 1 Doughnut, Appx ⅞ oz:25 g	98	1.2	4.7	1.2	–	–	12.9	20	0.05	0.05	0.43	–	–	–	0	10	0.5	–
1 Doughnut, Appx 2 oz:58 g	227	2.7	10.8	2.7	–	–	29.8	46.4	0.12	0.12	0.99	–	–	–	0	23.2	1.16	–
Doughnuts; Yeast, Glazed Doughnut, Appx 1½ oz:42 g	170	2.3	9.6	2.4	–	–	18.7	21	0.08	0.08	0.67	–	–	–	0	13.4	0.76	–
Doughnuts; Yeast, Plain Doughnut, Appx 1½ oz:42 g	174	2.7	11.2	2.8	–	–	15.8	25.2	0.1	0.09	0.8	–	–	–	0	16	0.88	–
Eclairs; W/Custard Fill and Choc-Ic Eclair: 100 g	239	6.2	13.6	4.4	–	–	23.2	340	0.04	0.16	0.1	–	–	–	0	80	0.7	–
Pie Crust, Baked, Veg-S 1 Pie Shell:180 g	900	11	60.1	14.9	–	–	78.8	0	0.54	0.4	5.04	–	–	–	0	25.2	4.5	–
Pie; Apple, Baked 1 Pie, 8-In Diam:550 g	1397	10.5	55.6	13.8	–	–	220	71.5	0.66	0.44	6.05	–	–	–	6.1	44	6.05	–

Food item & Measure	Cals	Protein (g)	All fat (g)	Sat/fat (g)	Pol/fat (g)	Mon/fat (g)	Carbo (g)	Vit A (IU)	Thiam (mg)	Ribo (mg)	Niac (mg)	Vit B₆ (mg)	Fol (mcg)	Vit B₁₂ (mcg)	Vit C (mg)	Ca (mg)	Iron (mg)	Zinc (mg)
Breads and Cakes cont.																		
Pie, Blackberry, Baked, Veg-S, Wo/Salt In Filling 1 Pie, 9-In Diam:945 g	2296	24.6	104	25.8	–	–	325.1	850.5	0.19	0.19	2.84	–	–	–	37.8	179.6	4.73	–
Pie; Cherry, Baked, Veg-S 1 Pie, 9-In Diam:945 g	2466	24.6	106.8	28.2	–	–	362.9	4158	0.19	0.19	4.73	–	–	–	0	132.3	2.84	–
Pie; Mince, Baked, Veg-S 1 Pie, 9-In Diam:945 g	2561	23.6	108.7	28.4	–	†	389.3	18.9	0.95	0.85	9.45	–	–	–	9.5	264.6	16.07	–
Pie; Pumpkin, Baked, Veg-S, Wo/Salt In Filling 1 Pie, 9-In Diam:910 g	1920	36.4	101.9	36	–	–	223	22477	0.27	0.91	4.55	–	–	–	0	464.1	4.55	–
Pie; Rhubarb, Baked, Veg-S, Wo/Salt In Filling 1 Pie, 9-In Diam:945 g	2391	23.6	101.1	25	–	–	361	472.5	0.19	0.38	2.84	–	–	–	28.4	604.8	6.62	–
Rice Pudding; W/Raisins 1 C:265 g	387	9.5	8.2	4.5	–	–	70.8	291.5	0.08	0.37	0.53	–	–	–	0	259.7	1.06	–
Yeast; Baker's, Dry, Active 1 oz:28 g	79	10.33	0.45	0	–	–	10.89	0	0.65	1.51	10.28	–	–	–	0	12.32	4.51	–
¼-oz Pkg:7 g	20	2.58	0.11	0	–	–	2.72	0	0.16	0.38	2.57	–	–	–	0	3.08	1.13	–
Yeast; Brewer's, Debittered 1 oz:28 g	79	10.86	0.28	0	–	–	10.75	0	4.37	1.2	10.61	–	–	–	0	58.8	4.84	–
1 Tbsp:8 g	23	3.1	0.08	0	–	–	3.07	0	1.25	0.34	3.03	–	–	–	0	16.8	1.38	–
DAIRY PRODUCTS																		
Butter 1 Pat:5 g	36	0	4.1	2.5	0.2	1.2	0	152.9	0	0	0	0	0.1	0.01	0	1.2	0.01	0
1 Slk Nt Wt 4 oz:113.4 g	813	1	92	57.3	3.4	26.6	0.1	3467.8	0.01	0.04	0.05	0	3.2	0.14	0	26.7	0.18	0.06
Cheese Pizza 1 Pizza:503 g	1122	61.3	25.7	12.3	3.9	7.9	163.7	3053.2	1.46	1.31	19.82	0.35	467.8	2.67	10.1	930.6	4.63	6.49
1 Slice:63 g	140	7.7	3.2	1.5	0.5	1	20.5	382.4	0.18	0.16	2.48	0.04	58.6	0.33	1.3	116.6	0.58	0.81
Cheese Souffle 1 C:95 g	207	9.4	16.3	8.2	–	–	5.9	760	0.05	0.23	0.19	–	–	–	0	191	0.95	–
1 Portion:110 g	240	10.9	18.8	9.5	–	–	6.8	880	0.06	0.26	0.22	–	–	–	0	221.1	1.1	–
Cheese Spread; Pasteurised Processed 1 Jar, Nt Wt 5 oz:142 g	412	23.3	30.2	18.9	0.9	8.8	12.4	1119	0.07	0.61	0.19	0.17	9.9	0.57	0	797.9	0.47	3.68
1 oz:28 g	81	4.6	5.9	3.7	0.2	1.7	2.4	220.6	0.01	0.12	0.04	0.03	2	0.11	0	157.3	0.09	0.73
Cheese; Blue 1 C, Crmbld, Not Pkd:135 g	477	28.9	38.8	25.2	1.1	10.5	3.2	973.4	0.04	0.52	1.37	0.22	49.1	1.64	0	712.3	0.42	3.59
1 oz:28 g	99	6	8.1	5.2	0.2	2.2	0.7	201.9	0.01	0.11	0.28	0.05	10.2	0.34	0	147.7	0.09	0.74
Cheese; Brie 1 oz:28 g	93	5.8	7.8	4.9	0.2	2.2	0.1	186.8	0.02	0.15	0.11	0.07	18.2	0.46	0	51.5	0.14	0.67
1 Pkg, Nt Wt 4.5 oz:128 g	427	26.6	35.4	22.3	1.1	10.3	0.6	853.8	0.09	0.67	0.49	0.3	83.2	2.11	0	235.5	0.64	3.05

Food item & Measure	Cals	Protein (g)	All fat (g)	Sat fat (g)	Pol fat (g)	Mon fat (g)	Carbo (g)	Vit A (IU)	Thiam (mg)	Ribo (mg)	Niac (mg)	Vit B6 (mg)	Fol (mcg)	Vit B12 (mcg)	Vit C (mg)	Ca (mg)	Iron (mg)	Zinc (mg)
Dairy Products cont.																		
Cheese; Camembert 1 oz:28 g	84	5.5	6.8	4.3	0.2	2	0.1	258.4	0.01	0.14	0.18	0.06	17.4	0.36	0	108.5	0.09	0.67
1 Wdg, Nt Wt 1.3 oz:38 g	114	7.5	9.2	5.8	0.2	2.7	0.2	350.7	0.01	0.19	0.24	0.09	23.6	0.49	0	147.3	0.13	0.9
Cheese; Cheddar 1 C, Grated, Not Pkd:113 g	455	28.1	37.5	23.8	1.1	10.6	1.5	1196.7	0.03	0.42	0.09	0.08	20.6	0.93	0	815.1	0.77	3.51
1 oz:28 g	113	7	9.3	5.9	0.3	2.6	0.4	296.5	0.01	0.11	0.02	0.02	5.1	0.23	0	202	0.19	0.87
Cheese; Cheshire 1 oz:28 g	108	6.5	8.6	5.5	0.2	2.4	1.3	275.8	0.01	0.08	0.02	0.02	5.1	0.23	0	180	0.06	0.78
Cheese; Cottage, Creamed, Large or Small Curd 1 C, Not Pkd:210 g	217	26.2	9.5	6	0.3	2.7	5.6	342.3	0.04	0.34	0.26	0.14	25.6	1.31	0	126	0.29	0.78
4 oz:113 g	117	14.1	5.1	3.2	0.2	1.5	3	184.2	0.02	0.18	0.14	0.08	13.8	0.7	0	67.8	0.16	0.42
Cheese; Cottage, Low fat (2%) 1 C, Not Pkd:226 g	203	31.1	4.4	2.8	0.1	1.2	8.2	158.2	0.05	0.42	0.33	0.17	29.6	1.61	0	154.8	0.36	0.95
4 oz:113 g	101	15.5	2.2	1.4	0.1	0.6	4.1	79.1	0.03	0.21	0.16	0.09	14.8	0.8	0	77.4	0.18	0.47
Cheese; Edam 1 oz:28 g	100	7	7.8	4.9	0.2	2.3	0.4	256.5	0.01	0.11	0.02	0.02	4.5	0.43	0	204.7	0.12	1.05
1 Pkg, Nt Wt 7 oz:198 g	706	49.5	55	34.8	1.3	16.1	2.8	1813.7	0.07	0.77	0.16	0.15	32.1	3.04	0	1447.4	0.87	7.43
Cheese; Feta 1 oz:28 g	74	4	6	4.2	0.2	1.3	1.2	125.2	0.04	0.24	0.28	0.12	9	0.47	0	137.9	0.18	0.81
Cheese; Gruyère 1 oz:28 g	116	8.4	9.1	5.3	0.5	2.8	0.1	341.3	0.02	0.08	0.03	0.02	2.9	0.45	0	283.1	0.05	1.09
1 Pkg, Nt Wt 6 oz:170 g	702	50.7	55	32.2	3	17.1	0.6	2072.3	0.1	0.47	0.18	0.14	17.7	2.72	0	1718.7	0.29	6.63
Cheese; Mozzarella, Whole Milk 1 oz:28 g	79	5.4	6.1	3.7	0.2	1.8	0.6	221.8	0	0.07	0.02	0.02	2	0.18	0	144.8	0.05	0.62
Cheese; Parmesan, Grated 1 oz:28 g	128	11.6	8.4	5.3	0.2	2.5	1.1	196.3	0.01	0.11	0.09	0.03	2.2	0.39	0	385.2	0.27	0.89
1 Tbsp:5 g	23	2.1	1.5	1	0	0.4	0.2	35.1	0	0.02	0.02	0.01	0.4	0.07	0	68.8	0.05	0.16
Cream; Fluid, Heavy Whipping 1 C or 2 C Whipped:238 g	821	4.9	88.1	54.8	3.3	25.4	6.6	3498.6	0.05	0.26	0.09	0.06	8.8	0.43	1.4	153.8	0.07	0.55
1 Tbsp:15 g	52	0.3	5.6	3.5	0.2	1.6	0.4	220.5	0	0.02	0.01	0	0.6	0.03	0.1	9.7	0	0.03
Cream; Fluid, Medium, 25% Fat 1 C:239 g	583	5.9	59.8	37.2	2.2	17.3	8.3	2251.4	0.07	0.33	0.12	0.07	5.5	0.52	1.7	215.6	0.1	0.62
1 Tbsp:15 g	37	0.4	3.8	2.3	0.1	1.1	0.5	141.3	0	0.02	0.01	0	0.4	0.03	0.1	13.5	0.01	0.04
Custard; Baked 1 C:265 g	305	14.3	14.6	6.8	–	–	29.4	927.5	0.11	0.5	0.27	–	–	–	1.3	296.8	1.06	–
Egg Substitute; Powder 0.35 oz:9.9 g	44	5.5	1.3	0.4	0.2	0.5	2.2	121.6	0.02	0.17	0.06	0.01	12.4	0.35	0.1	32.3	0.31	0.18
0.7 oz:19.8 g	88	11	2.6	0.8	0.3	1.1	4.3	243.1	0.04	0.35	0.11	0.03	24.8	0.7	0.2	64.6	0.63	0.36
Eggs; Chicken, White, Raw, Fresh and Frozen 1 C:243 g	122	25.6	0	–	–	–	2.5	0	0.01	1.1	0.22	0.01	7.3	0.49	0	14.6	0.07	0.02
1 Lrg Egg White:33.4 g	17	3.5	0	–	–	–	0.3	0	0	0.15	0.03	0	1	0.07	0	2	0.01	0

Dairy Products cont.

Food item & Measure	Cals	Protein (g)	All fat (g)	Sat/fat (g)	Pol/fat (g)	Mon/fat (g)	Carbo (g)	Vit A (IU)	Thiam (mg)	Ribo (mg)	Niac (mg)	Vit B_6 (mg)	Fol (mcg)	Vit B_{12} (mcg)	Vit C (mg)	Ca (mg)	Iron (mg)	Zinc (mg)
Eggs: Chicken, Whole, Fried																		
1 Lrg Egg:46 g	92	6.2	6.9	1.9	1.3	2.8	0.6	394.2	0.03	0.24	0.04	0.07	17.5	0.42	0	25.3	0.72	0.55
Eggs: Chicken, Whole, Hard-Boiled																		
1 C, Chopped:136 g	211	17.2	14.4	4.5	2	5.5	1.5	754.8	0.09	0.7	0.09	0.16	61.2	1.52	–	68	1.62	1.41
1 Lrg Egg:50 g	78	6.3	5.3	1.6	0.7	2	0.6	277.5	0.03	0.26	0.03	0.06	22.5	0.56	–	25	0.6	0.52
Eggs: Chicken, Whole, Omelette																		
1 Lrg Egg:59 g	90	6.1	6.8	1.9	1.3	2.7	0.6	385.9	0.03	0.24	0.03	0.06	17.1	0.41	0	24.8	0.7	0.54
Eggs: Chicken, Whole, Poached																		
1 Lrg Egg:50 g	75	6.2	5	1.5	0.7	1.9	0.6	316	0.02	0.22	0.03	0.06	17.5	0.4	0	24.5	0.72	0.55
Eggs: Chicken, Whole, Scrambled																		
1 C:220 g	365	24.4	26.9	8.1	4.7	10.5	4.8	1500.4	0.11	0.96	0.17	0.26	66	1.69	0.4	156.2	2.64	2.2
1 Egg:60 g	100	6.7	7.3	2.2	1.3	2.9	1.3	409.2	0.03	0.26	0.05	0.07	18	0.46	0.1	42.6	0.72	0.6
Eggs: Chicken, Whole, Dried																		
1 C, Sifted:85 g	505	39	35.5	10.7	4.6	14.2	4.1	1657.5	0.26	1	0.21	0.34	156.2	8.5	0	179.8	6.7	4.62
1 Tbsp:5 g	30	2.3	2.1	0.6	0.3	0.8	0.2	97.5	0.02	0.06	0.01	0.02	9.2	0.5	0	10.6	0.39	0.27
Eggs: Chicken, Whole, Raw, Fresh and Frozen 1 C:243 g	362	30.4	24.4	7.5	3.3	9.3	3	1543.1	0.15	1.23	0.18	0.34	114.2	2.43	0	119.1	3.5	2.67
1 Lrg Egg:50 g	75	6.3	5	1.6	0.7	1.9	0.6	317.5	0.03	0.25	0.04	0.07	23.5	0.5	0	24.5	0.72	0.55
Eggs: Chicken, Yolk, Raw, Fresh																		
1 C:243 g	870	40.7	75	23.2	10.2	28.5	4.3	4726.4	0.41	1.55	0.04	0.95	354.8	7.56	0	332.9	8.58	7.56
1 Lrg Egg Yolk:16.6 g	59	2.8	5.1	1.6	0.7	2	0.3	322.9	0.03	0.11	0	0.07	24.2	0.52	0	22.7	0.59	0.52
Eggs: Duck, Whole, Fresh, Raw 1 Egg:70 g	130	9	9.6	2.6	0.9	4.6	1	929.6	0.11	0.28	0.14	0.18	56	3.78	0	44.6	2.7	0.99
Eggs: Goose, Whole, Fresh, Raw																		
1 Egg:144 g	267	20	19.1	5.2	2.4	8.3	1.9	1843.2	0.21	0.55	0.27	0.34	108.9	7.34	0	86.7	5.24	1.92
Ice Cream: Vanilla, Regular, Appx 10% Fat																		
1 C (8 fl oz):133 g	269	4.8	14.3	8.9	0.5	4.1	31.7	542.6	0.05	0.33	0.13	0.06	2.8	0.63	0.7	175.7	0.12	1.41
½ Gal:1064 g	2153	38.4	114.6	71.3	4.3	33.1	253.8	4341.1	0.41	2.63	1.07	0.49	22.3	5	5.6	1405.5	0.96	11.28
Ice Cream: Vanilla, Rich, Appx 16% Fat																		
1 C (8 fl oz):148 g	349	4.1	23.7	14.7	0.9	6.8	32	896.9	0.04	0.28	0.12	0.05	2.4	0.54	0.6	151.1	0.1	1.21
½ Gal:1188 g	2805	33.2	190.1	118.3	7.1	54.9	256.5	7199.3	0.36	2.27	0.93	0.43	19	4.31	4.9	1213	0.83	9.74

Dairy Products cont.

Food item & Measure	Cals	Protein (g)	All fat (g)	Sat/fat (g)	Pol/fat (g)	Mon/fat (g)	Carbo (g)	Vit A (IU)	Thiam (mg)	Ribo (mg)	Niac (mg)	Vit B_6 (mg)	Fol (mcg)	Vit B_{12} (mcg)	Vit C (mg)	Ca (mg)	Iron (mg)	Zinc (mg)
Milk Shakes; Thick Vanilla 1 Cntnr, Nt Wt 11 oz:313 g	350	12.1	9.5	5.9	0.4	2.7	55.6	356.8	0.09	0.61	0.46	0.13	20.7	1.63	0	457.3	0.31	1.22
Milk; Cow, Cnd, Condensed, Sweetened 1 C:306 g	982	24.2	26.6	16.8	1	7.4	166.5	1003.7	0.28	1.27	0.64	0.16	34.3	1.36	8	867.5	0.58	2.88
1 fl oz:38.2 g	123	3	3.3	2.1	0.1	0.9	20.8	125.3	0.03	0.16	0.08	0.02	4.3	0.17	1	108.3	0.07	0.36
Milk; Cow, Dry, Skim, Non-Fat Solids, Instant, W/Added Vit A 1 C:68 g	244	23.9	0.5	0.3	0	0.1	35.5	1611.6	0.28	1.19	0.61	0.23	33.9	2.72	3.8	836.9	0.21	3
1.3 C or 3.2-oz Envl:91 g	326	31.9	0.7	0.4	0	0.2	47.5	2156.7	0.38	1.59	0.81	0.31	45.3	3.63	5.1	1120	0.28	4.01
Milk; Cow, Dry, Skim, Non-Fat Solids, Regular, Wo/Added Vit A 1 C:120 g	435	43.4	0.9	0.6	0	0.2	62.4	43.2	0.5	1.86	1.14	0.43	60.1	4.84	8.1	1508.3	0.38	4.9
⅓ C:30 g	109	10.9	0.2	0.2	0	0.1	15.6	10.8	0.12	0.47	0.29	0.11	15	1.21	2	377.1	0.1	1.22
Milk; Cow, Dry, Whole 1 C:128 g	635	33.7	34.2	21.4	0.9	10.1	49.2	1180.2	0.36	1.54	0.83	0.39	47.4	4.16	11.1	1167.9	0.6	4.28
⅓ C:32 g	159	8.4	8.6	5.4	0.2	2.5	12.3	295	0.09	0.39	0.21	0.1	11.8	1.04	2.8	292	0.15	1.07
Milk; Cow, Low fat (1%), Past and Raw, Fluid, 1 C:244 g	102	8	2.6	1.6	0.1	0.8	11.7	500.2	0.1	0.41	0.21	0.1	12.4	0.9	2.4	300.1	0.12	0.95
1 Qt:976 g	409	32.1	10.4	6.4	0.4	3	46.7	2000.8	0.38	1.63	0.85	0.42	49.8	3.59	9.5	1200.5	0.49	3.81
Milk; Cow, Malted, Beverage 1 C Mlk + ¾ oz Pdr:265 g	236	10.3	9.8	6	0.6	2.8	27.3	368.4	0.2	0.59	1.31	0.19	21.7	1.03	2.9	355.1	0.27	1.14
Milk; Cow, Skim, Past and Raw, Fluid, W/Nfms 1 C:245 g	90	8.8	0.6	0.4	0	0.2	12.3	499.8	0.1	0.43	0.22	0.11	13.2	0.95	2.5	316.3	0.12	1
1 Qt:980 g	361	35	2.5	1.6	0.1	0.6	49.2	1999.2	0.4	1.72	0.89	0.45	52.9	3.78	9.9	1265.2	0.49	4.02
Milk; Cow, Whole, Past and Raw, Fluid, 3.3% Fat 1 C:244 g	150	8	8.2	5.1	0.3	2.4	11.4	307.4	0.09	0.4	0.2	0.1	12.2	0.87	2.3	291.3	0.12	0.93
1 Qt:976 g	600	32.1	32.6	20.3	1.2	9.4	45.5	1229.8	0.37	1.58	0.82	0.41	48.8	3.48	9.2	1165.3	0.49	3.71
Milk; Goat, Fluid 1 C:244 g	168	8.7	10.1	6.5	0.4	2.7	10.9	451.4	0.12	0.34	0.68	0.11	1.5	0.16	3.2	325.7	0.12	0.73
1 Qt:976 g	672	34.8	40.4	26	1.5	10.8	43.4	1805.6	0.47	1.35	2.7	0.45	5.9	0.63	12.6	1303	0.49	2.93
Milk; Human, Mature, Fluid 1 C:246 g	171	2.5	10.8	4.9	1.2	4.1	17	592.9	0.03	0.09	0.44	0.03	12.8	0.11	12.3	79.2	0.07	0.42
1 fl oz:30.8 g	21	0.3	1.4	0.6	0.2	0.5	2.1	74.2	0	0.01	0.05	0	1.6	0.01	1.5	9.9	0.01	0.05

Food Item & Measure	Cals	Protein (g)	All fat (g)	Sat fat (g)	Pol fat (g)	Mon fat (g)	Carbo (g)	Vit A (IU)	Thiam (mg)	Ribo (mg)	Niac (mg)	Vit B6 (mg)	Fol (mcg)	Vit B12 (mcg)	Vit C (mg)	Ca (mg)	Iron (mg)	Zinc (mg)
Dairy Products cont.																		
Pancakes; Plain and Buttermilk, Made W/Egg and Milk 1 × 4 In–Diam:27 g	61	1.9	2	0.7	–	–	8.8	67.5	0.02	0.05	0.11	–	–	–	0.1	58.1	0.19	–
1 × 6-In Diam:73 g	164	5.3	5.3	1.9	–	–	23.7	182.5	0.04	0.12	0.29	–	–	–	0.4	157	0.51	–
Sundae, Strawberry 1 Sundae:153 g	268	6.3	7.9	3.7	1	2.7	44.7	221.9	0.06	0.28	0.9	0.08	18.4	0.64	2	160.7	0.32	0.66
Yoghurt; Fruit, Low fat, 10 Grams Protein																		
Per 8 oz 1 Cntnr, Nt Wt 8 oz:227 g	231	9.9	2.5	1.6	0.1	0.7	43.2	104.4	0.08	0.4	0.22	0.09	21.1	1.06	1.5	344.8	0.16	1.68
½ Cntnr, Nt Wt 4 oz:113 g	115	4.9	1.2	0.8	0	0.3	21.5	52	0.04	0.2	0.11	0.05	10.5	0.53	0.8	171.7	0.08	0.84
Yoghurt; Plain, Low fat, 12 Grams Protein																		
Per 8 oz 1 Cntnr, Nt Wt 8 oz:227 g	144	11.9	3.5	2.3	0.1	1	16	149.8	0.1	0.49	0.26	0.11	25.4	1.28	1.8	414.5	0.18	2.02
½ Cntnr, Nt Wt 4 oz:113 g	72	5.9	1.8	1.1	0.1	0.5	8	74.6	0.05	0.24	0.13	0.06	12.7	0.64	0.9	206.3	0.09	1.01
Yoghurt; Plain, Whole Milk, 8 Grams Protein																		
Per 8 oz 1 Cntnr, Nt Wt 8 oz:227 g	139	7.9	7.4	4.8	0.2	2	10.6	279.2	0.07	0.32	0.17	0.07	16.8	0.84	1.2	274	0.11	1.34
½ Cntnr, Nt Wt 4 oz:113 g	69	3.9	3.7	2.4	0.1	1	5.3	139	0.03	0.16	0.08	0.04	8.4	0.42	0.6	136.4	0.06	0.67
DRINKS																		
Apple Juice; Cnd Or Bottled, Unsweetened, Wo/Added Asc Acid																		
1 fl oz:31 g	15	0	0	0	0	0	3.6	0.3	0.01	0.01	0.03	0.01	0	0	0.3	2.2	0.11	0.01
1 C:248 g	117	0.2	0.3	0.1	0.1	0	29	2.5	0.05	0.04	0.25	0.07	0.3	0	2.2	17.4	0.92	0.07
Beer, Light 1 fl oz:29.5 g	8	0.1	0	0	0	0	0.4	0	0	0.01	0.12	0.07	1.2	0	0	1.5	0.1	0.01
12-fl oz Can:354 g	99	0.7	0	0	0	0	4.6	0	0.03	0.11	1.39	0.12	14.5	0.04	0	17.7	0.14	0.11
Beer, Regular 1 fl oz:29.7 g	12	0.1	0	0	0	0	1.1	0	0	0.01	0.13	0.01	1.8	0.01	0	1.5	0.01	0.01
12-fl oz Can:356 g	146	1.1	0	0	0	0	13.2	0	0.02	0.09	1.61	0.18	21.4	0.07	0	17.8	0.11	0.07
Bloody Mary 1 Cocktail (5 fl oz):148 g	115	0.7	0.2	0	0	0	4.9	507.6	0.05	0.03	0.64	0.11	19.7	0	20.4	10.4	0.55	0.13
1 fl oz:29.7 g	23	0.2	0	0	0	0	1	101.9	0.01	0.01	0.13	0.02	4	0	4.1	2.1	0.11	0.03
Carbonated Orange 1 fl oz:31.0 g	15	0	0	0	0	0	3.8	0	0	0	0	0	0	0	0	1.6	0.02	0.03
12-fl oz Can:372 g	179	0	0	0	0	0	45.8	0	0	0	0	0	0	0	0	18.6	0.22	0.37
Club Soda 1 fl oz:29.6 g	13	0	0	0	0	0	3.3	0	0	0	0	0	0	0	0	0.9	0.02	0.02
12-fl oz Can:355 g	153	0	0	0	0	0	39.8	0	0	0	0	0	0	0	0	10.7	0.28	0.25

Food item & Measure	Cals	Protein (g)	All fat (g)	Sat-fat (g)	Pol-fat (g)	Mon-fat (g)	Carbo (g)	Vit A (IU)	Thiam (mg)	Ribo (mg)	Niac (mg)	Vit B6 (mg)	Fol (mcg)	Vit B12 (mcg)	Vit C (mg)	Ca (mg)	Iron (mg)	Zinc (mg)
Drinks cont.																		
Cocoa Mix; No Added Nutrients, Powder 1-oz Pkt(3–4 Rd Tsp):28.4 g	103	3.1	1.1	0.7	0	0.4	22.5	4	0.03	0.16	0.17	0.03	0	0.37	0.5	92.6	0.34	0.41
Coffee Substitute; Cereal Grain Beverage, Powder 1 Tsp:2.3 g	9	0.1	0.1	0	0	0	1.9	0	0.01	0	0.39	0.02	0.6	0	0	1.2	0.11	0.01
Coffee Substitute; Cereal Grain Beverage, Prepared With Water 6 fl oz Wtr + 1 Tsp Pdr:180 g	9	0.2	0	0	0	0	1.8	0	0.01	0	0.39	0.02	0.5	0	0	5.4	0.11	0.05
Coffee, Brewed 1 fl oz:29.6 g	1	0	0	0	0	0	0.1	0	0	0	0.07	0	0	0	0	0.6	0.01	0.01
6 fl oz:177 g	4	0.2	0	0	0	0	0.7	0	0	0	0.39	0	0.2	0	0	3.5	0.09	0.04
Coffee; Instant, Decaffeinated, Powder, Prepared With Water 6 fl oz Wtr +1 Rd Tsp:179 g	4	0.2	0	0	0	0	0.7	0	0	0.03	0.5	0	0	0	0	5.4	0.07	0.05
Coffee; Instant, Regular 6 fl oz Wtr +1 Rd Tsp:179 g	4	0.2	0	0	0	0	0.7	0	0	0	0.51	0	0	0	0	5.4	0.09	0.05
Cola 1 fl oz:30.8 g	13	0	0	0	—	—	3.2	0	0	0	0	0	0	0	0	0.9	0.01	0
12-fl oz Can:370 g	152	0	0	0	—	—	38.5	0	0	0	0	0	0	0	0	11.1	0.11	0.04
Cola, Low-Cal 1 fl oz:29.6	0	0	0	0	0	0	0	0	0	0	0	0	0	0	0	1.2	0.11	0.01
12-fl oz Can:355 g	0	0	0	0	—	—	0.4	0	0	0	0	0	0	0	0	14.2	0.14	0.18
Daiquiri, Cnd 1 fl oz:30.5 g	38	0	0	0	—	—	4.8	0.6	0	0	0	0	0.2	0	0.4	0	0	0.01
6.8-fl oz Can:207 g	259	0	0	0	—	—	32.5	4.1	0	0	0.03	0.01	1.7	0	2.7	0	0.02	0.06
Distilled Alcohol, All (Gin, Rum, Vodka, Whiskey) 80% Proof 1 fl oz:27.8 g	64	0	0	0	0	0	0	0	0	0	0	0	0	0	0	0	0	0
1.5-fl oz Jigger:42 g	97	0	0	0	0	0	0	0	0	0	0.01	0	0	0	0	0	0.02	0.02
Gin and Tonic 1 Cocktail (7.5 fl oz):225 g	171	0	0	0	0	0	15.8	2.3	0	0	0.02	0	1.1	0	0.9	4.5	0.05	0.18
1 fl oz:30.0 g	23	0	0	0	0	0	2.1	0.3	0	0	0	0	0.2	0	0.1	0.6	0.01	0.02
Grapefruit Juice; Cnd; Unsweetened 1 fl oz:30.9 g	12	0.2	0	0	0	0	2.8	2.2	0.01	0.01	0.07	0.01	3.2	0	9	2.2	0.06	0.03
1 C:247 g	94	1.3	0.3	0	0.1	0	22.1	17.3	0.1	0.05	0.57	0.05	25.7	0	72.1	17.3	0.49	0.22
Lemon Juice; Raw 1 Tbsp:15.2 g	4	0.1	0	—	—	—	1.3	3	0	0	0.02	0.01	2	0	7	1.1	0	0.01
1 C:244 g	61	0.9	0	—	—	—	21.1	48.8	0.07	0.02	0.24	0.12	31.5	0	112.2	17.1	0.07	0.12

234 COMPOSITION OF VEGETARIAN FOODS

Food Item & Measure	Cals	Protein (g)	All fat (g)	Sat/fat (g)	Pol/fat (g)	Mon/fat (g)	Carbo (g)	Vit A (IU)	Thiam (mg)	Ribo (mg)	Niac (mg)	Vit B_6 (mg)	Fol (mcg)	Vit B_{12} (mcg)	Vit C (mg)	Ca (mg)	Iron (mg)	Zinc (mg)
Drinks cont.																		
Lime Juice; Raw 1 Tbsp:15.4 g	4	0.1	0	0	0	0	1.4	1.5	0	0	0.02	0.01	1.3	0	4.5	1.4	0	0.01
1 C:246 g	66	1.1	0.3	0	0.1	0	22.2	24.6	0.05	0.02	0.25	0.11	20.2	0	72.1	22.1	0.07	0.15
Lime Juice; Cnd or Bottled, Unsweetened																		
1 Tbsp:15.4 g	10	0.1	0.1	0	0	0	2.6	0	0	0.01	0.09	0	1.2	0	11	0.8	0.05	0.01
1 C:246 g	162	.2	1.1	0.1	0.2	0.1	40.7	0	0.03	0.16	1.48	0.07	19.4	0	175.9	12.3	0.76	0.17
Malt Beverage 1 fl oz:30.0 g	3	0.1	0	0	0	0	0.4	0	0	0.01	0.14	0.02	1.8	0.01	0	2.1	0.04	0
12-fl oz Can:360 g	32	1.1	0	0	0	0	5	0	0.02	0.09	1.63	0.18	21.6	0.07	0	25.2	0.04	0.04
Orange Drink; Cnd 1 fl oz:31.0 g	16	0	0	0	0	0	4	5.6	0	0	0.01	0	0.7	0	10.6	1.9	0.09	0.03
6-fl oz Glass:186 g	95	0	0	0	0	0	24	33.5	0.01	0.01	0.06	0.02	4.1	0	63.4	11.2	0.52	0.17
Orange Juice; Raw Juice																		
from 1 Fruit:86 g	39	0.6	0.2	0	0	0	8.9	172	0.08	0.03	0.34	0.03	26.1	0	43	9.5	0.17	0.04
1 C:248 g	112	1.7	0.5	0.1	0.1	0.1	25.8	496	0.22	0.07	0.99	0.1	75.1	0	124	27.3	0.5	0.12
Orange Juice; Cnd, Unsweetened																		
1 fl oz:31.1 g	13	0.2	0	0	0	0	3.1	54.4	0.02	0.01	0.1	0.03	5.6	0	10.7	2.5	0.14	0.02
1 C:249 g	105	1.5	0.4	0	0.1	0.1	24.5	435.8	0.15	0.07	0.78	0.22	45.1	0	85.7	19.9	1.1	0.17
Pina Colada 1 Cocktail (4.5 fl oz):141 g	262	0.6	2.7	1.2	0.5	0.2	39.9	2.8	0.04	0.02	0.17	0.06	14.4	0	6.6	11.3	0.31	0.18
1 fl oz:31.4 g	58	0.1	0.6	0.3	0.1	0.1	8.9	0.6	0.01	0	0.04	0.01	3.2	0	1.5	2.5	0.07	0.04
Pineapple Juice; Cnd, Unsweetened, Wo/Added Asc Acid 1 fl oz:31.3 g	18	0.1	0	0	0	0	4.3	1.6	0.02	0.01	0.08	0.03	7.2	0	3.4	5.3	0.08	0.03
1 C:250 g	140	0.8	0.2	0	0.1	0	34.5	12.5	0.14	0.06	0.64	0.24	57.8	0	26.8	42.5	0.65	0.28
Prune Juice; Cnd 1 fl oz:32.0 g	23	0.2	0	0	0	0	5.6	1	0.01	0.02	0.25	0.07	0.1	0	1.3	3.8	0.38	0.07
1 C:256 g	182	1.6	0.1	0	0	0.1	44.7	7.7	0.04	0.18	2.01	0.56	1	0	10.5	30.7	3.02	0.54
Tea; Brewed 1 fl oz:29.6 g	0	0	0	0	0	0	0.1	0	0	0	0	0	1.5	0	0	0	0	0
6 fl oz:178 g	2	0	0	0	0	0	0.5	0	0	0.02	0	0	9.3	0	0	0	0.02	0.02
Tea; Herb, Camomile, Brewed																		
1 fl oz:29.6 g	0	0	0	0	0	0	0.1	5.9	0	0	0	0	0.2	0	0	0.6	0.02	0.01
6 fl oz:178 g	2	0	0	0	0	0	0.4	35.6	0.02	0.01	0	0	1.1	0	0	3.6	0.14	0.07
Tea; Herb, Other Than Camomile, Brewed																		
1 fl oz:29.6 g	0	0	0	0	0	0	0.1	0	0	0	0	0	0.2	0	0	0.6	0.02	0.01
6 fl oz:178 g	2	0	0	0	0	0	0.4	0	0.02	0.01	0	0	1.1	0	0	3.6	0.14	0.07

Food item & Measure	Cals	Protein (g)	All fat (g)	Sat/fat (g)	Pol/fat (g)	Mon/fat (g)	Carbo (g)	Vit A (IU)	Thiam (mg)	Ribo (mg)	Niac (mg)	Vit B6 (mg)	Fol (mcg)	Vit B12 (mcg)	Vit C (mg)	Ca (mg)	Iron (mg)	Zinc (mg)
Drinks cont.																		
Tequila Sunrise 1 Cocktail (5.5 fl oz):172 g	189	0.5	0.2	0	0	0	14.8	166.8	0.07	0.03	0.33	0.09	18.2	0	33.2	10.3	0.48	0.1
1 fl oz:31.2 g	34	0.1	0	0	0	0	2.7	30.3	0.01	0	0.06	0.02	3.3	0	6	1.9	0.09	0.02
Tomato Juice; Cnd, W/Salt Added ½ C:122 g	21	0.9	0.1	0	0	0	5.2	678.3	0.06	0.04	0.82	0.14	24.3	0	22.3	11	0.71	0.17
6 fl oz:182 g	31	1.4	0.1	0	0	0	7.7	1011.9	0.09	0.06	1.22	0.2	36.2	0	33.3	16.4	1.06	0.25
Tonic Water 1 fl oz:30.5 g	10	0	0	0	0	0	2.7	0	0	0	0	0	0	0	0	0.3	0.04	0.03
12-fl oz Can:366 g	124	0	0	0	0	0	32.2	0	0	0	0	0	0	0	0	3.7	0.04	0.37
Vegetable Juice Cocktail; Cnd ½ C:121 g	23	0.8	0.1	0	0.1	0	5.5	1415.7	0.05	0.03	0.88	0.17	25.5	0	33.5	13.3	0.51	0.24
6 fl oz:182 g	35	1.2	0.2	0	0.1	0	8.3	2129.4	0.08	0.05	1.32	0.25	38.4	0	50.4	20	0.76	0.36
Water; Bottled Mineral 1 C:237 g	0	0	0	0	0	0	0	0	0	0	0	0	0	0	0	33.2	0	0
6.5-fl oz Bottle:192 g	0	0	0	0	0	0	0	0	0	0	0	0	0	0	0	26.9	0	0
Wine; Dessert, Sweet 1 fl oz:30.0 g	46	0.1	0	0	0	0	3.5	0	0	0.01	0.06	0	0.1	0	0	2.4	0.07	0.02
2-fl oz Glass:59 g	90	0.1	0	0	0	0	7	0	0.01	0.01	0.13	0	0.2	0	0	4.7	0.14	0.04
Wine; Table, All 1 fl oz:29.5 g	21	0.1	0	0	0	0	0.4	0	0	0.02	0.02	0.01	0.3	0	0	2.4	0.12	0.02
3.5-fl oz Glass:103 g	72	0.2	0	0	0	0	1.4	0	0	0.02	0.08	0.02	1.1	0.01	0	8.2	0.42	0.07
Wine; Table, Red 1 fl oz:29.5 g	21	0.1	0	0	0	0	0.5	0	0	0.01	0.02	0.01	0.6	0	0	2.4	0.13	0.03
3.5-fl oz Glass:103 g	74	0.2	0	0	0	0	1.8	0	0.01	0.03	0.08	0.04	2.1	0.01	0	8.2	0.44	0.09
FRUIT																		
Apples: Dehydrated (Low Moisture), Sulphured, Uncooked 1 C:60 g	208	0.8	0.4	0.1	0.1	0	56.1	48.6	0.03	0.08	0.41	0.17	0.6	0	1.3	11.4	1.2	0.17
½ C:30 g	104	0.4	0.2	0	0.1	0	28.1	24.3	0.01	0.04	0.2	0.08	0.3	0	0.7	5.7	0.6	0.09
Apples: Dried, Sulphured, Stewed, W/Added Sugar 1 C:280 g	232	0.6	0.2	0	0.1	0	58	44.8	0.02	0.05	0.34	0.13	0	0	2.5	8.4	0.87	0.11
½ C:140 g	116	0.3	0.1	0	0	0	29	22.4	0.01	0.03	0.17	0.07	0	0	1.3	4.2	0.43	0.06
Apples: Dried, Sulphured, Stewed, Wo/Added Sugar 1 C:255 g	145	0.6	0.2	0	0.1	0	39.1	43.4	0.02	0.05	0.33	0.13	0	0	2.6	7.7	0.84	0.13
½ C:128 g	73	0.3	0.1	0	0	0	19.6	21.8	0.01	0.02	0.17	0.06	0	0	1.3	3.8	0.42	0.06
Apples: Dried, Sulphured, Uncooked 1 C:86 g	209	0.8	0.3	0.1	0.1	0	56.7	0	0	0.14	0.8	0.11	0	0	3.4	12	1.2	0.17
10 Rings:64 g	156	0.6	0.2	0	0.1	0	42.2	0	0	0.1	0.59	0.08	0	0	2.5	9	0.9	0.13

Food Item & Measure	Cals	Protein (g)	All fat (g)	Sat fat (g)	Pol/fat (g)	Mon/fat (g)	Carbo (g)	Vit A (IU)	Thiam (mg)	Ribo (mg)	Niac (mg)	Vit B6 (mg)	Fol (mcg)	Vit B12 (mcg)	Vit C (mg)	Ca (mg)	Iron (mg)	Zinc (mg)
Fruit cont.																		
Apples; Raw, W/Skin 1 C Slices:110 g	65	0.2	0.4	0.1	0.1	0	16.8	58.3	0.02	0.02	0.08	0.05	3.1	0	6.3	7.7	0.2	0.04
1 Fruit, 3/lb, Wo/Rt:138 g	81	0.3	0.5	0.1	0.1	0	21.1	73.1	0.02	0.02	0.11	0.07	3.9	0	7.9	9.7	0.25	0.06
Apples, Wo/Skin, Boiled																		
1 C Slices:171 g	91	0.4	0.6	0.1	0.2	0	23.3	75.2	0.03	0.02	0.16	0.08	1	0	0.3	8.6	0.32	0.07
½ C Slices:86 g	46	0.2	0.3	0.1	0.1	0	11.7	37.8	0.01	0.01	0.08	0.04	0.5	0	0.2	4.3	0.16	0.03
Apples, Wo/Skin, Microwaved																		
1 C Slices:170 g	95	0.5	0.7	0.1	0.2	0	24.5	68	0.03	0.02	0.1	0.08	1	0	0.5	8.5	0.29	0.07
½ C Slices:85 g	48	0.2	0.4	0.1	0.1	0	12.3	34	0.01	0.01	0.05	0.04	0.5	0	0.3	4.3	0.14	0.03
Apple Sauce; Cnd, Unsweetened, Wo/Added																		
Asc Acid 1 C:244 g	105	0.4	0.1	0	0	0	27.6	70.8	0.03	0.06	0.46	0.06	1.5	0	2.9	7.3	0.29	0.07
½ C:122 g	52	0.2	0.1	0	0	0	13.8	35.4	0.02	0.03	0.23	0.03	0.7	0	1.5	3.7	0.15	0.04
Apricots; Cnd, Water Pack, W/Skin, Sol and																		
Liq 1 C Halves:243 g	66	1.7	0.4	0	0.1	0.2	15.5	3142	0.05	0.06	0.96	0.13	4.1	0	8.3	19.4	0.78	0.27
3 Hvs, 1-¾ Tbsp Lq:84 g	23	0.6	0.1	0	0	0.1	5.4	1086.1	0.02	0.02	0.33	0.05	1.4	0	2.9	6.7	0.27	0.09
Apricots; Dried, Sulphured, Stewed,																		
Wo/Added Sugar 1 C Halves:250 g	213	3.3	0.4	0	0.1	0.2	54.8	5907.5	0.02	0.08	2.36	0.29	0	0	4	40	4.18	0.65
½ C Halves:125 g	106	1.6	0.2	0	0	0.1	27.4	2953.8	0.01	0.04	1.18	0.14	0	0	2	20	2.09	0.33
Apricots; Dried, Sulphured, Uncooked																		
1 C Halves:130 g	309	4.8	0.6	0	0.1	0.3	80.3	9412	0.01	0.2	3.9	0.2	13.4	0	3.1	58.5	6.11	0.96
10 Halves:35 g	83	1.3	0.2	0	0.1	0.1	21.6	2534	0	0.05	1.05	0.05	3.6	0	0.8	15.8	1.65	0.26
Apricots; Raw 1 C Halves:155 g	74	2.2	0.6	0	0.1	0.3	17.2	4048.6	0.05	0.06	0.93	0.08	13.3	0	15.5	21.7	0.84	0.4
3 Fruits, 12/lb, Wo/Rt:106 g	51	1.5	0.4	0	0.1	0.2	11.8	2768.7	0.03	0.04	0.64	0.06	9.1	0	10.6	14.8	0.57	0.28
Avocados; Raw, All Commer Varieties																		
1 C Puree:230 g	370	4.6	35.2	5.6	4.5	22.1	17	1407.6	0.25	0.28	4.42	0.64	142.4	0	18.2	25.3	2.35	0.97
1 Fruit, Wo/Rt:201 g	324	4	30.8	4.9	3.9	19.3	14.9	1230.1	0.22	0.25	3.86	0.56	124.4	0	15.9	22.1	2.05	0.84
Bananas; Dehydrated 1 C:100 g	346	3.9	1.8	0.7	0.3	0.2	88.3	305	0.18	0.24	2.8	–	–	0	7	22	1.15	0.61
1 Tbsp:6.2 g	21	0.2	0.1	0	0	0	5.5	18.9	0.01	0.01	0.17	–	–	0	0.4	1.4	0.07	0.04
Blackberries; Cnd, Heavy Syrup, Sol and Liq																		
1 C:256 g	236	3.4	0.4	–	–	–	59.1	560.6	0.07	0.1	0.73	0.09	67.8	0	7.2	53.8	1.66	0.46
½ C:128 g	118	1.7	0.2	–	–	–	29.6	280.3	0.03	0.05	0.37	0.05	33.9	0	3.6	26.9	0.83	0.23

Fruit cont.

Food Item & Measure	Cals	Protein (g)	All fat (g)	Sat/fat (g)	Pol/fat (g)	Mon/fat (g)	Carbo (g)	Vit A (IU)	Thiam (mg)	Ribo (mg)	Niac (mg)	Vit B6 (mg)	Fol (mcg)	Vit B12 (mcg)	Vit C (mg)	Ca (mg)	Iron (mg)	Zinc (mg)
Blackberries; Raw 1 C:144 g	75	1	0.6	–	–	–	18.4	237.6	0.04	0.06	0.58	0.08	49	0	30.2	46.1	0.82	0.39
½ C:72 g	37	0.5	0.3	–	–	–	9.2	118.8	0.02	0.03	0.29	0.04	24.5	0	15.1	23	0.41	0.19
Cherries; Sweet, Raw 1 C:145 g	104	1.7	1.4	0.3	0.4	–	24	310.3	0.07	0.09	0.58	0.05	6.1	0	10.2	21.8	0.57	0.09
10 Fruits, Wo/Rt:68 g	49	0.8	0.7	0.2	0.2	0.4	11.3	145.5	0.03	0.04	0.27	0.02	2.9	0	4.8	10.2	0.27	0.04
Currants; European Black, Raw 1 C:112 g	71	1.6	0.5	0.2	0.2	0.1	17.2	257.6	0.06	0.06	0.34	0.07	–	0	202.7	61.6	1.72	0.3
½ C:56 g	35	0.8	0.2	0	0.1	0	8.6	128.8	0.03	0.03	0.17	0.04	–	0	101.4	30.8	0.86	0.15
Dates; Domestic, Natural and Dry																		
1 C Chopped:178 g	490	3.5	0.8	–	–	–	130.9	89	0.16	0.18	3.92	0.34	22.4	0	0	57	2.05	0.52
10 Fruits:83 g	228	1.6	0.4	–	–	–	61	41.5	0.07	0.08	1.83	0.16	10.5	0	0	26.6	0.95	0.24
Figs; Dried, Stewed 1 C:259 g	280	3.3	1.3	0.3	0.6	0.3	71.4	411.8	0.03	0.28	1.66	0.34	2.6	0	11.4	158	2.43	0.54
½ C:130 g	140	1.7	0.6	0.1	0.3	0.1	35.8	206.7	0.01	0.14	0.83	0.17	1.3	0	5.7	79.3	1.22	0.27
Figs; Dried, Uncooked 1 C:199 g	507	6.1	2.3	0.5	1.1	0.5	130.1	264.7	0.14	0.18	1.38	0.45	14.9	0	1.6	286.6	4.44	1.01
10 Fruits, Wo/Rt:187 g	477	5.7	2.2	0.4	1.1	0.5	122.2	248.7	0.13	0.16	1.3	0.42	14	0	1.5	269.3	4.17	0.95
Fruit Salad; (Peach, Pear, Aprct, Pnappl and Cherr), Cnd, Water Pack, Sol and Liq 1 C:245 g	74	0.9	0.2	0	0.1	0	19.3	1078	0.04	0.05	0.92	0.08	6.4	0	4.7	17.2	0.74	0.2
½ C:122 g	37	0.4	0.1	0	0	0	9.6	536.8	0.02	0.03	0.46	0.04	3.2	0	2.3	8.5	0.37	0.1
Gooseberries; Raw 1 C:150 g	66	1.3	0.9	0.1	0.5	0.1	15.3	435	0.06	0.05	0.45	0.12		0	41.6	37.5	0.47	0.18
1 C Sections W/Juice:230 g	74	1.5	0.2	0	0.1	0	18.6	285.2	0.08	0.05	0.58	0.1	23.5	0	79.1	27.6	0.21	0.16
Grapefruit; Raw ½ Frt, 3-7/8-In Diam:120 g	38	0.8	0.1	0	0	0	9.7	148.8	0.04	0.02	0.3	0.05	12.2	0	41.3	14.4	0.11	0.08
Grapefruit; Sections, Cnd, Juice Pack, Sol and Liq 1 C:249 g	92	1.7	0.2	0	0.1	0	22.9	0	0.07	0.04	0.62	0.05	21.9	0	84.4	37.4	0.52	0.2
½ C:124 g	46	0.9	0.1	0	0	0	11.4	0	0.04	0.02	0.31	0.02	10.9	0	42	18.6	0.26	0.1
Grapes; Raw 1 C, Wo/Rf:160 g	114	1.1	0.9	0.3	0.3	0	28.4	116.8	0.15	0.09	0.48	0.18	6.2	0	17.3	17.6	0.42	0.08
10 Fruits, Wo/Rf:50 g	36	0.3	0.3	0.1	0.1	0	8.9	36.5	0.05	0.03	0.15	0.06	2	0	5.4	5.5	0.13	0.03
Guavas; Raw 1 C:165 g	84	1.4	1	0.3	0.4	0.1	19.6	1306.8	0.08	0.08	1.98	0.24	–	0	302.8	33	0.51	0.38
1 Fruit, Wo/Rf:90 g	46	0.7	0.5	0.2	0.2	0.1	10.7	712.8	0.05	0.05	1.08	0.13	–	0	165.2	18	0.28	0.21
Kiwifruit (Chinese Gooseberries); Fresh, Raw																		
1 Lge Frt, Wo/Skin:91 g	56	0.9	0.4	–	–	–	13.5	159.3	0.02	0.05	0.46	–	–	0	89.2	23.7	0.37	–
1 Med Frt, Wo/Skin:76 g	46	0.8	0.3	–	–	–	11.3	133	0.02	0.04	0.38	–	–	0	74.5	19.8	0.31	–

Food item & Measure	Cals	Protein (g)	All fat (g)	Sat/fat (g)	Pol/fat (g)	Mon/fat (g)	Carbo (g)	Vit A (IU)	Thiam (mg)	Ribo (mg)	Niac (mg)	Vit B$_6$ (mg)	Fol (mcg)	Vit B$_{12}$ (mcg)	Vit C (mg)	Ca (mg)	Iron (mg)	Zinc (mg)
Fruit cont.																		
Kumquats; Raw 1 Fruit, Wo/Rf:19 g	12	0.2	0	–	–	–	3.1	57.4	0.02	0.02	0.1	–	–	0	7.1	8.4	0.07	0.02
Lemons; Raw, W/Peel																		
1 Med Frt, Wo/Rf:108 g	22	1.3	0.3	0	0.1	0	11.6	32.4	0.05	0.04	0.22	0.12	–	0	83.2	65.9	0.76	0.11
1 Wedge, Wo/Rf:27 g	5	0.3	0.1	0	0	0	2.9	8.1	0.01	0.01	0.05	0.03	–	0	20.8	16.5	0.19	0.03
Lemons; Raw, Wo/Peel 1 Lge Frt, Wo/Rf:84 g	24	0.9	0.3	0	0.1	0	7.8	24.4	0.03	0.02	0.08	0.07	8.9	0	44.5	21.8	0.5	0.05
1 Med Frt, Wo/Rf:58 g	17	0.6	0.2	0	0.1	0	5.4	16.8	0.02	0.01	0.06	0.05	6.2	0	30.7	15.1	0.35	0.03
Limes; Raw 1 Fruit, 2-In Diam:67 g	20	0.5	0.1	0	0	0	7.1	6.7	0.02	0.01	0.13	0.03	5.5	0	19.5	22.1	0.4	0.07
Mangos; Raw 1 C Slices:165 g	107	0.8	0.5	0.1	0.1	0.2	28.1	6425.1	0.1	0.09	0.96	0.22	–	0	45.7	16.5	0.21	0.07
1 Fruit, Wo/Rf:207 g	135	1.1	0.6	0.1	0.1	0.2	35.2	8060.6	0.12	0.12	1.21	0.28	–	0	57.3	20.7	0.27	0.08
Melons; Cantaloup, Raw																		
1 C Cubed Pieces:160 g	56	1.4	0.5	–	–	–	13.4	5158.4	0.06	0.03	0.92	0.18	27.2	0	67.5	17.6	0.34	0.26
½ Frt, 5-In Diam:267 g	93	2.4	0.8	–	–	–	22.3	8608.1	0.1	0.06	1.53	0.31	45.4	0	112.7	29.4	0.56	0.43
Melons; Honeydew, Raw																		
1 C Cubed Pieces:170 g	60	0.8	0.2	–	–	–	15.6	68	0.13	0.03	1.02	0.1	–	0	42.2	10.2	0.12	–
1 Frt, 7x2 In:129 g	45	0.6	0.1	–	–	–	11.8	51.6	0.1	0.02	0.77	0.08	–	0	32	7.7	0.09	–
Nectarines; Raw 1 C Slices:138 g	68	1.3	0.6	–	–	–	16.3	1015.7	0.02	0.06	1.37	0.03	5.1	0	7.5	6.9	0.21	0.12
1 Fruit, Wo/Rf:136 g	67	1.3	0.6	–	–	–	16	1001	0.02	0.06	1.35	0.03	5	0	7.3	6.8	0.2	0.12
Olives, Ripe, Cnd (Small-Extra Large)																		
1 Large:4.4 g	5	0	0.5	0.1	0	0.4	0.3	17.7	0	0	0	0	0	0	0	3.9	0.15	0.01
1 Small:3.2 g	4	0	0.3	0.1	0	0.3	0.2	12.9	0	0	0	0	0	0	0	2.8	0.11	0.01
Oranges; Raw, All Commer Varieties																		
1 C Sect. Wo/Membrane:180 g	85	1.7	0.2	0	0.1	0	21.2	369	0.16	0.07	0.51	0.11	54.5	0	95.8	72	0.18	0.13
1 Frt, 2⅝-In Diam:131 g	62	1.2	0.2	0	0	0	15.4	268.6	0.11	0.05	0.37	0.08	39.7	0	69.7	52.4	0.13	0.09
Papayas; Raw 1 C Cubed Pieces:140 g	55	0.9	0.2	0.1	0	0.1	13.7	397.6	0.04	0.04	0.47	0.03	53.2	0	86.5	33.6	0.14	0.1
1 Frt, 3½-In Diam:304 g	119	1.9	0.4	0.1	0.1	0.1	29.8	863.4	0.08	0.1	1.03	0.06	115.5	0	187.9	73	0.3	0.21
Passion-Fruit (Granadilla); Purple, Raw																		
1 Fruit, Wo/Rf:18 g	17	0.4	0.1	–	–	–	4.2	126	0	0.02	0.27	–	–	0	5.4	2.2	0.29	–
Peaches; Cnd, Light Syrup Pack, Sol and																		
Liq 1 C Halves or Slices:251 g	134	1.1	0.1	0	0	0	36.1	877.9	0.02	0.06	1.47	0.05	8.2	0	6	7.4	0.89	0.22
1 Half, 1¾ Tbsp Lq:81 g	44	0.4	0	0	0	0	11.8	286.7	0.01	0.02	0.48	0.02	2.7	0	1.9	2.4	0.29	0.07

Fruit cont.

Food Item & Measure	Cals	Protein (g)	All fat (g)	Sat/fat (g)	Pol/fat (g)	Mon/fat (g)	Carbo (g)	Vit A (IU)	Thiam (mg)	Ribo (mg)	Niac (mg)	Vit B6 (mg)	Fol (mcg)	Vit B12 (mcg)	Vit C (mg)	Ca (mg)	Iron (mg)	Zinc (mg)
Peaches; Dried, Sulphured, Stewed, Wo/Added Sugar 1 C Halves:258 g	199	3	0.7	0.1	0.3	0.2	50.8	508.3	0.01	0.05	3.92	0.1	0.3	0	9.6	23.2	3.38	0.46
½ C Halves:129 g	99	1.5	0.3	0	0.2	0.1	25.4	254.1	0.01	0.03	1.96	0.05	0.1	0	4.8	11.6	1.69	0.23
Peaches; Dried, Sulphured, Uncooked 1 C Halves:160 g	382	5.8	1.2	0.1	0.6	0.4	98.1	3460.8	0	0.34	7	0.11	0.5	0	7.7	44.8	6.5	0.91
10 Halves:130 g	311	4.7	1	0.1	0.5	0.4	79.7	2811.9	0	0.28	5.69	0.09	0.4	0	6.2	36.4	5.28	0.74
Peaches; Raw 1 C Slices:170 g	73	1.2	0.2	0	0.1	0.1	18.9	909.5	0.03	0.07	1.68	0.03	5.8	0	11.2	8.5	0.19	0.24
1 Fruit, 4/lb:87 g	37	0.6	0.1	0	0	0	9.7	465.5	0.01	0.04	0.86	0.02	3	0	5.7	4.4	0.1	0.12
Pears; Cnd, Light Syrup Pack, Sol and Liq 1 C Halves:251 g	143	0.5	0.1	0	0	0	38.1	0	0.03	0.04	0.39	0.04	3	0	1.8	12.6	0.7	0.2
1 Half, 1¾ Tbsp Lq:79 g	45	0.2	0	0	0	0	12	0	0.01	0.01	0.12	0.01	1	0	0.6	4	0.22	0.06
Pears; Dried, Sulphured, Stewed, Wo/Added Sugar 1 C Halves:255 g	324	2.3	0.8	0	0.2	0.2	86.2	107.1	0.01	0.05	0.9	0.09	0	0	10.2	40.8	2.6	0.48
½ C Halves:128 g	163	1.2	0.4	0	0.1	0.1	43.3	53.8	0.01	0.03	0.45	0.04	0	0	5.1	20.5	1.31	0.24
Pears; Dried, Sulphured, Uncooked 1 C Halves:180 g	472	3.4	1.1	0.1	0.3	0.2	125.5	5.4	0.01	0.26	2.47	0.13	0	0	12.6	61.2	3.78	0.7
10 Halves:175 g	459	3.3	1.1	0.1	0.3	0.2	122	5.3	0.01	0.25	2.4	0.13	0	0	12.3	59.5	3.68	0.68
Pears; Raw 1 C Slices:165 g	97	0.6	0.7	0	0.2	0.1	24.9	33	0.03	0.07	0.17	0.03	12.1	0	6.6	18.2	0.41	0.2
1 Fruit, 2.5 lb:166 g	98	0.7	0.7	0	0.2	0.1	25.1	33.2	0.03	0.07	0.17	0.03	12.1	0	6.6	18.3	0.42	0.2
Persimmons; Japanese, Raw 1 Frt, 2½-In Diam:168 g	118	1	0.3	–	–	–	31.2	3640.6	0.05	0.03	0.17	–	12.6	0	12.6	13.4	0.25	0.18
Pineapple; Cnd, Light Syrup Pack, Sol and Liq 1 C:252 g	131	0.9	0.3	0	0.1	0	33.9	37.8	0.23	0.06	0.74	0.19	11.8	0	18.9	35.3	0.98	0.3
1 Slc, 1¼ Tbsp Liq:58 g	30	0.2	0.1	0	0	0	7.8	8.7	0.05	0.01	0.17	0.04	2.7	0	4.4	8.1	0.23	0.07
Pineapple; Raw 1 C Diced Pieces:155 g	76	0.6	0.7	0.1	0.2	0.1	19.2	35.7	0.14	0.06	0.65	0.13	16.4	0	23.9	10.9	0.57	0.12
1 Slice, 3½-In Diam:84 g	41	0.3	0.4	0	0.1	0	10.4	19.3	0.08	0.03	0.35	0.07	8.9	0	12.9	5.9	0.31	0.07
Plantain; Cooked 1 C Slices:154 g	179	1.2	0.3	–	–	–	48	1399.9	0.07	0.08	1.16	0.37	40	0	16.8	3.1	0.89	0.2
½ C Slices:77 g	89	0.6	0.1	–	–	–	24	699.9	0.04	0.04	0.58	0.18	20	0	8.4	1.5	0.45	0.1
Plantain; Raw 1 C Slices:148 g	181	1.9	0.6	–	–	–	47.2	1668	0.08	0.08	1.02	0.44	32.6	0	27.2	4.4	0.89	0.21
1 Fruit, Wo/Rt:179 g	218	2.3	0.7	–	–	–	57.1	2017.3	0.09	0.1	1.23	0.54	39.4	0	32.9	5.4	1.07	0.25

Fruit cont.

Food item & Measure	Cals	Protein (g)	All fat (g)	Sat/fat (g)	Pol/fat (g)	Mon/fat (g)	Carbo (g)	Vit A (IU)	Thiam (mg)	Ribo (mg)	Niac (mg)	Vit B_6 (mg)	Fol (mcg)	Vit B_{12} (mcg)	Vit C (mg)	Ca (mg)	Iron (mg)	Zinc (mg)
Plums; Purple, Cnd Light Syrup Pack, Sol and Liq 1 C:252 g	159	0.9	0.3	0	0.1	0.2	41	665.3	0.04	0.1	0.75	0.07	6.6	0	1	22.7	2.17	0.2
3 Frts, 2½ Tbsp Lq:133 g	84	0.5	0.1	0	0	0.1	21.7	351.1	0.02	0.05	0.4	0.04	3.5	0	0.5	12	1.14	0.11
Plums; Raw 1 C Slices:165 g	91	1.3	1	0.1	0.2	0.7	21.5	533	0.07	0.16	0.83	0.13	3.6	0	15.7	6.6	0.17	0.17
1 Fruit:66 g	36	0.5	0.4	0	0.1	0.3	8.6	213.2	0.03	0.06	0.33	0.05	1.5	0	6.3	2.6	0.07	0.07
Pomegranate; Raw																		
1 Fruit:154 g	105	1.5	0.5	–	–	–	26.4	0	0.05	0.05	0.46	0.16	–	0	9.4	4.6	0.46	–
1 Fruit, Wo/Rf:103 g	42	0.8	0.5	–	–	–	9.9	52.5	0.01	0.06	0.47	–	–	0	14.4	57.7	0.31	–
Prunes; Cnd, Heavy Syrup Pack, Sol and Liq																		
1 C:234 g	246	2	0.5	0	0.1	0.3	65.1	1865	0.08	0.29	2.03	0.48	0.2	0	6.6	39.8	0.96	0.44
5 Fruits, 2 Tbsp Liq:86 g	90	0.8	0.2	0	0	0.1	23.9	685.4	0.03	0.1	0.74	0.17	0.1	0	2.4	14.6	0.35	0.16
Prunes; Dried, Stewed, Wo/Added Sugar																		
1 C, Wo/Stones:212 g	227	2.5	0.5	0	0.1	0.3	59.5	648.7	0.05	0.21	1.53	0.46	0.2	0	6.2	48.8	2.35	0.51
½ C, Wo/Stones:106 g	113	1.2	0.2	0	0.1	0.2	29.8	324.4	0.03	0.11	0.77	0.23	0.1	0	3.1	24.4	1.18	0.25
Prunes; Dried, Uncooked																		
1 C, Wo/Stones:161 g	385	4.2	0.8	0.1	0.2	0.6	101	3199.1	0.13	0.26	3.16	0.43	6	0	5.3	82.1	3.99	0.85
10 Fruits, Wo/Stones:84 g	201	2.2	0.4	0	0.1	0.3	52.7	1669.1	0.07	0.14	1.65	0.22	3.1	0	2.8	42.8	2.08	0.45
Quinces; Raw 1 Fruit, Wo/Rf:92 g	52	0.4	0.1	0	0.1	0	14.1	36.8	0.02	0.03	0.18	0.04	–	0	13.8	10.1	0.64	–
Raisins; Seedless 1 C Not Packed:145 g	435	4.7	0.7	0.2	0.2	0	114.7	11.6	0.23	0.13	1.19	0.36	4.8	0	4.8	71.1	3.02	0.39
1 C Packed:165 g	495	5.3	0.8	0.3	0.2	0	130.6	13.2	0.26	0.15	1.35	0.41	5.5	0	5.5	80.9	3.43	0.45
Raspberries; Cnd, Heavy Syrup Pack, Sol and Liq 1 C:256 g	233	2.1	0.3	0	0.2	0	59.8	84.5	0.05	0.08	1.13	0.11	26.9	0	22.3	28.2	1.08	0.41
½ C:128 g	116	1.1	0.2	0	0.1	0	29.9	42.2	0.03	0.04	0.57	0.05	13.4	0	11.1	14.1	0.54	0.2
Raspberries; Raw 1 C:123 g	60	1.1	0.7	0	0.4	0.1	14.2	159.9	0.04	0.11	1.11	0.07	32	0	30.8	27.1	0.7	0.57
1 Pint, Wo/Rf:312 g	153	2.8	1.7	0.1	1	0.2	36.1	405.6	0.09	0.28	2.81	0.18	81.1	0	78	68.6	1.78	1.44
Rhubarb; Cooked, W/Sugar 1 C:240 g	278	0.9	0.1	–	–	–	74.9	165.6	0.04	0.06	0.48	0.05	12.7	0	7.9	348	0.5	0.19
½ C:120 g	139	0.5	0.1	–	–	–	37.4	82.8	0.02	0.03	0.24	0.02	6.4	0	4	174	0.25	0.1
Strawberries; Cnd, Heavy Syrup Pack, Sol and Liq 1 C:254 g	234	1.4	0.7	0	0.3	0.1	59.8	66	0.05	0.09	0.14	0.12	71.1	0	80.5	33	1.24	0.23
½ C:127 g	117	0.7	0.3	0	0.2	0.1	29.9	33	0.03	0.04	0.07	0.06	35.6	0	40.3	16.5	0.62	0.11

Fruit cont.

Food item & Measure	Cals	Protein (g)	All fat (g)	Sat/fat (g)	Pol/fat (g)	Mon/fat (g)	Carbo (g)	Vit A (IU)	Thiam (mg)	Ribo (mg)	Niac (mg)	Vit B6 (mg)	Fol (mcg)	Vit B12 (mcg)	Vit C (mg)	Ca (mg)	Iron (mg)	Zinc (mg)
Strawberries; Raw 1 C:149 g	45	0.9	0.6	0	0.3	0.1	10.5	40.2	0.03	0.1	0.34	0.09	26.4	0	84.5	20.9	0.57	0.19
1 Pint, Wo/Rt:320 g	96	2	1.2	0.1	0.6	0.2	22.5	86.4	0.06	0.21	0.74	0.19	56.6	0	181.4	44.8	1.22	0.42
Tangerines (Mandarin Oranges); Cnd, Juice Pack 1 C:249 g	92	1.5	0.1	0	0	0	23.8	2121.5	0.2	0.07	1.11	0.1	11.5	0	85.2	27.4	0.67	1.27
½ C:124 g	46	0.8	0	0	0	0	11.9	1056.5	0.1	0.04	0.55	0.05	5.7	0	42.4	13.6	0.33	0.63
Tangerines (Mandarin Oranges); Raw 1 C Sect, Wo/Membrane:195 g	86	1.2	0.4	0	0.1	0.1	21.8	1794	0.2	0.04	0.31	0.13	39.8	0	60.1	27.3	0.2	0.47
1 Frt, 2⅜-In Diam:84 g	37	0.5	0.2	0	0	0	9.4	772.8	0.09	0.02	0.13	0.06	17.1	0	25.9	11.8	0.08	0.2
Watermelon; Raw 1 C Diced Pieces:160 g	51	1	0.7	-	-	-	11.5	585.6	0.13	0.03	0.32	0.23	3.5	0	15.4	12.8	0.27	0.11
1 Fruit, 10-In Diam:482 g	154	3	2.1	-	-	-	34.6	1764.1	0.39	0.1	0.96	0.69	10.6	0	46.3	38.6	0.82	0.34

GRAINS AND GRAIN PRODUCTS

Food item & Measure	Cals	Protein (g)	All fat (g)	Sat/fat (g)	Pol/fat (g)	Mon/fat (g)	Carbo (g)	Vit A (IU)	Thiam (mg)	Ribo (mg)	Niac (mg)	Vit B6 (mg)	Fol (mcg)	Vit B12 (mcg)	Vit C (mg)	Ca (mg)	Iron (mg)	Zinc (mg)
Barley, Pearled, Cooked 1 C:157 g	193	3.6	0.7	0.2	0.3	0.1	44.3	11	0.13	0.1	3.24	0.18	25.1	0	0	17.3	2.09	1.29
½ C:79 g	97	1.8	0.4	0.1	0.2	0.1	22.3	5.5	0.07	0.05	1.63	0.09	12.6	0	0	8.7	1.05	0.65
Buckwheat Flour, Whole-Groat 1 C:120 g	402	15.1	3.7	0.8	1.1	1.1	84.7	0	0.5	0.23	7.38	0.7	64.8	0	0	49.2	4.87	3.74
½ C:60 g	201	7.6	1.9	0.4	0.6	0.6	42.4	0	0.25	0.11	3.69	0.35	32.4	0	0	24.6	2.44	1.87
Buckwheat Groats, Roasted, Cooked 1 C:198 g	182	6.7	1.2	0.3	0.4	0.4	39.5	0	0.08	0.08	1.86	0.15	27.7	0	0	13.9	1.58	1.21
½ C:99 g	91	3.4	0.6	0.1	0.2	0.2	19.7	0	0.04	0.04	0.93	0.08	13.9	0	0	6.9	0.79	0.6
Bulghur, Cooked 1 C:182 g	151	5.6	0.4	0.1	0.2	0.1	33.8	0	0.1	0.05	1.82	0.15	32.8	0	0	18.2	1.75	1.04
½ C:91 g	76	2.8	0.2	0	0.1	0	16.9	0	0.05	0.03	0.91	0.08	16.4	0	0	9.1	0.87	0.52
Cereals, Ready-To-Eat; Corn Flakes 1 oz:28.4 g	110	2.3	0.1	-	-	-	24.5	1252.2	0.37	0.43	5	0.51	100.3	0	15.1	0.9	1.79	0.08
¾-oz Box:21.3 g	83	1.7	0.1	-	-	-	18.3	939.1	0.28	0.32	3.75	0.38	75.2	0	11.3	0.6	1.34	0.06
Cereals, Ready-To-Eat; Honey and Nut Corn Flakes (Corn) 1 oz:28.4 g	113	1.8	1.6	-	-	-	23.3	1252.2	0.37	0.43	5	0.51	100.3	0	15.1	3.4	1.79	0.11
⅞ oz Box:24.8 g	99	1.6	1.4	-	-	-	20.4	1093.4	0.32	0.37	4.36	0.45	87.5	0	13.1	3	1.56	0.09
Cereals, Ready-To-Eat; Muesli, Homemade (Oats, Wheat Germ) 1 C:122 g	436	14.3	0.5	5.8	17.2	9.4	100	5379	1.59	1.83	21.47	2.2	430.7	6.47	0	11.6	5.29	2.68
1 oz:28.4 g	101	3.3	0.1	1.4	4	2.2	23.3	1252.2	0.37	0.43	5	0.51	100.3	1.51	0	2.7	1.23	0.62

242 COMPOSITION OF VEGETARIAN FOODS

Food item & Measure	Cals	Protein (g)	All fat (g)	Sat fat (g)	Pol/fat (g)	Mon/fat (g)	Carbo (g)	Vit A (IU)	Thiam (mg)	Ribo (mg)	Niac (mg)	Vit B6 (mg)	Fol (mcg)	Vit B12 (mcg)	Vit C (mg)	Ca (mg)	Iron (mg)	Zinc (mg)
Grains and Grain Products cont.																		
Cereals, Ready-To-Eat; Raisin Bran,																		
Kellogg's (Wheat) 1⅓-oz Box:35.4 g	110	3.8	0.7	–	–	–	26.7	1199.4	0.35	0.42	4.81	0.5	95.9	1.45	0	12.4	16.07	3.61
1.3 oz:36.9 g	115	4	0.7	–	–	–	27.9	1250.2	0.37	0.44	5.02	0.52	100	1.51	0	12.9	16.75	3.76
Cereals, Ready-To-Eat; Rice Krispies (Rice)																		
1 oz:28.4 g	112	1.9	0.2	–	–	–	24.8	1252.2	0.37	0.43	5	0.51	100.3	0	15.1	4	1.79	0.48
⅝-oz Box:17.7 g	70	1.2	0.1	–	–	–	15.5	780.4	0.23	0.27	3.12	0.32	62.5	0	9.4	2.5	1.12	0.3
Cereals, Ready-To-Eat; Special K																		
(Rice, Wheat) 1 oz:28.4 g	111	5.6	0.1	–	–	–	21.3	1252.2	0.37	0.43	5	0.51	100.3	0.01	15.1	8.2	4.52	3.75
⅝-oz Box:17.7 g	69	3.5	0.1	–	–	–	13.3	780.4	0.23	0.27	3.12	0.32	62.5	0.01	9.4	5.1	2.81	2.34
Cereals, Ready-To-Eat; Sugar Frosted Flakes,																		
Kellogg's (Corn) 1 C:35 g	133	1.8	0.1	–	–	–	31.7	1543.2	0.46	0.53	6.16	0.63	123.6	0	18.6	1.4	2.21	0.05
1 oz:28.4 g	108	1.4	0.1	–	–	–	25.7	1252.2	0.37	0.43	5	0.51	100.3	0	15.1	1.1	1.79	0.04
Cereals, Ready-To-Eat; Wheat, Shredded, Large Biscuit (Wheat)																		
1 Rect Bisct:23.6 g	83	2.6	0.3	–	–	–	18.8	0	0.07	0.07	1.08	0.06	11.8	0	0	9.7	0.74	0.59
2 Round Bisct:37.8 g	133	4.1	0.5	–	–	–	30.2	0	0.11	0.11	1.73	0.1	18.9	0	0	15.5	1.19	0.95
Cereals, Ready-To-Eat; Wheat, Shredded, Small Biscuit (Wheat) 1 oz:28.4 g	102	3.1	0.6	–	–	–	22.7	0	0.07	0.08	1.49	0.07	14.2	0	0	10.8	1.2	0.94
⅞ oz Box:24.8 g	89	2.7	0.6	–	–	–	19.8	0	0.06	0.07	1.3	0.06	12.4	0	0	9.4	1.05	0.82
Cereals, Oats, Instant, Fort, Plain, Dry 1 Pkt:28.4 g	105	4.4	1.7	–	–	–	18.2	1516.6	0.53	0.29	5.5	0.74	150.8	0	0	163.6	6.33	0.87
Cereals, Oats, Instant, Fort, Plain, Prep W/Water 1 Pkt Prepared:177 g	104	4.4	1.8	–	–	–	18.1	1509.8	0.53	0.28	5.47	0.74	150.5	0	0	162.8	6.3	0.87
Cereals, Oats, Reg & Quick & Instant, Wo/Fort, Ckd W/Water, Wo/Slt																		
1 C:234 g	145	6.1	2.3	0.4	0.9	0.8	25.3	37.4	0.26	0.05	0.3	0.05	9.4	0	–	18.7	1.59	1.15
¾ C:175 g	109	4.6	1.8	0.3	0.7	0.6	18.9	28	0.19	0.04	0.23	0.04	7	0	–	14	1.19	0.86
Cereals, Oats, Reg & Quick & Instant, Wo/Fort, Dry 1 C:81 g	311	13	5.1	0.9	1.9	1.6	54.3	81.8	0.59	0.11	0.63	0.1	25.9	0	0	42.1	3.4	2.49
¾ C:27 g	104	4.3	1.7	0.3	0.6	0.5	18.1	27.3	0.2	0.04	0.21	0.03	8.6	0	0	14	1.13	0.83

Food Item & Measure	Cals	Protein (g)	All fat (g)	Sat/fat (g)	Pol/fat (g)	Mon/fat (g)	Carbo (g)	Vit A (IU)	Thiam (mg)	Ribo (mg)	Niac (mg)	Vit B6 (mg)	Fol (mcg)	Vit B12 (mcg)	Vit C (mg)	Ca (mg)	Iron (mg)	Zinc (mg)
Grains and Grain Products cont.																		
Cornmeal, Degermed, Unenriched, White																		
1 C:138 g	505	11.7	2.3	0.3	1	0.6	107.2	0	0.19	0.07	1.38	0.35	66.2	0	0	6.9	1.52	0.99
¼ C:34 g	124	2.9	0.6	0.1	0.2	0.1	26.4	0	0.05	0.02	0.34	0.09	16.3	0	0	1.7	0.37	0.24
Cornflour 1 C:128 g	488	0.3	0.1	0	0	0	116.8	0	0	0	0	0	0	0	0	2.6	0.6	0.08
¼ C:43 g	164	0.1	0	0	0	0	39.3	0	0	0	0	0	0	0	0	0.9	0.2	0.03
Couscous, Cooked 1 C:179 g	200	6.8	0.3	0.1	0.1	0	41.6	0	0.11	0.05	1.76	0.09	26.9	0	0	14.3	0.68	0.47
½ C:90 g	101	3.4	0.1	0	0.1	0	20.9	0	0.06	0.02	0.88	0.05	13.5	0	0	7.2	0.34	0.23
Macaroni and Cheese; Cnd																		
1 Cn, 15 oz:430 g	409	16.8	17.2	7.6	-	-	46	473	0.22	0.43	1.72	-	-	-	0.4	356.9	1.72	-
1 C:240 g	228	9.4	9.6	4.2	-	-	25.7	264	0.12	0.24	0.96	-	-	-	0.2	199.2	0.96	-
Macaroni, Cooked, Enriched 1 C:140 g	197	6.7	0.9	0.1	0.4	0.1	39.7	0	0.29	0.14	2.34	0.05	9.8	0	0	9.8	1.96	0.74
½ C:70 g	99	3.3	0.5	0.1	0.2	0.1	19.8	0	0.14	0.07	1.17	0.02	4.9	0	0	4.9	0.98	0.37
Macaroni, Cooked, Unenriched 1 C:140 g	197	6.7	0.9	0.1	0.4	0.1	39.7	-	0.03	0.03	0.56	0.05	9.8	0	0	9.8	0.7	0.74
½ C:70 g	99	3.3	0.5	0.1	0.2	0.1	19.8	-	0.01	0.01	0.28	0.02	4.9	0	0	4.9	0.35	0.37
Macaroni, Dry, Enriched 1 C:105 g	390	13.4	1.7	0.2	0.7	0.2	78.4	0	1.08	0.47	7.88	0.11	18.9	0	0	18.9	4.05	1.27
2 oz:57 g	211	7.3	0.9	0.1	0.4	0.1	42.6	0	0.59	0.25	4.28	0.06	10.3	0	0	10.3	2.2	0.69
Macaroni, Dry, Unenriched 1 C:105 g	390	13.4	1.7	0.2	0.7	0.2	78.4	0	0.09	0.06	1.79	0.11	18.9	0	0	18.9	1.37	1.27
2 oz:57 g	211	7.3	0.9	0.1	0.4	0.1	42.6	0	0.05	0.03	0.97	0.06	10.3	0	0	10.3	0.74	0.69
Macaroni; Whole-Wheat, Cooked																		
1 C:140 g	174	7.5	0.8	0.1	0.3	0.1	37.2	0	0.15	0.06	0.99	0.11	7	0	0	21	1.48	1.13
½ C:70 g	87	3.7	0.4	0.1	0.2	0.1	18.6	0	0.08	0.03	0.49	0.06	3.5	0	0	10.5	0.74	0.57
Macaroni, Whole-Wheat, Dry 1 C:105 g	365	15.4	1.5	0.3	0.6	0.2	78.8	0	0.51	0.15	5.39	0.23	59.9	0	0	42	3.81	2.49
2 oz:57 g	198	8.3	0.8	0.2	0.3	0.1	42.8	0	0.28	0.08	2.92	0.13	32.5	0	0	22.8	2.07	1.35
Millet, Cooked 1 C:240 g	286	8.4	2.4	0.4	1.2	0.4	56.8	0	0.25	0.2	3.19	0.26	45.6	0	0	7.2	1.51	2.18
½ C:120 g	143	4.2	1.2	0.2	0.6	0.2	28.4	0	0.13	0.1	1.6	0.13	22.8	0	0	3.6	0.76	1.09
Noodles; Chinese, Chow Mein 1 C:45 g	237	3.8	13.8	2	7.8	3.5	25.9	38.3	0.26	0.19	2.68	0.05	9.9	0	0	9	2.13	0.63
1.5 oz:43 g	227	3.6	13.2	1.9	7.5	3.3	24.7	36.6	0.25	0.18	2.56	0.05	9.5	0	0	8.6	2.03	0.6
Noodles; Egg, Cooked, Enriched																		
1 C:160 g	213	7.6	2.4	0.5	0.7	0.7	39.7	32	0.3	0.13	2.38	0.06	11.2	0.14	0	19.2	2.54	0.99
½ C:80 g	106	3.8	1.2	0.3	0.3	0.3	19.9	16	0.15	0.07	1.19	0.03	5.6	0.07	0	9.6	1.27	0.5

Grains and Grain Products cont.

Food Item & Measure	Cals	Protein (g)	All fat (g)	Sat/fat (g)	Pol/fat (g)	Mon/fat (g)	Carbo (g)	Vit A (IU)	Thiam (mg)	Ribo (mg)	Niac (mg)	Vit B6 (mg)	Fol (mcg)	Vit B12 (mcg)	Vit C (mg)	Ca (mg)	Iron (mg)	Zinc (mg)
Noodles; Egg, Cooked, Unenriched, With Added Salt 1 C:160 g	213	7.6	2.4	0.5	0.7	0.7	39.7	32	0.05	0.03	0.64	0.06	11.2	0.14	0	19.2	0.96	0.99
½ C:80 g	106	3.8	1.2	0.3	0.3	0.3	19.9	16	0.02	0.02	0.32	0.03	5.6	0.07	0	9.6	0.48	0.5
Noodles; Egg, Cooked, Unenriched, Without Added Salt 1 C:160 g	213	7.6	2.4	0.5	0.7	0.7	39.7	32	0.05	0.03	0.64	0.06	11.2	0.14	0	19.2	0.96	0.99
½ C:80 g	106	3.8	1.2	0.3	0.3	0.3	19.9	16	0.02	0.02	0.32	0.03	5.6	0.07	0	9.6	0.48	0.5
Oat Bran, Cooked 1 C:219 g	88	7	1.9	0.4	0.7	0.6	25.1	0	0.35	0.07	0.16	0.05	13.1	0	0	21.9	1.93	1.16
½ C:110 g	44	3.5	1	0.2	0.4	0.3	12.6	0	0.18	0.04	0.16	0.03	6.6	0	0	11	0.97	0.58
Oat Bran, Raw ½ C:47 g	116	8.1	3.3	0.6	1.3	1.1	31.1	0	0.55	0.1	0.44	0.08	24.4	0	0	27.3	2.54	1.46
⅓ C:31 g	76	5.4	2.2	0.4	0.9	0.7	20.5	0	0.36	0.07	0.29	0.05	16.1	0	0	18	1.68	0.96
Pasta; Fresh-Refrigerated, Plain, Cooked 2 oz:57 g	75	2.9	0.6	0.1	0.2	0.1	14.2	11.4	0.12	0.09	0.57	0.02	4	0.08	0	3.4	0.65	0.32
Pasta; Homemade with Egg, Cooked 2 oz:57 g	74	3	1	0.2	0.3	0.3	13.4	33.1	0.1	0.1	0.72	0.02	10.8	0.06	0	5.7	0.66	0.25
Pasta; Homemade without Egg, Cooked 2 oz:57 g	71	2.5	0.6	0.1	0.3	0.1	14.3	0	0.1	0.08	0.77	0.02	9.7	0	0	3.4	0.64	0.21
Pizza; with Cheese Topping 1 Pizza, 13½-In Dia:520 g	1227	62.4	43.2	16.9	–	–	147.2	3276	1.14	1.61	11.96	–	–	–	41.6	1149.2	11.44	–
1 S:65 g	142	7.2	5	2	–	–	17	378	0.13	0.19	1.38	–	–	–	4.8	132.6	1.32	–
Popcorn; Popped, Plain 1 C:6 g	23	0.8	0.3	0	–	–	4.6	0	0.02	0.01	0.13	–	–	–	0	0.7	0.16	–
	386	12.7	5	0.5	–	–	76.7	0	0.37	0.12	2.2	–	–	–	0	11	2.7	–
Popcorn; Popped, W/Butter and Salt added 1 C:9 g	41	0.9	2	0.9	–	–	5.3	0	0.03	0.01	0.15	–	–	–	0	0.7	0.19	–
	456	9.8	21.8	10.4	–	–	59.1	0	0.37	0.09	1.7	–	–	–	0	8	2.1	–
Pretzels 7½-oz Pkg Dutch:213 g	332	9.4	0.4	0	–	–	78.6	979.8	0.04	0.3	1.07	–	–	–	4.3	46.9	2.77	–
1 Pretzel, Dutch-Type:16 g	25	0.7	0	0	–	–	5.9	73.6	0	0.02	0.08	–	–	–	0.3	3.5	0.21	–
Quinoa 1 C:170 g	636	22.3	9.9	1	4	2.6	117.1	0	0.34	0.67	4.98	0.38	83.3	0	0	102	15.73	5.61
½ C:85 g	318	11.1	4.9	0.5	2	1.3	58.6	0	0.17	0.34	2.49	0.19	41.7	0	0	51	7.86	2.81
Rice Flour, White 1 C:158 g	578	9.4	2.2	0.6	0.6	0.7	126.6	0	0.22	0.03	4.09	0.69	6.3	0	0	15.8	0.55	1.26
½ C:79 g	289	4.7	1.1	0.3	0.3	0.4	63.3	0	0.11	0.02	2.05	0.34	3.2	0	0	7.9	0.28	0.63

Grains and Grain Products cont.

Food item & Measure	Cals	Protein (g)	All fat (g)	Sat.fat (g)	Pol.fat (g)	Mon.fat (g)	Carbo (g)	Vit A (IU)	Thiam (mg)	Ribo (mg)	Niac (mg)	Vit B_6 (mg)	Fol (mcg)	Vit B_{12} (mcg)	Vit C (mg)	Ca (mg)	Iron (mg)	Zinc (mg)
Rice; Brown, Long-Grain, Cooked 1 C:195 g	216	5	1.8	0.4	0.6	0.6	44.8	0	0.19	0.05	2.98	0.28	7.8	0	0	19.5	0.82	1.23
½ C:98 g	109	2.5	0.9	0.2	0.3	0.3	22.5	0	0.09	0.02	1.5	0.14	3.9	0	0	9.8	0.41	0.62
Rice; Brown, Medium-Grain, Cooked 1 C:195 g	218	4.5	1.6	0.3	0.6	0.6	45.8	0	0.2	0.02	2.59	0.29	7.8	0	0	19.5	1.03	1.21
½ C:98 g	110	2.3	0.8	0.2	0.3	0.3	23	0	0.1	0.01	1.3	0.15	3.9	0	0	.98	0.52	0.61
Rice; White, Long-Grain, Parboiled, Cooked, Unenriched 1 C:175 g	200	4	0.5	0.1	0.1	0.2	43.3	–	0.04	0.03	2.45	0.03	7	0	0	33.3	0.35	0.54
½ C:88 g	100	2	0.2	0.1	0.1	0.1	21.8	–	0.02	0.02	1.23	0.02	3.5	0	0	16.7	0.18	0.27
Rice; White, Long-Grain, Parboiled, Dry, Unenriched 1 C:185 g	686	12.6	1	0.3	0.3	0.3	151.2	–	0.19	0.13	6.72	0.65	31.5	0	0	111	2.78	1.78
½ C:92 g	341	6.3	0.5	0.1	0.1	0.2	75.2	–	0.09	0.06	3.34	0.32	15.6	0	0	55.2	1.38	0.88
Rice; White, Long-Grain, Regular, Cooked, Unenriched, With Salt 1 C:205 g	264	5.5	0.6	0.2	0.2	0.2	57.2	–	0.04	0.03	0.82	0.19	6.2	0	0	22.6	0.41	0.94
½ C:102 g	132	2.7	0.3	0.1	0.1	0.1	28.5	–	0.02	0.01	0.41	0.09	3.1	0	0	11.2	0.2	0.47
Rice; White, Long-Grain, Regular, Cooked, Unenriched, Without Salt 1 C:205 g	264	5.5	0.6	0.2	0.2	0.2	57.2	–	0.04	0.03	0.82	0.19	6.2	0	0	22.6	0.41	0.94
½ C:102 g	132	2.7	0.3	0.1	0.1	0.1	28.5	–	0.02	0.01	0.41	0.09	3.1	0	0	11.2	0.2	0.47
Rice; White, Long-Grain, Regular, Raw, Unenriched 1 C:185 g	675	13.2	1.2	0.3	0.3	0.4	147.9	–	0.13	0.09	2.96	0.3	14.8	0	0	51.8	1.48	2.02
½ C:92 g	336	6.6	0.6	0.2	0.2	0.2	73.6	–	0.06	0.05	1.47	0.15	7.4	0	0	25.8	0.74	1
Rice; White, Medium-Grain, Cooked, Enriched 1 C:205 g	267	4.9	0.4	0.1	0.1	0.1	58.6	0	0.34	0.03	3.76	0.1	4.1	0	0	6.2	3.05	0.86
½ C:102 g	133	2.4	0.2	0.1	0.1	0.1	29.2	0	0.17	0.02	1.87	0.05	2	0	0	3.1	1.52	0.43
Rice; White, Medium-Grain, Cooked, Unenriched 1 C:205 g	267	4.9	0.4	0.1	0.1	0.1	58.6	0	0.04	0.03	0.82	0.1	4.1	0	0	6.2	0.41	0.86
½ C:102 g	133	2.4	0.2	0.1	0.1	0.1	29.2	0	0.02	0.02	0.41	0.05	2	0	0	3.1	0.2	0.43
Rice; White, Medium-Grain, Raw, Unenriched 1 C:195 g	702	12.9	1.1	0.3	0.4	0.4	154.7	–	0.14	0.09	3.12	0.28	17.6	0	0	17.6	1.56	2.26
½ C:98 g	353	6.5	0.6	0.2	0.2	0.2	77.8	–	0.07	0.05	1.57	0.14	8.8	0	0	8.8	0.78	1.14

Food Item & Measure	Cals	Protein (g)	All fat (g)	Sat/fat (g)	Pol/fat (g)	Mon/fat (g)	Carbo (g)	Vit A (IU)	Thiam (mg)	Ribo (mg)	Niac (mg)	Vit B6 (mg)	Fol (mcg)	Vit B12 (mcg)	Vit C (mg)	Ca (mg)	Iron (mg)	Zinc (mg)
Grains and Grain Products cont.																		
Rice; White, Short-Grain, Cooked, Enriched																		
1 C:205 g	267	4.8	0.4	0.1	0.1	0.1	58.9	0	0.34	0.03	3.06	0.12	4.1	0	0	2.1	2.99	0.82
½ C:102 g	133	2.4	0.2	0.1	0.1	0.1	29.3	0	0.17	0.02	1.52	0.06	2	0	0	1	1.49	0.41
Rice; White, Short-Grain, Cooked, Unenriched 1 C:205 g	738	26	2.2	0.3	0.9	0.3	149.3	-	0.57	0.16	6.79	0.21	147.6	0	0	34.9	2.52	2.15
½ C:102 g	367	12.9	1.1	0.2	0.4	0.1	74.3	-	0.29	0.08	3.38	0.11	73.4	0	0	17.3	1.25	1.07
Rice; White, Short-Grain, Raw, Unenriched																		
1 C:200 g	716	13	1	0.3	0.3	0.3	158.3	-	0.14	0.1	3.2	0.34	12	0	0	6	1.6	2.2
½ C:100 g	358	6.5	0.5	0.1	0.1	0.2	79.2	-	0.07	0.05	1.6	0.17	6	0	0	3	0.8	1.1
Rye Flour, Dark 1 C:128 g	415	18	3.4	0.4	1.5	0.4	88	0	0.4	0.32	5.47	0.57	76.8	0	0	71.7	8.26	7.19
½ C:64 g	207	9	1.7	0.2	0.8	0.2	44	0	0.2	0.16	2.73	0.28	38.4	0	0	35.8	4.13	3.6
Rye Flour, Light 1 C:102 g	374	8.6	1.4	0.2	0.6	0.2	81.8	0	0.34	0.09	0.82	0.24	22.4	0	0	21.4	1.84	1.79
½ C:51 g	187	4.3	0.7	0.1	0.3	0.1	40.9	0	0.17	0.05	0.41	0.12	11.2	0	0	10.7	0.92	0.89
Rye Flour, Medium 1 C:102 g	361	9.6	1.8	0.2	0.8	0.2	79	0	0.29	0.12	1.76	0.27	19.4	0	0	24.5	2.16	2.03
½ C:51 g	181	4.8	0.9	0.1	0.4	0.1	39.5	0	0.15	0.06	0.88	0.14	9.7	0	0	12.2	1.08	1.01
Semolina, Enriched 1 C:167 g	601	21.2	1.8	0.3	0.7	0.2	121.6	0	1.35	0.95	10	0.17	120.2	0	0	28.4	7.28	1.75
½ C:84 g	302	10.7	0.9	0.1	0.4	0.1	61.2	0	0.68	0.48	5.03	0.09	60.5	0	0	14.3	3.66	0.88
Sorghum 1 C:192 g	651	21.7	6.3	0.9	2.6	1.9	143.3	0	0.46	0.27	5.62	-	-	0	0	53.8	8.45	-
½ C:96 g	325	10.9	3.2	0.4	1.3	1	71.6	0	0.23	0.14	2.81	-	-	0	0	26.9	4.22	-
Soy Flour; Defatted, Crude Protein Basis																		
1 C Stirred:100 g	327	51.5	1.2	0.1	0.5	0.2	33.9	40	0.7	0.25	2.61	0.57	305.4	0	0	241	9.24	2.46
½ C Stirred:50 g	164	25.7	0.6	0.1	0.3	0.1	17	20	0.35	0.13	1.31	0.29	152.7	0	0	120.5	4.62	1.23
Soy Flour; Full-Fat, Raw, Crude Protein Basis																		
1 C Stirred:85 g	369	32.1	17.6	2.5	9.9	3.9	27.1	102	0.49	0.99	3.67	0.39	293.3	0	0	175.1	5.41	3.33
½ C Stirred:42 g	182	15.9	8.7	1.3	4.9	1.9	13.4	50.4	0.24	0.49	1.81	0.19	144.9	0	0	86.5	2.68	1.65
Soy Flour; Low-Fat, Crude Protein Basis																		
1 C Stirred:88 g	325	44.8	5.9	0.9	3.3	1.3	29.6	35.2	0.33	0.25	1.9	0.46	360.8	0	0	165.4	5.27	1.04
½ C Stirred:44 g	162	22.4	3	0.4	1.7	0.7	14.8	17.6	0.17	0.13	0.95	0.23	180.4	0	0	82.7	2.64	0.52

Grains and Grain Products cont.

Food Item & Measure	Cals	Protein (g)	All fat (g)	Sat/fat (g)	Pol/fat (g)	Mon/fat (g)	Carbo (g)	Vit A (IU)	Thiam (mg)	Ribo (mg)	Niac (mg)	Vit B6 (mg)	Fol (mcg)	Vit B12 (mcg)	Vit C (mg)	Ca (mg)	Iron (mg)	Zinc (mg)
Spaghetti; Cooked, Enriched, Without Added Salt 1 C:140 g	197	6.7	0.9	0.1	0.4	0.1	39.7	--	0.29	0.14	2.34	0.05	9.8	0	0	9.8	1.96	0.74
½ C:70 g	99	3.3	0.5	0.1	0.2	0.1	19.8	--	0.14	0.07	1.17	0.02	4.9	0	0	4.9	0.98	0.37
Spaghetti; Cooked, Unenriched, With Added Salt 1 C:140 g	197	6.7	0.9	0.1	0.4	0.1	39.7	--	0.03	0.03	0.56	0.05	9.8	0	0	9.8	0.7	0.74
½ C:70 g	99	3.3	0.5	0.1	0.2	0.1	19.8	--	0.01	0.01	0.28	0.02	4.9	0	0	4.9	0.35	0.37
Spaghetti; Cooked, Unenriched, Without Added Salt 1 C:140 g	197	6.7	0.9	0.1	0.4	0.1	39.7	--	0.03	0.03	0.56	0.05	9.8	0	0	9.8	0.7	0.74
½ C:70 g	99	3.3	0.5	0.1	0.2	0.1	19.8	--	0.01	0.01	0.28	0.02	4.9	0	0	4.9	0.35	0.37
Spaghetti; Dry, Enriched 8 oz:227 g	842	29	3.6	0.5	1.5	0.4	169.6	0	2.34	1.01	17.04	0.24	40.9	0	0	40.9	8.76	2.75
2 oz:57 g	211	7.3	0.9	0.1	0.4	0.1	42.6	0	0.59	0.25	4.28	0.06	10.3	0	0	10.3	2.2	0.69
Spaghetti; In Tomato Sauce W/Cheese, Cnd 15¼-oz Cn, 1-¾ C:432 g	328	9.5	2.6	0	--	--	66.5	1598.4	0.6	0.48	7.78	--	--	--	17.3	69.1	4.75	--
1 C:250 g	190	5.5	1.5	0	--	--	38.5	925	0.35	0.28	4.5	--	--	--	10	40	2.75	--
Spaghetti; Whole-Wheat, Cooked 1 C:140 g	174	7.5	0.8	0.1	0.3	0.1	37.2	0	0.15	0.06	0.99	0.11	7	0	0	21	1.48	1.13
½ C:70 g	87	3.7	0.4	0.1	0.2	0.1	18.6	0	0.08	0.03	0.49	0.06	3.5	0	0	10.5	0.74	0.57
Spaghetti; Whole-Wheat, Dry 8 oz:227 g	790	33.2	3.2	0.6	1.3	0.4	170.3	0	1.11	0.32	11.65	0.51	129.4	0	0	90.8	8.24	5.38
2 oz:57 g	198	8.3	0.8	0.2	0.3	0.1	42.8	0	0.28	0.08	2.92	0.13	32.5	0	0	22.8	2.07	1.35
Tapioca; Pearl, Dry 1 C:152 g	518	0.3	0	--	--	--	134.8	0	0.01	0	0	0.01	6.1	0	0	30.4	2.4	0.18
⅓ C:51 g	174	0.1	0	--	--	--	45.2	0	0	0	0	0	2	0	0	10.2	0.81	0.06
Welsh Rarebit 1 C:232 g	415	18.8	31.6	17.3	--	--	14.6	1229.6	0.09	0.53	0.23	--	--	--	0	582.3	0.7	--
Wheat Bran, Crude ½ C:30 g	65	4.7	1.3	0.2	0.7	0.2	19.4	0	0.16	0.17	4.07	0.39	23.7	0	0	21.9	3.17	2.18
2 Tbsp:7 g	15	1.1	0.3	0	0.2	0	4.5	0	0.04	0.04	0.95	0.09	5.5	0	0	5.1	0.74	0.51
Wheat Flour, White, Plain, Enriched, Bleached 1 C:125 g	455	12.9	1.2	0.2	0.5	0.1	95.4	0	0.98	0.62	7.38	0.06	32.5	0	0	18.8	5.8	0.88
½ C:62 g	226	6.4	0.6	0.1	0.3	0.1	47.3	0	0.49	0.31	3.66	0.03	16.1	0	0	9.3	2.88	0.43
Wheat Flour, White, Plain, Enriched, Unbleached 1 C:125 g	455	12.9	1.2	0.2	0.5	0.1	95.4	0	0.98	0.62	7.38	0.06	32.5	0	0	18.8	5.8	0.88
½ C:62 g	226	6.4	0.6	0.1	0.3	0.1	47.3	0	0.49	0.31	3.66	0.03	16.1	0	0	9.3	2.88	0.43

Food Item & Measure	Cals	Protein (g)	All fat (g)	Sat/fat (g)	Pol/fat (g)	Mon/fat (g)	Carbo (g)	Vit A (IU)	Thiam (mg)	Ribo (mg)	Niac (mg)	Vit B_6 (mg)	Fol (mcg)	Vit B_{12} (mcg)	Vit C (mg)	Ca (mg)	Iron (mg)	Zinc (mg)
Grains and Grain Products cont.																		
Wheat Flour; White, Self-Raising,																		
Enriched 1 C:125 g	443	12.4	1.2	0.2	0.5	0.1	92.8	0	0.84	0.52	7.29	0.06	52.5	0	0	422.5	5.84	0.78
½ C:62 g	219	6.1	0.6	0.1	0.3	0.1	46	0	0.42	0.26	3.62	0.03	26	0	0	209.6	2.9	0.38
Wheat Flour; White, Plain, Unenriched																		
1 C:125 g	455	12.9	1.2	0.2	0.5	0.1	95.4	–	0.15	0.05	1.56	0.06	32.5	0	0	18.8	1.46	0.88
½ C:62 g	226	6.4	0.6	0.1	0.3	0.1	47.3	–	0.07	0.02	0.78	0.03	16.1	0	0	9.3	0.73	0.43
Wheat Flour; White, Bread, Enriched																		
1 C:137 g	495	16.4	2.3	0.3	1	0.2	99.4	0	1.11	0.7	10.35	0.05	39.7	0	0	20.6	6.04	1.16
½ C:69 g	249	8.3	1.2	0.2	0.5	0.1	50.1	0	0.56	0.35	5.21	0.03	20	0	0	10.4	3.04	0.59
Wheat Flour; White, Cake, Enriched																		
1 C:109 g	441	10.5	11.6	4.5	1.7	5	73.2	0	0.8	0.54	6.34	0.04	25.7	0	0	223.5	7.68	0.7
½ C:54 g	219	5.2	5.7	2.2	0.8	2.5	36.3	0	0.4	0.27	3.14	0.02	12.7	0	0	110.7	3.81	0.35
Wheat Flour; Whole-Grain 1 C:120 g	407	16.4	2.2	0.4	0.9	0.3	87.1	0	0.54	0.26	7.64	0.41	52.8	0	0	40.8	4.66	3.52
½ C:60 g	203	8.2	1.1	0.2	0.5	0.1	43.5	0	0.27	0.13	3.82	0.2	26.4	0	0	20.4	2.33	1.76
Wheat Germ, Crude 1 C	360	23.2	9.7	1.7	6	1.4	51.8	0	1.88	0.5	6.81	1.3	281	0	0	39	6.26	12.29
¼ C:29 g	104	6.7	2.8	0.5	1.7	0.4	15	0	0.55	0.14	1.98	0.38	81.5	0	0	11.3	1.82	3.56
Wheat, Sprouted 1 C:108 g	214	8.1	1.4	0.2	0.6	0.2	45.9	0	0.24	0.17	3.33	0.29	41	0	2.8	30.2	2.31	1.78
1/3 C:36 g	71	2.7	0.5	0.1	0.2	0.1	15.3	0	0.08	0.06	1.11	0.1	13.7	0	0.9	10.1	0.77	0.59
Wild Rice, Cooked 1 C:164 g	166	6.5	0.6	0.1	0.4	0.1	35	0	0.09	0.14	2.11	0.22	42.6	0	0	4.9	0.98	2.2
½ C:82 g	83	3.3	0.3	0	0.2	0	17.5	0	0.04	0.07	1.06	0.11	21.3	0	0	2.5	0.49	1.1
SPICES AND HERBS																		
Allspice, Ground 1 Tbsp:6.0 g	20	1.1	1	0.2	0.2	0.6	3	32.4	0.01	0	0.17	–	–	0	2.4	38.7	2.22	0.32
1 Tsp:1.9 g	6	0.3	0.3	0.1	0.1	0.2	1	10.3	0	0	0.05	–	–	0	0.7	12.3	0.7	0.1
Basil, Ground 1 Tbsp:4.5 g	11	0.7	0.2	–	–	–	2.7	421.9	0.01	0.01	0.31	–	–	0	2.8	95.1	1.89	0.26
1 Tsp:1.4 g	4	0.2	0.1	–	–	–	0.9	131.3	0.01	0	0.1	–	–	0	0.9	29.6	0.59	0.08
Bay Leaf, Crumbled 1 Tbsp:1.8 g	6	0.1	0.2	0	0	0	1.4	111.3	0	0.01	0.04	–	–	0	0.8	15	0.77	0.07
1 Tsp:0.6 g	2	0.1	0.1	0	0	0	0.5	37.1	0	0	0.01	–	–	0	0.3	5	0.26	0.02
Caraway Seed 1 Tbsp:6.7 g	22	1.3	1	0.2	0.5	0.3	3.3	24.3	0.03	0.03	0.24	–	–	0	–	46.2	1.09	0.37
1 Tsp:2.1 g	7	0.4	0.3	0	0.1	0.2	1.1	7.6	0.01	0.01	0.08	–	–	0	–	14.5	0.34	0.12

Spices and Herbs cont.

Food item & Measure	Cals	Protein (g)	All fat (g)	Sat/fat (g)	Pol/fat (g)	Mon/fat (g)	Carbo (g)	Vit A (IU)	Thiam (mg)	Ribo (mg)	Niac (mg)	Vit B$_6$ (mg)	Fol (mcg)	Vit B$_{12}$ (mcg)	Vit C (mg)	Ca (mg)	Iron (mg)	Zinc (mg)
Cardamom, Ground 1 Tbsp:5.8 g	18	0.6	0.4	0	0	0.1	4	0	0.01	0.01	0.06	–	–	0	–	22.2	0.81	0.43
1 Tsp:2.0 g	6	0.2	0.1	0	0	0	1.4	0	0	0	0.02	–	–	0	–	7.7	0.28	0.15
Celery Seed 1 Tbsp:6.5 g	25	1.2	1.6	0.1	0.2	1	2.7	3.4	–	–	–	–	–	0	1.1	114.8	2.92	0.45
1 Tsp:2.0 g	8	0.4	0.5	0	0.1	0.3	0.8	1	–	–	–	–	–	0	0.3	35.3	0.9	0.14
Chervil, Dried 1 Tbsp:1.9 g	4	0.4	0.1	–	–	–	0.9	–	–	–	–	0.02	–	0	–	25.6	0.61	0.17
1 Tsp:0.6 g	1	0.1	0	–	–	–	0.3	–	–	–	–	0.01	–	0	–	8.1	0.19	0.05
Chilli Powder 1 Tbsp:7.5 g	24	0.9	1.3	–	–	–	4.1	2619.5	0.03	0.06	0.59	–	–	0	4.8	20.9	1.07	0.2
1 Tsp:2.6 g	8	0.3	0.4	–	–	–	1.4	908.1	0.01	0.02	0.21	–	–	0	1.7	7.2	0.37	0.07
Cinnamon, Ground 1 Tbsp:6.8 g	18	0.3	0.2	0	0	0	5.4	17.7	0.01	0.01	0.09	–	–	0	1.9	83.5	2.59	0.13
1 Tsp:2.3 g	6	0.1	0.1	0	0	0	1.8	6	0	0	0.03	–	–	0	0.7	28.3	0.88	0.05
Cloves, Ground 1 Tbsp:6.6 g	21	0.4	1.3	0.3	–	–	1.3	4	0.01	0.02	0.1	–	–	0	5.3	42.6	0.57	0.07
1 Tsp:2.1 g	7	0.1	0.4	0.1	–	–	0.9	11.1	0	0.01	0.03	–	–	0	1.7	13.6	0.18	0.02
Coriander Leaf, Dried 1 Tbsp:1.8 g	5	0.4	0.1	–	–	–	0.9	–	0.02	0.03	0.19	–	–	0	10.3	22.5	0.77	–
1 Tsp:0.6 g	2	0.1	0	–	–	–	0.3	–	0.01	0.01	0.06	–	–	0	3.4	7.5	0.26	–
Coriander Seed 1 Tbsp:5.0 g	19	0.9	1.1	0.1	0.1	0.7	2.2	63.5	0.03	0.02	0.23	–	0	0	0.4	46.5	3.32	0.24
1 Tsp:1.8 g	7	0.3	0.4	0	0	0.2	0.8	22.9	0.01	0.01	0.08	–	0	0	0.1	16.8	1.19	0.09
Curry Powder 1 Tbsp:6.3 g	20	0.8	0.9	–	–	–	3.7	62.1	0.02	0.02	0.22	–	–	0	0.7	30.1	1.86	0.26
1 Tsp:2.0 g	7	0.3	0.3	–	–	–	1.2	19.7	0.01	0.01	0.07	–	–	0	0.2	9.6	0.59	0.08
Dill Seed 1 Tbsp:6.6 g	20	1.1	1	0.1	0.1	0.6	3.6	3.5	0.03	0.02	0.19	–	–	0	–	100	1.08	0.34
1 Tsp:2.1 g	6	0.3	0.3	0	0	0.2	1.2	1.1	0.01	0.01	0.06	–	–	0	–	31.8	0.34	0.11
Dill Weed, Dried 1 Tbsp:3.1 g	8	0.6	0.1	–	–	–	1.7	–	0.01	0.01	0.09	0.05	–	0	–	55.3	1.51	0.1
1 Tsp:1.0 g	3	0.2	0	–	–	–	0.6	–	0	0	0.03	0.01	–	0	–	17.8	0.49	0.03
Fennel Seed 1 Tbsp:5.8 g	20	0.9	0.9	0.1	0	0.6	3	7.8	0.02	0.02	0.35	–	–	0	–	69.4	1.08	0.21
1 Tsp:2.0 g	7	0.3	0.3	0	0	0.2	1.1	2.7	0.01	0.01	0.12	–	–	0	–	23.9	0.37	0.07
Fenugreek Seed 1 Tbsp:11.1 g	36	2.6	0.7	–	–	–	6.5	–	0.04	0.04	0.18	–	6.3	0	0.3	19.5	3.72	0.28
1 Tsp:3.7 g	12	0.9	0.2	–	–	–	2.2	–	0.01	0.01	0.06	–	2.1	0	0.1	6.5	1.24	0.09
Garlic Powder 1 Tbsp:8.4 g	28	1.4	0.1	–	–	–	6.1	0	0.04	0.01	0.06	–	–	0	–	6.7	0.23	0.22
1 Tsp:2.8 g	9	0.5	0	–	–	–	2	0	0.01	0	0.02	–	–	0	–	2.2	0.08	0.07
Ginger Root, Crystallised, Candied 1 lb:454 g	1544	1.4	0.9	0	–	–	395.4	454	0.91	1.82	31.78	–	–	–	181.6	1044.2	95.34	–

Spices and Herbs cont.

Food item & Measure	Cals	Protein (g)	All fat (g)	Sat fat (g)	Pol fat (g)	Mon fat (g)	Carbo (g)	Vit A (IU)	Thiam (mg)	Ribo (mg)	Niac (mg)	Vit B6 (mg)	Fol (mcg)	Vit B12 (mcg)	Vit C (mg)	Ca (mg)	Iron (mg)	Zinc (mg)
Ginger, Ground 1 Tbsp:5.4 g	19	0.5	0.3	0.1	0.1	0.1	3.8	7.9	0	0.01	0.28	-	-	0	-	6.3	0.62	0.25
1 Tsp:1.8 g	6	0.2	0.1	0	0	0	1.3	2.7	0	0	0.09	-	-	0	-	2.1	0.21	0.08
Horseradish, Prepared 1 Tbsp:15 g	6	0.2	0	0	-	-	1.4	0	0	0	0	-	-	-	0	9.2	0.14	-
1 Tsp:5.0 g	2	0.1	0	0	-	-	0.5	0	0	0	0	-	-	-	0	3.1	0.05	-
Mace, Ground 1 Tbsp:5.3 g	25	0.4	1.7	0.5	0.2	0.6	2.7	42.4	0.02	0.02	0.07	-	-	0	-	13.4	0.74	0.12
1 Tsp:1.7 g	8	0.1	0.6	0.2	0.1	0.2	0.9	13.6	0.01	0.01	0.02	-	-	0	-	4.3	0.24	0.04
Marjoram, Dried 1 Tbsp:1.7 g	5	0.2	0.1	-	-	-	1	137.2	-	0.01	0.07	-	-	0	0.9	33.8	1.41	0.06
1 Tsp:0.6 g	2	0.1	0	-	-	-	0.4	48.4	-	0	0.02	-	-	0	0.3	11.9	0.5	0.02
Mustard Seed, Yellow 1 Tbsp:11.2 g	53	2.8	3.2	0.2	0.6	2.2	3.9	6.9	0.06	0.04	0.88	-	-	0	-	58.4	1.12	0.64
1 Tsp:3.3 g	15	0.8	1	0.1	0.2	0.7	1.2	2.1	0.02	0.01	0.26	-	-	0	-	17.2	0.33	0.19
Nutmeg, Grated 1 Tbsp:7.0 g	37	0.4	2.5	1.8	0.2	0.2	3.5	7.1	0.02	0.01	0.09	-	-	0	-	12.9	0.21	0.15
1 Tsp:2.2 g	12	0.1	0.8	0.6	0	0.1	1.1	2.2	0.01	0	0.03	-	-	0	-	4.1	0.07	0.05
Onion Powder 1 Tbsp:6.5 g	23	0.7	0.1	-	-	-	5.2	0	0.03	0	0.04	-	-	0	1	23.6	0.17	0.15
1 Tsp:2.1 g	7	0.2	0	-	-	-	1.7	0	0.01	0	0.01	-	-	0	0.3	7.6	0.05	0.05
Oregano, Grated 1 Tbsp:4.5 g	14	0.5	0.5	0.1	0.2	0	2.9	310.6	0.02	-	0.28	-	-	0	-	70.9	1.98	0.2
1 Tsp:1.5 g	5	0.2	0.2	0	0.1	0	1	103.6	0.01	-	0.09	-	-	0	-	23.6	0.66	0.07
Paprika 1 Tbsp:6.9 g	20	1	0.9	0.1	0.6	0.1	3.9	4181.7	0.04	0.12	1.06	-	-	0	4.9	12.2	1.63	0.28
1 Tsp:2.1 g	6	0.3	0.3	0	0.2	0	1.2	1272.7	0.01	0.04	0.32	-	-	0	1.5	3.7	0.5	0.09
Parsley, Dried 1 Tbsp:1.3 g	4	0.3	0.1	-	-	-	0.7	303.4	0.01	0.02	0.1	0.01	-	0	1.6	19.1	1.27	0.06
1 Tsp:0.3 g	1	0.1	0	-	-	-	0.2	70	0	0	0.02	0	-	0	0.4	4.4	0.29	0.01
Pepper, Black 1 Tbsp:6.4 g	16	0.7	0.2	0.1	0.1	0.1	4.2	12.2	0.01	0.02	0.07	-	-	0	-	27.9	1.85	0.09
1 Tsp:2.1 g	5	0.2	0.1	0	0	0	1.4	4	0	0.01	0.02	-	-	0	-	9.2	0.61	0.03
Pepper, Red or Cayenne 1 Tbsp:5.3 g	16	0.6	0.8	0.2	0.4	0.2	3.1	2205.3	0.02	0.05	0.46	-	-	0	4.1	7.9	0.41	0.13
1 Tsp:1.8 g	6	0.2	0.3	0.1	0.2	0.1	1.1	749	0.01	0.02	0.16	-	-	0	1.4	2.7	0.14	0.04
Pepper, White 1 Tbsp:7.1 g	38	1.3	3.2	0.4	2.2	0.5	1.7	0	0.06	0.01	0.07	0.03	-	0	-	102.8	0.67	0.73
1 Tsp:2.4 g	13	0.4	1.1	0.1	0.7	0.2	0.6	0	0.02	0	0.02	0.01	-	0	-	34.8	0.23	0.25
Rosemary, Dried 1 Tbsp:3.3 g	11	0.2	0.5	-	-	-	2.1	103.2	0.02	0.01	0.03	-	-	0	2	42.2	0.97	0.11
1 Tsp:1.2 g	4	0.1	0.2	-	-	-	0.8	37.5	0.01	0	0.01	-	-	0	0.7	15.4	0.35	0.04
Saffron 1 Tbsp:2.1 g	7	0.2	0.1	-	-	-	1.4	-	-	-	-	-	-	0	-	2.3	0.23	-
1 Tsp:0.7 g	2	0.1	0	-	-	-	0.5	-	-	-	-	-	-	0	-	0.8	0.08	-

Food item & Measure	Cals	Protein (g)	All fat (g)	Sat fat (g)	Pol/fat (g)	Mon/fat (g)	Carbo (g)	Vit A (IU)	Thiam (mg)	Ribo (mg)	Niac (mg)	Vit B6 (mg)	Fol (mcg)	Vit B12 (mcg)	Vit C (mg)	Ca (mg)	Iron (mg)	Zinc (mg)
Spices and Herbs cont.																		
Sage, Dried 1 Tbsp:2.0 g	6	0.2	0.3	0.1	0	0	1.2	118	0.02	0.01	0.11	-	-	0	0.7	33	0.56	0.09
1 Tsp:0.7 g	2	0.1	0.1	0.1	0	0	0.4	41.3	0.01	0	0.04	-	-	0	0.2	11.6	0.2	0.03
Savory, Dried 1 Tbsp:4.4 g	12	0.3	0.3	-	-	-	3	225.7	0.02	-	0.18	-	-	0	-	93.8	1.67	0.19
1 Tsp:1.4 g	4	0.1	0.1	-	-	-	1	71.8	0.01	-	0.06	0.01	-	0	-	29.8	0.53	0.06
Sesame Seeds, Dry 1 Tbsp:8.0 g	47	2.1	4.4	-	-	-	0.8	5.3	0.06	0.01	0.37	-	0	0	0	10.5	0.62	0.82
1 Tsp:2.7 g	16	0.7	1.5	-	-	-	0.3	1.8	0.02	0	0.13	0	0	0	0	3.5	0.21	0.28
Tarragon, Dried 1 Tbsp:4.8 g	14	1.1	0.4	-	-	-	2.4	201.6	0.01	0.06	0.43	-	-	0	-	54.7	1.55	0.19
1 Tsp:1.6 g	5	0.4	0.1	-	-	-	0.8	67.2	0	0.02	0.14	-	-	0	-	18.2	0.52	0.06
Thyme, Dried 1 Tbsp:4.3 g	12	0.4	0.3	0.1	0.1	0	2.8	163.4	0.02	0.02	0.21	-	-	0	-	81.3	5.31	0.27
1 Tsp:1.4 g	4	0.1	0.1	0	0	0	0.9	53.2	0.01	0.01	0.07	-	-	0	-	26.5	1.73	0.09
Turmeric, Ground 1 Tbsp:6.8 g	24	0.5	0.7	-	-	-	4.4	0	0.01	0.02	0.35	-	-	0	1.8	12.4	2.82	0.3
1 Tsp:2.2 g	8	0.2	0.2	-	-	-	1.4	0	0	0.01	0.11	-	0	0	0.6	4	0.91	0.1
NUTS AND SEEDS																		
Almond Paste																		
1 C Firmly Packed:227 g	1012	26.9	61.7	5.9	13	40.1	98.9	0	0.47	1.67	6.56	0.22	126.2	0	1.1	522.1	7.17	5.86
1 oz:28.4 g	127	3.4	7.7	0.7	1.6	5	12.4	0	0.06	0.21	0.82	0.03	15.8	0	0.1	65.3	0.9	0.73
Almonds, Dried, Blanched																		
1 C Whole Kernels:145 g	850	29.6	76.2	7.2	16	49.5	26.9	0	0.23	0.98	4.59	0.15	55.7	0	0.9	358.2	5.26	4.58
1 oz:28.4 g	166	5.8	14.9	1.4	3.1	9.7	5.3	0	0.05	0.19	0.9	0.03	10.9	0	0.2	70.2	1.03	0.9
Almonds, Dried, Unblanched																		
1 C Whole Kernels:142 g	836	28.3	74.1	7	15.6	48.1	29	0	0.3	1.11	4.77	0.16	83.4	0	0.9	377.7	5.2	4.15
1 oz (24 Whl Kernels):28.4 g	167	5.7	14.8	1.4	3.1	9.6	5.8	0	0.06	0.22	0.95	0.03	16.7	0	0.2	75.5	1.04	0.83
Almonds, Dry-Roasted, Unblanched, Wo/Salt Added																		
1 C Whole Kernels:138 g	810	22.5	71.2	6.8	14.9	46.2	33.4	0	0.18	0.83	3.89	0.1	88	0	1	389.2	5.24	6.76
1 oz:28.4 g	167	4.6	14.7	1.4	3.1	9.5	6.9	0	0.04	0.17	0.8	0.02	18.1	0	0.2	80.1	1.08	1.39
Almonds, Ground, Full-Fat																		
1 C Not Packed:65 g	385	12.9	33.6	3.2	7.1	21.8	14.5	0	0.13	0.77	1.53	0.07	38.6	0	0.3	141.7	1.82	0.14
1 oz:28.4 g	168	5.6	14.7	1.4	3.1	9.5	6.4	0	0.06	0.34	0.67	0.03	16.8	0	0.1	61.9	0.8	0.06

Food Item & Measure	Cals	Protein (g)	All fat (g)	Sat/fat (g)	Pol/fat (g)	Mon/fat (g)	Carbo (g)	Vit A (IU)	Thiam (mg)	Ribo (mg)	Niac (mg)	Vit B6 (mg)	Fol (mcg)	Vit B12 (mcg)	Vit C (mg)	Ca (mg)	Iron (mg)	Zinc (mg)
Nuts and Seeds cont.																		
Almonds, Ground, Partially Defatted																		
1 C Not Packed:65 g	255	24.4	10.4	1	2.2	6.8	20.7	0	0.1	0.42	1.97	0.06	23.9	0	0.4	154.1	2.26	1.97
1 oz:28.4 g	112	10.6	4.5	0.4	1	3	9	0	0.04	0.18	0.86	0.03	10.5	0	0.2	67.3	0.99	0.86
Almonds, Oil-Roasted, Blanched, Wo/Salt Added 1 C Whole Kernels:142 g	870	27	80.3	7.6	16.9	52.1	25.6	0	0.11	0.4	5.54	0.13	90.2	0	1.4	275.5	7.53	2.02
1 oz(24 Whl Kernels):28.4 g	174	5.4	16.1	1.5	3.4	10.4	5.1	0	0.02	0.08	1.11	0.03	18	0	0.3	55.1	1.51	0.4
Almonds, Oil-Roasted, Unblanched, Wo/Salt Added																		
1 C Whole Kernels:157 g	878	29	81.9	7.8	17.2	53.2	22.6	0	0.18	1.41	4.97	0.12	90.6	0	1	332.3	5.44	6.96
1 oz (22 Whl Kernels):28.4 g	176	5.8	16.4	1.6	3.4	10.6	4.5	0	0.04	0.28	0.99	0.02	18.1	0	0.2	66.5	1.09	1.39
Almonds, Toasted, Unblanched																		
1 oz:28.4 g	167	5.8	14.4	1.4	3	9.4	6.5	0	0.04	0.17	0.8	0.02	18.2	0	0.2	80.4	1.4	1.4
Brazil Nuts, Dried, Unblanched																		
1 C (32 Kernels):140 g	918	20.1	92.7	22.6	33.8	32.2	17.9	0	1.4	0.17	2.27	0.35	5.6	0	1	246.4	4.76	6.43
1 oz (6-8 Kernels):28.4 g	186	4.1	18.8	4.6	6.9	6.5	3.6	0	0.28	0.03	0.46	0.07	1.1	0	0.2	50	0.97	1.3
Cashew Butter, Plain, Wo/Salt Added																		
1 oz:28.4 g	167	5	14	2.8	2.4	8.3	7.8	0	0.09	0.05	0.45	0.07	19.4	0	0	12.2	1.43	1.47
1 Tbsp:16 g	94	2.8	7.9	1.6	1.3	4.7	4.4	0	0.05	0.03	0.26	0.04	10.9	0	0	6.9	0.8	0.83
Cashew Nuts, Dry-Roasted, Wo/Salt Added																		
1 C Wholes & Halves:137 g	786	21	63.5	12.6	10.7	37.4	44.8	0	0.27	0.27	1.92	0.35	94.8	0	0	61.7	8.22	7.67
1 oz:28.4 g	163	4.4	13.2	2.6	2.2	7.8	9.3	0	0.06	0.06	0.4	0.07	19.7	0	0	12.8	1.7	1.59
Cashew Nuts, Oil-Roasted, Wo/Salt Added																		
1 C Wholes & Halves:130 g	749	21	62.7	12.4	10.6	36.9	37.1	0	0.55	0.23	2.34	0.33	88	0	0	53.3	5.33	6.18
1 oz:28.4 g	164	4.6	13.7	2.7	2.3	8.1	8.1	0	0.12	0.05	0.51	0.07	19.2	0	0	11.6	1.16	1.35
Chestnuts, Boiled and Steamed 1 oz:28.4 g	37	0.6	0.4	0.1	0.2	0.1	7.9	4.8	0.04	0.03	0.21	0.07	10.9	0	7.6	13.1	0.49	0.07
Chestnuts, Roasted 1 C:143 g	350	4.5	3.2	0.6	1.2	1.1	75.7	34.3	0.35	0.25	1.92	0.71	100.1	0	37.2	41.5	1.3	0.82
Chestnuts, Raw, Peeled 1 oz:28.4 g	70	0.9	0.6	0.1	0.3	0.2	15	6.8	0.07	0.05	0.38	0.14	19.9	0	7.4	8.2	0.26	0.16
Coconut Meat, Dried (Desiccated), Not Sweetened 1 oz:28.4 g	187	2	18.3	16.3	0.2	0.8	6.9	0	0.02	0.03	0.17	0.09	2.6	0	0.4	7.4	0.94	0.57

Nuts and Seeds cont.

Food Item & Measure	Cals	Protein (g)	All fat (g)	Sat/fat (g)	Pol/fat (g)	Mon/fat (g)	Carbo (g)	Vit A (IU)	Thiam (mg)	Ribo (mg)	Niac (mg)	Vit B6 (mg)	Fol (mcg)	Vit B12 (mcg)	Vit C (mg)	Ca (mg)	Iron (mg)	Zinc (mg)
Coconut Meat, Raw																		
1 C Shredded or Grated:80 g	-	-	-	0	0	-	-	0	0	0	-	0	-	0	-	-	-	-
1 Piece (2x2x½-in):45 g	159	1.5	15.1	13.4	0.2	0.6	6.9	0	0.03	0.01	0.24	0.02	11.9	0	1.5	6.3	1.09	0.5
Coconut Milk, Raw (Liquid Expressed from Grated Meat and Water)																		
1 C:240 g	552	5.5	57.2	50.7	0.6	2.4	13.3	0	0.06	0	1.82	0.08	38.6	0	6.7	38.4	3.94	1.61
1 Tbsp:15 g	35	0.3	3.6	3.2	0	0.2	0.8	0	0	0	0.11	0	2.4	0	0.4	2.4	0.25	0.1
Coconut Water (Liquid from Coconuts)																		
1 C:240 g	46	1.7	0.5	0.4	0	0	8.9	0	0.07	0.14	0.19	0.08	6	0	5.8	57.6	0.7	0.24
1 Tbsp:15 g	3	0.1	0	0	0	0	0.6	0	0	0.01	0.01	0	0.4	0	0.4	3.6	0.04	0.02
Filberts or Hazelnuts, Dried, Blanched																		
1 oz:28.4 g	191	3.6	19.1	1.4	1.8	15	4.5	19.6	0.15	0.03	0.33	0.18	21.2	0	0.3	55.4	0.96	0.71
Filberts or Hazelnuts, Dried, Unblanched																		
1 C Chopped Kernels:115 g	727	15	72	5.3	6.9	56.5	17.6	77.1	0.58	0.13	1.31	0.7	82.6	0	1.2	216.2	3.76	2.76
1 oz:28.4 g	179	3.7	17.8	1.3	1.7	13.9	4.4	19	0.14	0.03	0.32	0.17	20.4	0	0.3	53.4	0.93	0.68
Filberts or Hazelnuts, Dry-Roasted, Unblanched, Wo/Salt Added 1 oz:28.4 g	188	2.8	18.8	1.4	1.8	14.8	5.1	19.6	0.06	0.06	0.79	0.18	21.2	0	0.3	55.4	0.96	0.71
Filberts or Hazelnuts, Oil-Roasted, Unblanched, Wo/Salt Added 1 oz:28.4 g	187	4.1	18.1	1.3	1.7	14.2	5.4	19.9	0.06	0.06	0.79	0.18	21.3	0	0.3	55.7	0.97	0.71
Ginkgo Nuts, Cnd 1 C:155 g	172	3.6	2.5	0.5	0.9	0.9	34.3	522.4	0.21	0.08	5.62	0.31	50.7	0	14.1	6.2	0.45	0.33
1 oz (14 Med Krnls):28.4 g	32	0.7	0.5	0.1	0.2	0.2	6.3	95.7	0.04	0.02	1.03	0.06	9.3	0	2.6	1.1	0.08	0.06
Macadamia Nuts, Dried 1 C:134 g	941	11.1	98.8	14.8	1.7	78	18.4	0	0.47	0.15	2.87	0.26	21	0	0	93.8	3.23	2.29
1 oz:28.4 g	199	2.4	20.9	3.1	0.4	16.5	3.9	0	0.1	0.03	0.61	0.06	4.5	0	0	19.9	0.68	0.49
Mixed Nuts, With Peanuts, Dry-Roasted, Wo/Salt Added 1 C:137 g	814	23.7	70.5	9.5	14.8	43	34.7	20.6	0.27	0.27	6.44	0.41	69.1	0	0.6	95.9	5.07	5.21
1 oz:28.4 g	169	4.9	14.6	2	3.1	8.9	7.2	4.3	0.06	0.06	1.33	0.08	14.3	0	0.1	19.9	1.05	1.08
Mixed Nuts, With Peanuts, Oil-Roasted, Wo/Salt Added																		
1 C:142 g	876	23.8	80	12.4	18.9	45	30.4	27	0.71	0.32	7.19	0.34	117.9	0	0.7	153.4	4.56	7.21
1 oz:28.4 g	175	4.8	16	2.5	3.8	9	6.1	5.4	0.14	0.06	1.44	0.07	23.6	0	0.1	30.7	0.91	1.44

Nuts and Seeds cont.

Food Item & Measure	Cals	Protein (g)	All fat (g)	Sat/fat (g)	Pol/fat (g)	Mon/fat (g)	Carbo (g)	Vit A (IU)	Thiam (mg)	Ribo (mg)	Niac (mg)	Vit B6 (mg)	Fol (mcg)	Vit B12 (mcg)	Vit C (mg)	Ca (mg)	Iron (mg)	Zinc (mg)
Peanut Butter, Crunchy Style, W/Salt																		
1 C:258 g	1520	62.1	128.9	24.7	37	60.8	55.7	0	0.32	0.29	35.32	1.16	237.4	0	0	105.8	4.9	7.17
2 Tbsp:32 g	188	7.7	16	3.1	4.6	7.5	6.9	0	0.04	0.04	4.38	0.14	29.4	0	0	13.1	0.61	0.89
Peanut Butter; Smooth Style, W/Salt																		
1 C:258 g	1517	63.4	129	24.7	37.1	60.8	53.4	0	0.35	0.26	33.77	0.98	201.8	0	0	87.7	4.31	6.48
2 Tbsp:32 g	188	7.9	16	3.1	4.6	7.6	6.6	0	0.04	0.03	4.19	0.12	25	0	0	10.9	0.53	0.8
Peanuts: All Types, Raw 1 oz:28 g	159	7.2	13.8	1.9	4.4	6.8	4.5	0	0.18	0.04	3.38	0.1	67.1	0	0	25.8	1.28	0.92
Pecans, Dried 1 C (Halves):108 g	720	8.4	73.1	5.9	18.1	45.5	19.7	138.2	0.92	0.14	0.96	0.2	42.3	0	2.2	38.9	2.3	5.91
1 oz (20 Halves):28.4 g	189	2.2	19.2	1.5	4.8	12	5.2	36.4	0.24	0.04	0.25	0.05	11.1	0	0.6	10.2	0.6	1.55
Pecans, Dry-Roasted, Wo/Salt Added 1 oz:28.4 g	187	2.3	18.4	1.5	4.5	11.4	6.3	37.8	0.09	0.03	0.26	0.06	11.6	0	0.6	9.9	0.62	1.61
Pecans, Oil-Roasted, Wo/Salt Added																		
1 C:110 g	754	7.7	78.3	6.3	19.4	48.8	17.7	141.9	0.34	0.11	0.98	0.21	43.3	0	2.2	37.4	2.32	6.05
1 oz (15 Halves):28.4 g	195	2	20.2	1.6	5	12.6	4.6	36.6	0.09	0.03	0.25	0.05	11.2	0	0.6	9.7	0.6	1.56
Pine Nuts (Pignolia), Dried 1 oz:28.4 g	146	6.8	14.4	2.2	6.1	5.4	4	8.2	0.23	0.05	1.01	0.03	16.3	0	0.5	7.4	2.61	1.21
1 Tbsp:10 g	52	2.4	5.1	0.8	2.1	1.9	1.4	2.9	0.08	0.02	0.36	0.01	5.7	0	0.2	2.6	0.92	0.43
Pine Nuts (Pinyon), Dried 1 oz:28.4 g	161	3.3	17.3	2.7	7.3	6.5	5.5	8.2	0.35	0.06	1.24	0.03	16.4	0	0.6	2.3	0.87	1.22
10 Kernels:1 g	6	0.1	0.6	0.1	0.3	0.2	0.2	0.3	0.01	0	0.04	0	0.6	0	0	0.1	0.03	0.04
Pistachio Nuts, Dried 1 C:128 g	739	26.3	61.9	7.8	9.4	41.8	31.8	298.2	1.05	0.22	1.38	0.32	74.2	0	9.2	172.8	8.68	1.72
1 oz (47 Kernels):28.4 g	164	5.8	13.7	1.7	2.1	9.3	7.1	66.2	0.23	0.05	0.31	0.07	16.5	0	2	38.3	1.93	0.38
Pistachio Nuts, Dry Roasted, Wo/Salt Added																		
1 C:128 g	776	19.1	67.6	8.6	10.2	45.6	35.2	304.6	0.54	0.31	1.8	0.33	75.7	0	9.3	89.6	4.06	1.74
1 oz:28.4 g	172	4.2	15	1.9	2.3	10.1	7.8	67.6	0.12	0.07	0.4	0.07	16.8	0	2.1	19.9	0.9	0.39
Pumpkin Seed Kernels, Dried 1 C:138 g	747	33.9	63.3	12	28.9	19.7	24.6	524.4	0.29	0.44	2.41	0.31	79.4	0	2.6	59.3	20.66	10.29
1 oz (142 Kernels):28.4 g	154	7	13	2.5	5.9	4.1	5.1	107.9	0.06	0.09	0.5	0.06	16.3	0	0.5	12.2	4.25	2.12
Pumpkin Seed Kernels, Roasted, Wo/Salt																		
1 C:227 g	1185	74.8	95.6	18.1	43.6	29.7	30.5	862.6	0.48	0.72	3.95	0.2	130.3	0	4.1	97.6	33.91	16.89
1 oz:28.4 g	148	9.4	12	2.3	5.5	3.7	3.8	107.9	0.06	0.09	0.49	0.03	16.3	0	0.5	12.2	4.24	2.11

Food Item & Measure	Cals	Protein (g)	All fat (g)	Sat/fat (g)	Pol/fat (g)	Mon/fat (g)	Carbo (g)	Vit A (IU)	Thiam (mg)	Ribo (mg)	Niac (mg)	Vit B_6 (mg)	Fol (mcg)	Vit B_{12} (mcg)	Vit C (mg)	Ca (mg)	Iron (mg)	Zinc (mg)
Nuts and Seeds cont.																		
Sesame Butter (Tahini), from Raw and Stone Ground Kernels 1 oz:28.4 g	162	5.1	13.6	1.9	6	5.2	7.4	19	0.36	0.14	1.68	0.04	27.8	0	0	119.3	0.71	1.32
1 Tbsp:15 g	86	2.7	7.2	1	3.2	2.7	3.9	10.1	0.19	0.08	0.89	0.02	14.7	0	0	63	0.38	0.7
Sesame Butter (Tahini), from Rstd and Tstd Krnls (Most Common Type) 1 oz:28.4 g	169	4.8	15.3	2.1	6.7	5.8	6	19	0.35	0.13	1.55	0.04	27.8	0	0	121	2.54	1.31
1 Tbsp:15 g	89	2.6	8.1	1.1	3.5	3.1	3.2	10.1	0.18	0.07	0.82	0.02	14.7	0	0	63.9	1.34	0.69
Sesame Seeds, Whole, Dried 1 C:144 g	825	25.5	71.5	10	31.4	27	33.8	13	1.14	0.36	6.5	1.14	139.3	0	0	1404	20.95	11.16
1 Tbsp:9 g	52	1.6	4.5	0.6	2	1.7	2.1	0.8	0.07	0.02	0.41	0.07	8.7	0	0	87.8	1.31	0.7
Sesame Seeds, Whole, Roasted and Toasted 1 oz:28.4 g	160	4.8	13.6	1.9	6	5.2	7.3	2.6	0.23	0.07	1.3	0.23	27.9	0	0	280.9	4.19	2.03
Sunflower Seed Kernels, Dried 1 C:144 g	821	32.8	71.4	7.5	47.1	13.6	27	72	3.3	0.36	6.48	1.11	327.5	0	2	167	9.75	7.29
1 oz:28.4 g	162	6.5	14.1	1.5	9.3	2.7	5.3	14.2	0.65	0.07	1.28	0.22	64.6	0	0.4	32.9	1.92	1.44
Sunflower Seed Kernels, Oil-Roasted, Wo/Salt 1 C:135 g	830	28.8	77.6	8.1	51.2	14.8	19.9	68.9	0.43	0.38	5.58	1.07	315.9	0	1.9	75.6	9.05	7.03
1 oz:28.4 g	175	6.1	16.3	1.7	10.8	3.1	4.2	14.5	0.09	0.08	1.17	0.22	66.5	0	0.4	15.9	1.9	1.48
Walnuts, English or Persian, Dried 1 C Pieces or Chips:120 g	770	17.2	74.2	6.7	47	17	22	148.8	0.46	0.18	1.25	0.67	79.2	0	3.8	112.8	2.93	3.28
1 oz (14 Halves):28.4 g	182	4.1	17.6	1.6	11.1	4	5.2	35.2	0.11	0.04	0.3	0.16	18.7	0	0.9	26.7	0.69	0.78
OILS, SPREADS AND DRESSINGS																		
Ketchup 1 Packet:6 g	6	0.1	0	0	0	0	1.6	61	0.01	0	0.08	0.01	0.9	0	0.9	1.1	0.04	0.01
1 Tbsp:15 g	16	0.2	0.1	0	0	0	4.1	152.4	0.01	0.01	0.21	0.03	2.3	0	2.3	2.9	0.11	0.03
Ketchup, Low Salt 1 Packet:6 g	6	0.1	0	0	0	0	1.6	61	0.01	0	0.08	0.01	0.9	0	0.9	1.1	0.04	0.01
1 Tbsp:15 g	16	0.2	0.1	0	0	0	4.1	152.4	0.01	0.01	0.21	0.03	2.3	0	2.3	2.9	0.11	0.03
Margarine; Regular, Hard, Sflwr and Sybn (Hydr) 1 Stick:113.4 g	815	1	91.3	15.7	35.6	36	1	3750.1	0.01	0.04	0.03	0.01	1.3	0.11	0.2	33.9	0	0
1 Tsp:4.7 g	34	0	3.8	0.7	1.5	1.5	0	155.4	0	0	0	0	0.1	0	0	1.4	0	0
Margarine; Regular, Hard, Sybn (Hydr) 1 Stick:113.4 g	815	1	91.3	18.9	23.7	44.6	1	3750.1	0.01	0.04	0.03	0.01	1.3	0.11	0.2	33.9	0	0
1 Tsp:4.7 g	34	0	3.8	0.8	1	1.9	0	155.4	0	0	0	0	0.1	0	0	1.4	0	0

Oils, Spreads and Dressings cont.

Food Item & Measure	Cals	Protein (g)	All fat (g)	Sat/fat (g)	Pol/fat (g)	Mon/fat (g)	Carbo (g)	Vit A (IU)	Thiam (mg)	Ribo (mg)	Niac (mg)	Vit B_6 (mg)	Fol (mcg)	Vit B_{12} (mcg)	Vit C (mg)	Ca (mg)	Iron (mg)	Zinc (mg)
Margarine; Regular, Unspecified Oils, W/Salt																		
1 Stick:113.4 g	815	1	91.3	17.9	28.8	40.6	1	3750.1	0.01	0.04	0.03	0.01	1.3	0.11	0.2	33.9	0.07	0
1 Tsp:4.7 g	34	0	3.8	0.7	1.2	1.7	0	155.4	0	0	0	0	0.1	0	0	1.4	0	0
Margarine; Soft, Sthwr (Hydr and Reg) 1 C:227 g	1626	1.8	182.5	20.9	101	52.7	1.1	7506.9	0.02	0.07	0.05	0.02	2.4	0.19	0.3	60.2	0.82	0.13
1 Tsp:4.7 g	34	0	3.8	0.4	2.1	1.1	0	155.4	0	0	0	0	0.1	0	0	1.3	0.05	0.01
Margarine; Soft, Sybn (Hydr and Reg), Wo/Salt 1 C:227 g	1626	1.8	182.3	30.7	60.8	82.6	2	7506.9	0.02	0.07	0.05	0.02	2.4	0.19	0.3	60.2	0	0
1 Tsp:4.7 g	34	0	3.8	0.6	1.3	1.7	0	155.4	0	0	0	0	0.1	0	0	1.3	0	0
Margarine; Soft, Unspecified Oils, W/Salt 1 C:227 g	1626	1.8	182.5	31.3	78.5	64.7	1.1	7506.9	0.02	0.07	0.05	0.02	2.4	0.19	0.3	60.2	0	0
1 Tsp:4.7 g	34	0	3.8	0.7	1.6	1.3	0	155.4	0	0	0	0	0.1	0	0	1.3	0	0
Oil; Olive, Salad or Cooking 1 C:216 g	1909	0	216	29.2	18.1	159.2	0	0	0	0	0	0	0	0	0	0.4	0.82	0.13
1 Tbsp:13.5 g	119	0	13.5	1.8	1.1	10	0	0	0	0	0	0	0	0	0	0	0.05	0.01
Oil; Sesame, Salad or Cooking 1 C:218 g	1927	0	218	31	90.9	86.6	0	0	0	0	0	0	0	0	0	0	0	0
1 Tbsp:13.6 g	120	0	13.6	1.9	5.7	5.4	0	0	0	0	0	0	0	0	0	0	0	0
Oil; Soybean, Salad or Cooking 1 C:218 g	1927	0	218	31.4	126.2	50.8	0	0	0	0	0	0	0	0	0	0.1	0.04	0
1 Tbsp:13.6 g	120	0	13.6	2	7.9	3.2	0	0	0	0	0	0	0	0	0	0	0	0
Oil; Vegetable, Sunflower, Linoleic (Less Than 60%) 1 C:218 g	1927	0	218	22	87.4	99	0	0	0	0	0	0	0	0	0	0.4	0.07	0
1 Tbsp:13.6 g	120	0	13.6	1.4	5.5	6.2	0	0	0	0	0	0	0	0	0	0	0	0
Oil; Wheat Germ 1 C:218 g	1927	0	218	41	134.5	32.9	0	0	0	0	0	0	0	0	0	0	0	0
1 Tbsp:13.6 g	120	0	13.6	2.6	8.4	2.1	0	0	0	0	0	0	0	0	0	0	0	0
Peanut Butter; Crunchy Style, W/Salt 1 C:258 g	1520	62.1	128.9	24.7	37	60.8	55.7	0	0.32	0.29	35.32	1.16	237.4	0	0	105.8	4.9	7.17
2 Tbsp:32 g	188	7.7	16	3.1	4.6	7.5	6.9	0	0.04	0.04	4.38	0.14	29.4	0	0	13.1	0.61	0.89
Peanut Butter; Smooth Style, W/Salt 1 C:258 g	1517	63.4	129	24.7	37.1	60.8	53.4	0	0.35	0.26	33.77	0.98	201.8	0	0	87.7	4.31	6.48
2 Tbsp:32 g	188	7.9	16	3.1	4.6	7.6	6.6	0	0.04	0.03	4.19	0.12	25	0	0	10.9	0.53	0.8
Pickle, Cucumber, Dill 1 Slice:6 g	1	0	0	0	0	0	0.3	19.7	0	0	0	0	0.1	0	0.1	0.5	0.03	0.01
1 Medium:65 g	12	0.4	0.1	0	0.1	0	2.7	213.9	0.01	0.02	0.04	0.01	0.7	0	1.2	5.9	0.34	0.09

Oils, Spreads and Dressings cont.

Food item & Measure	Cals	Protein (g)	All fat (g)	Sat/fat (g)	Pol/fat (g)	Mon/fat (g)	Carbo (g)	Vit A (IU)	Thiam (mg)	Ribo (mg)	Niac (mg)	Vit B6 (mg)	Fol (mcg)	Vit B12 (mcg)	Vit C (mg)	Ca (mg)	Iron (mg)	Zinc (mg)
Pickle, Cucumber, Sweet 1 Slice:6 g	7	0	0	0	0	0	1.9	7.6	0	0	0.01	0	0.1	0	0.1	0.2	0.04	0
1 Medium:35 g	41	0.1	0.1	0	0	0	11.1	44.1	0	0.01	0.06	0.01	0.4	0	0.4	1.4	0.21	0.03
Pickle, Cucumber, Sour 1 Slice:7 g	1	0	0	0	0	0	0.2	10.2	0	0	0	0	0.1	0	0.1	0	0.03	0
1 Medium:35 g	4	0.1	0.1	0	0	0	0.8	50.8	0	0	0	0	0.3	0	0.4	0	0.14	0.01
Pickle, Cucumber, Sweet, Low Salt																		
1 Slice:6 g	7	0	0	0	0	0	1.9	7.6	0	0	0.01	0	0.1	0	0.1	0.2	0.04	0
1 Medium:35 g	41	0.1	0.1	0	0	0	11.1	44.1	0	0.01	0.06	0.01	0.4	0	0.4	1.4	0.21	0.03
Salad Dressing; French, Diet, Low-Fat, 5 Cal/Tsp, W/Salt 1 C:260 g	349	0.5	15.1	2.1	8.8	3.6	56.4	0	0	0	0	0	0	0	0	28.6	1.04	0.47
1 Tbsp:16.3 g	22	0	1	0.1	0.6	0.2	3.5	0	0	0	0	0	0	0	0	1.8	0.07	0.03
Salad Dressing; Italian, Commer, Diet, 2 Cal/Tsp, W/Salt 1 C:240 g	253	0.2	23.5	3.1	14.4	4.8	11.8	0	0	0	0	0	0	0	0	4.8	0.48	0.26
1 Tbsp:15.0 g	16	0	1.5	0.2	0.9	0.3	0.7	0	0	0	0	0	0	0	0	0.3	0.03	0.02
Salad Dressing; Mayonnaise-Type, Diet, Low-Cal, 8 Cal/Tsp, W/Salt																		
1 C:250 g	340	2.8	31.8	5.8	–	–	12	550	0.03	0.08	0	–	–	–	0	45	0.5	–
1 Tbsp:16 g	22	0.2	2	0.4	–	–	0.8	35.2	0	0	0	–	–	–	0	2.9	0.03	–
Salad Dressing; Mayonnaise-Type, Regular, W/Salt 1 C:235 g	916	2.1	78.5	11.5	42.3	21.2	56.2	517	0.03	0.06	0.01	0.04	14.8	0.49	0	32.9	0.47	0.42
1 Tbsp:14.7 g	57	0.1	4.9	0.7	2.7	1.3	3.5	32.3	0	0	0	0	0.9	0.03	0	2.1	0.03	0.03
Salad Dressing; Mayonnaise-Type, Regular, Wo/Salt 1 C:235 g	1022	2.4	99.4	18	–	–	33.8	517	0.02	0.07	0	–	–	–	0	32.9	0.47	–
1 Tbsp:15 g	65	0.2	6.4	1.2	–	–	2.2	33	0	0	0	–	–	–	0	2.1	0.03	–
Salad Dressing; Mayonnaise, Imitation, Soybean 1 C:240 g	556	0.7	46.1	7.9	25.4	10.8	38.4	0	0	0	0	0	0	0	0	0	0	0.26
1 Tbsp:15.0 g	35	0.1	2.9	0.5	1.6	0.7	2.4	0	0	0	0	0	0	0	0	0	0	0.02
Salad Dressing; Thousand Island, Commer, Regular, W/Salt 1 C:250 g	943	2.3	89.3	15	49.5	20.8	38	800	0.03	0.06	0.01	0.04	15.7	0.52	0	27.5	1.5	0.35
1 Tbsp:15.6 g	59	0.1	5.6	0.9	3.1	1.3	2.4	49.9	0	0	0	0	1	0.03	0	1.7	0.09	0.02

258 COMPOSITION OF VEGETARIAN FOODS

Food Item & Measure	Cals	Protein (g)	All fat (g)	Sat fat (g)	Pol fat (g)	Mon fat (g)	Carbo (g)	Vit A (IU)	Thiam (mg)	Ribo (mg)	Niac (mg)	Vit B₆ (mg)	Fol (mcg)	Vit B₁₂ (mcg)	Vit C (mg)	Ca (mg)	Iron (mg)	Zinc (mg)
Oils, Spreads and Dressings cont.																		
Salad Dressing; Thousand Island, Commer, Regular, Wo/Salt 1 C:250 g	1255	2	125.5	22.2	-	-	-	1	800	0.05	0.08	0.5	-	-	-	7.5	27.5	1.5	-
1 Tbsp:16 g	80	0.1	8	1.4	-	-	-	0.1	51.2	0	0.03	0.03	-	-	0.5	1.8	0.1	-
Salad Dressing; Thousand Island, Diet, Low-Cal, 10 Cal/Tsp, W/Salt 1 C:245 g	389	2	26.2	3.9	15.2	5.9	39.7	784	0.03	0.05	0.01	0.04	13.7	0.45	0	27	1.47	0.39
1 Tbsp:15.3 g	24	0.1	1.6	0.2	1	0.4	2.5	49	0	0	0	0	0.9	0.03	0	1.7	0.09	0.02
Salad Dressing; Thousand Island, Diet, Low-Cal, 10 Cal/Tsp, Wo/Slt 1 C:245 g	441	2.2	33.6	5.9	-	-	38.2	784	0.05	0.07	0.49	-	-	-	7.4	27	1.47	-
1 Tbsp:15 g	27	0.1	2.1	0.4	-	-	2.3	48	0	0	0.03	-	-	-	0.5	1.7	0.09	-
Sandwich Spread; W/Chopped Pickle, Regular, Unspecified Oils 1 C:245 g	953	2.2	83.3	12.5	49	18.1	54.9	0	0	0	0	0	0	0	0	0	0	0
1 Tbsp:15.3 g	60	0.1	5.2	0.8	3.1	1.1	3.4	0	0	0	0	0	0	0	0	0	0	0
Sauce; Barbecue 1 C (8 fl oz):250 g	188	4.5	4.5	0.7	1.7	1.9	32	2170	0.08	0.05	2.25	0.19	10	0	17.5	47.5	2.25	0.5
Sauce; Cheese, Dehydrated, Dry 1 Pkt:35.2 g	158	8	9	4.2	1.3	3	11.9	83.8	0.06	0.17	0.11	0.04	4.6	0.28	0.4	279.8	0.15	0.88
Sauce; Curry, Dehydrated, Dry 1 C (8 fl oz):28.3 g	121	2.7	6.5	1	2.5	2.8	14.3	28.9	0.03	0.06	0.23	0.01	2.6	0.11	0.6	50.1	0.88	0.25
1 Pkt:35.4 g	151	3.3	8.2	1.2	3.1	3.5	17.9	36.1	0.04	0.07	0.28	0.02	3.2	0.14	0.7	62.7	1.1	0.32
Sauce; Mushroom, Dehydrated, Dry 1 C (8 fl oz):22.7 g	79	3.3	2.2	0.3	0.8	0.9	12.4	0	0.02	0.11	1.04	0.02	6.1	0	0	2	0.23	0.2
1 Pkt:28.35 g	99	4.1	2.7	0.4	1	1.2	15.5	0	0.03	0.14	1.3	0.03	7.7	0	0	2.6	0.28	0.26
Sauce; Soy 1 Tbsp:18 g	10	0.9	0	0	0	0	1.5	0	0.01	0.02	0.6	0.03	2.8	0	0	3.1	0.36	0.07
1 fl oz:36 g	19	1.9	0	0	0	0	3.1	0	0.02	0.05	1.21	0.06	5.6	0	0	6.1	0.73	0.13
Sauce; Spaghetti, Dehydrated, Dry 1 Serving:10 g	28	0.6	0.1	0.1	0	0	6.4	0.6	0.01	0.06	0.23	0.06	3	0.05	0.2	17.2	0.27	0.02
1 Pkt:42 g	118	2.5	0.4	0.3	0	0.1	27	2.5	0.04	0.24	0.97	0.25	12.6	0.21	0.8	72.2	1.12	0.1
Sauce; Sweet and Sour, Dehydrated, Dry 1 C (8 fl oz):75.6 g	294	0.8	0.1	0	0.1	0	72.7	0	0.02	0.1	0.83	0.38	2.3	0	0	40.8	1.62	0.09
1 Pkt:56.7 g	221	0.6	0.1	0	0	0	54.5	0	0.01	0.07	0.62	0.28	1.7	0	0	30.6	1.21	0.07
Sauce; Tartar, Dietary, Low-Cal, 10 Cal/Tsp 1 Tbsp:14 g	31	0.1	3.1	0.6	-	-	0.9	30.8	0	0	0	-	-	-	0.1	2.5	0.13	-

Food item & Measure	Cals	Protein (g)	All fat (g)	Sat/fat (g)	Pol/fat (g)	Mon/fat (g)	Carbo (g)	Vit A (IU)	Thiam (mg)	Ribo (mg)	Niac (mg)	Vit B6 (mg)	Fol (mcg)	Vit B12 (mcg)	Vit C (mg)	Ca (mg)	Iron (mg)	Zinc (mg)
Oils, Spreads and Dressings cont.																		
Sauce: Tartar, Regular																		
1 C:230 g	1221	3.2	132.9	25.3	–	–	9.7	506	0.02	0.07	0	–	–	–	2.3	41.4	2.07	–
1 Tbsp:14 g	74	0.2	8.1	1.5	–	–	0.6	30.8	0	0	0	–	–	–	0.1	2.5	0.13	–
Sauce: White, Dehydrated, Dry																		
1 C (8 fl oz):19.8 g	92	2.2	5.3	1.3	1.4	2.4	10	1.4	0.04	0.08	0.14	0.02	2.8	0.16	0.6	133.5	0.1	0.22
Sauce: White, Medium																		
1 C:250 g	405	9.8	31.3	17.2	–	–	22	1150	0.13	0.43	0.75	–	–	–	1.8	287.5	0.5	–
Sauce: White, Thick 1 C:250 g	495	10	39	21.4	–	–	27.5	1425	0.18	0.45	1.25	–	–	–	1.5	267.5	0.75	–
Sauce: White, Thin 1 C:250 g	303	9.8	21.8	12	–	–	18	800	0.1	0.45	0.5	–	–	–	1.8	305	0.25	–
Vinegar: Cider 1 C:240 g	34	0	0	0	–	–	14.2	0	0	0	0	–	–	–	0	14.4	1.44	–
1 Tbsp:15 g	2	0	0	0	–	–	0.9	0	0	0	0	–	–	–	0	0.9	0.09	–
Vinegar: Distilled 1 C:240 g	29	0	0	0	–	–	12	0	0	0	0	–	–	–	0	0	0	–
1 Tbsp:15 g	2	0	0	0	–	–	0.8	0	0	0	0	–	–	–	0	0	0	–
SOUPS																		
Cream of Asparagus, Cnd, Condensed,																		
Commer 1 Can (10.75 oz):305 g	210	5.6	9.9	2.5	4.5	2.3	26	1082.8	0.13	0.19	1.89	0.03	58	0.12	6.7	70.2	1.95	2.14
1 C (8 fl oz):251 g	173	4.6	8.2	2.1	3.7	1.9	21.4	891.1	0.11	0.16	1.56	0.03	47.7	0.1	5.5	57.7	1.61	1.76
Cream of Asparagus, Dehydrated, Dry Mix																		
1 C (8 fl oz):14.2 g	52	2	1.5	0.2	0.6	0.7	7.9	239.7	0.03	0.04	0.43	0.01	6.5	0.03	0.7	19.5	0.54	0.6
1 Pkt:63.8 g	234	8.8	6.9	1	2.6	3	35.6	1076.9	0.13	0.19	1.91	0.03	29.4	0.13	3.3	87.4	2.42	2.68
Cream of Celery, Cnd, Condensed, Commer																		
1 C (8 fl oz):251 g	181	3.3	11.2	2.8	5	2.6	17.7	612.4	0.06	0.1	0.67	0.03	4.8	0.1	0.5	80.3	1.26	0.3
1 Can (10.75 oz):305 g	220	4	13.6	3.4	6.1	3.1	21.4	744.2	0.07	0.12	0.81	0.03	5.8	0.12	0.6	97.6	1.53	0.37
Cream of Mushroom, Cnd, Condensed,																		
Commer 1 C (8 fl oz):251 g	259	4	19	5.2	8.9	3.6	18.6	0	0.06	0.17	1.62	0.03	7.5	0.25	2.3	65.3	1.05	1.19
1 Can (10.75 oz):305 g	314	4.9	23.1	6.3	10.8	4.4	22.6	0	0.07	0.2	1.97	0.03	9.2	0.31	2.8	79.3	1.28	1.44
Cream of Onion, Cnd, Condensed, Commer																		
1 C (8 fl oz):251 g	221	5.5	10.5	2.9	2.9	4.2	26.1	592.4	0.1	0.15	1	0.05	14.3	0.1	2.5	67.8	1.26	0.3
1 Can (10.75 oz):305 g	268	6.7	12.8	3.6	3.5	5.1	31.7	719.8	0.12	0.18	1.22	0.06	17.4	0.12	3.1	82.4	1.53	0.37

Food Item & Measure	Cals	Protein (g)	All fat (g)	Sat fat (g)	Pol fat (g)	Mon fat (g)	Carbo (g)	Vit A (IU)	Thiam (mg)	Ribo (mg)	Niac (mg)	Vit B6 (mg)	Fol (mcg)	Vit B12 (mcg)	Vit C (mg)	Ca (mg)	Iron (mg)	Zinc (mg)
Soups cont.																		
Cream of Vegetable, Dehydrated, Dry																		
1 C (8 fl oz):23.6 g	105	1.9	5.7	1.4	1.5	2.5	35.4	1.22	0.11	0.52	0.02	7.1	0.12	3.9	31.6	0.61	0.38	
1 Pkt:17.7 g	79	1.4	4.3	1.1	1.1	1.9	9.2	26.6	0.92	0.08	0.39	0.02	5.3	0.09	2.9	23.7	0.46	0.28
Gazpacho, Cnd, Ready-To-Serve																		
1 C (8 fl oz):244 g	56	8.7	2.2	0.3	1.3	0.5	0.8	200.1	0.05	0.02	0.33	0.15	9.8	0	3.2	24.4	0.98	0.24
1 Can (13 oz):369 g	85	13.1	3.4	0.4	2	0.8	1.2	302.6	0.07	0.04	1.4	0.22	14.8	0	4.8	36.9	1.48	0.37
Minestrone, Cnd, Chunky, Ready-To-Serve																		
1 C (8 fl oz):240 g	127	5.1	2.8	1.5	0.3	0.9	20.7	4351.2	0.06	0.12	1.18	0.24	31.2	0	4.8	60	1.78	1.44
1 Can (19 oz):539 g	286	11.5	6.3	3.3	0.6	2.1	46.6	9772.1	0.12	0.26	2.65	0.54	70.1	0	10.8	134.8	3.99	3.23
Minestrone, Dehydrated, Dry																		
1 C (8 fl oz):17.3 g	62	3.5	1.4	0.6	0.1	0.6	9.3	229.4	0.05	0.03	0.8	0.09	13.7	0	0.9	29.1	0.78	0.62
1 Pkt:78 g	279	15.6	6.1	2.9	0.4	2.6	41.8	1034.3	0.23	0.16	3.59	0.39	61.6	0	3.9	131	3.51	2.81
Mushroom, Dehydrated, Dry																		
1 Instnt Pkt:16.7 g	74	1.7	3.7	0.6	1.2	1.7	8.6	5	0.22	0.09	0.38	0.01	3	0.1	0.8	51.1	0.42	0.07
1 Reg Pkt:74.4 g	328	7.6	16.6	2.8	5.3	7.7	38.1	22.3	0.96	0.38	1.71	0.04	13.4	0.45	3.7	227.7	1.86	0.3
Onion Mix, Dehydrated, Dry Form																		
1 Pkt (6 fl oz):7.1 g	21	0.8	0.4	0.1	0.1	0.3	3.8	1.4	0.02	0.04	0.36	0.01	1.2	0	0.2	10	0.11	0.04
1 Pkt:39 g	115	4.5	2.3	0.5	0.3	1.4	20.9	7.8	0.11	0.24	1.99	0.04	6.3	0	0.9	55	0.58	0.23
Onion, Cnd, Condensed, Commer																		
1 C (8 fl oz):246 g	113	7.5	3.5	0.5	1.3	1.5	16.4	0	0.07	0.05	1.21	0.1	30.5	0	2.5	54.1	1.35	1.23
1 Can (10.5 oz):298 g	137	9.1	4.2	0.6	1.6	1.8	19.9	0	0.08	0.06	1.46	0.12	37	0	3	65.6	1.64	1.49
Pea, Green, Cnd, Condensed, Commer																		
1 C (8 fl oz):263 g	329	17.2	5.9	2.8	0.8	2	53.1	402.4	0.22	0.14	2.48	0.11	3.7	0	3.4	55.2	3.89	3.42
1 Can (11.25 oz):319 g	399	20.9	7.1	3.4	0.9	2.4	64.4	488.1	0.26	0.17	3.01	0.13	4.5	0	4.2	67	4.72	4.15
Tomato, Cnd, Condensed, Commer																		
1 C (8 fl oz):251 g	171	4.1	3.8	0.7	1.9	0.9	33.2	1393.1	0.18	0.1	2.84	0.23	29.4	0	133	27.6	3.51	0.49
1 Can (10.75 oz):305 g	207	5	4.7	0.9	2.3	1	40.3	1692.8	0.21	0.12	3.45	0.27	35.7	0	161.7	33.6	4.27	0.59
Tomato, Dehydrated, Dry																		
1 C (8 fl oz):22.7 g	82	2	1.9	0.9	0.2	0.7	15.5	664.7	0.05	0.04	0.62	0.08	5.3	0.07	3.7	43.1	0.34	0.17
1 Pkt (6 fl oz):21.3 g	77	1.8	1.8	0.8	0.2	0.7	14.6	623.7	0.05	0.04	0.59	0.07	5	0.06	3.5	40.5	0.32	0.16

Food Item & Measure	Cals	Protein (g)	All fat (g)	Sat fat (g)	Pol/fat (g)	Mon/fat (g)	Carbo (g)	Vit A (IU)	Thiam (mg)	Ribo (mg)	Niac (mg)	Vit B6 (mg)	Fol (mcg)	Vit B12 (mcg)	Vit C (mg)	Ca (mg)	Iron (mg)	Zinc (mg)
Soups cont.																		
Vegetable, Cnd, Chunky, Ready-To-Serve, Commer 1 C (8 fl oz);240 g	122	3.5	3.7	0.6	1.4	1.6	26.5	5877.6	0.07	0.06	1.2	0.19	16.6	0	6	55.2	1.63	3.12
1 Can (19 oz);539 g	275	7.9	8.3	1.2	3.1	3.6	42.7	13200.1	0.16	0.15	2.7	0.43	37.2	0	13.5	124	3.67	7.01
SWEET THINGS																		
Butterscotch 1 oz;28 g	111	0	1	0.5	—	—	26.5	39.2	0	0	0	—	—	—	0	4.8	0.39	—
Caramels; Plain or Chocolate 1 oz;28 g	112	1.1	2.9	1.6	—	—	21.5	2.8	0.01	0.05	0.06	—	—	—	0	41.4	0.39	—
Caramels; Plain or Chocolate with Nuts																		
1 oz;28 g	120	1.3	4.6	1.6	—	—	19.7	5.6	0.03	0.05	0.06	—	—	—	0	39.2	0.42	—
Chocolate-Coated Peanuts 1 C;170 g	954	27.9	70.2	18.3	—	—	66.5	—	0.63	0.31	12.58	—	—	—	0	197.2	2.55	—
1 oz;28 g	157	4.6	11.6	3	—	—	11	0	0.1	0.05	2.07	—	—	—	0	32.5	0.42	—
Chocolate-Coated Raisins 1 C;190 g	808	10.3	32.5	18.1	—	—	134	285	0.15	0.4	0.76	—	—	—	0	288.8	4.75	—
1 oz;28 g	119	1.5	4.8	2.7	—	—	19.7	42	0.02	0.06	0.11	—	—	—	0	42.6	0.7	—
Chocolate-Coated Vanilla Creams 1 oz;28 g	122	1.1	4.8	1.4	—	—	19.7	0	0.01	0.02	0.03	—	—	—	0	35.8	0.17	—
Chocolate-Coated Almonds 1 C;165 g	939	20.3	72.1	12.2	—	—	65.3	0	0.2	0.87	2.81	—	—	—	0	335	4.62	—
1 oz;28 g	159	3.4	12.2	2.1	—	—	11.1	0	0.03	0.15	0.48	—	—	—	0	56.8	0.78	—
Chocolate; Bittersweet 1 oz;28 g	134	2.2	11.1	6.2	—	—	13.1	11.2	0.01	0.05	0.28	—	—	—	0	16.2	1.4	—
Chocolate; Semisweet																		
1 C or 6-oz Pkg;170 g	862	7.1	60.7	34	—	—	96.9	34	0.02	0.14	0.85	—	—	—	0	51	4.42	—
1 oz;28 g	142	1.2	10	5.6	—	—	16	5.6	0	0.02	0.14	—	—	—	0	8.4	0.73	—
Chocolate; Sweet 1 oz;28 g	148	1.2	9.8	5.5	—	—	16.2	2.8	0.01	0.04	0.08	—	—	—	0	26.3	0.39	—
Chocolate; Bitter Or Baking																		
1 C, Grated;132 g	667	14.1	70	39.2	—	—	38.2	79.2	0.07	0.32	1.98	—	—	—	0	103	8.84	—
1 oz;28 g	141	.3	14.8	8.3	—	—	8.1	16.8	0.01	0.07	0.42	—	—	—	0	21.8	1.88	—
Fondant 1 C Candy Corn;200 g	728	0.2	4	1	—	—	179.2	0	0	0	0	—	—	—	0	28	2.2	—
1 C Mints, Uncoated;110 g	400	0.1	2.2	0.6	—	—	98.6	0	0	0	0	—	—	—	0	15.4	1.21	—
Fudge; Chocolate 1 Cu In;21 g	84	0.6	2.6	0.9	—	—	15.8	0	0.01	0.02	0.04	—	—	—	0	16.2	0.21	—
1 oz;28 g	112	0.8	3.4	1.2	—	—	21	0	0.01	0.03	0.06	—	—	—	0	21.6	0.28	—
Fudge; Chocolate, W/Nuts 1 Cu In;21 g	89	0.8	3.7	0.9	—	—	14.5	0	0.01	0.02	0.06	—	—	—	0	16.6	0.25	—
1 oz;28 g	119	1.1	4.9	1.2	—	—	19.3	0	0.01	0.03	0.08	—	—	—	0	22.1	0.34	—

262 COMPOSITION OF VEGETARIAN FOODS

Food item & Measure	Cals	Protein (g)	All fat (g)	Sat/fat (g)	Pol/fat (g)	Mon/fat (g)	Carbo (g)	Vit A (IU)	Thiam (mg)	Ribo (mg)	Niac (mg)	Vit B_6 (mg)	Fol (mcg)	Vit B_{12} (mcg)	Vit C (mg)	Ca (mg)	Iron (mg)	Zinc (mg)
Sweet Things cont.																		
Fudge, Vanilla 1 Cu In:21 g	84	0.6	2.3	0.6	–	–	15.7	0	0	0.03	0.02	–	–	–	0	23.5	0.11	–
1 oz:28 g	111	0.8	3.1	0.8	–	–	20.9	0	0.01	0.04	0.03	–	–	–	0	31.4	0.14	–
Fudge, Vanilla, W/Nuts 1 Cu In:21 g	89	0.9	3.4	0.6	–	–	14.5	0	0.01	0.03	0.02	–	–	–	0	23.3	0.17	–
1 oz:28 g	119	1.2	4.6	0.8	–	–	19.3	0	0.01	0.04	0.03	–	–	–	0	31.1	0.22	–
Gum Drops, Starch Jelly Pcs 1 oz:28 g	108	0	0.3	0	–	–	27.2	0	0	0	0	–	–	–	0	5.9	0.53	–
Honey; Strained or Extracted 1 C:339 g	1031	1	0	0	–	–	279	0	0.02	0.14	1.02	–	–	–	3.4	17	1.7	–
1 Tbsp:21 g	64	0.1	0	0	–	–	17.3	0	0	0.01	0.06	–	–	–	0.2	1.1	0.11	–
Jams and Preserves; (Red Cherry or Strawberry) 1 Glass or Jar,																		
10 oz:284 g	772	1.7	0.3	0	–	–	198.8	28.4	0.03	0.09	0.57	–	–	–	42.6	56.8	2.84	–
1 Tbsp:20 g	54	0.1	0	0	–	–	14	2	0	0.01	0.04	–	–	–	3	4	0.2	–
Jams and Preserves (Other than Red Cherry or Strawberry) 1 Glass or Jar,																		
10 oz:284 g	772	1.7	0.3	0	–	–	198.8	28.4	0.03	0.09	0.57	–	–	–	5.7	56.8	2.84	–
1 Tbsp:20 g	54	0.1	0	0	–	–	14	2	0	0.01	0.04	–	–	–	0.4	4	0.2	–
Jelly Beans 1 C (Appx 75):220 g	807	0	1.1	0	–	–	204.8	0	0	0	0	–	–	–	0	26.4	2.42	–
1 oz (Appx 10):28 g	103	0	0.1	0	–	–	26.1	0	0	0	0	–	–	–	0	3.4	0.31	–
Marshmallows 1 Lrg, Reg Type:7.2 g	23	0.1	0	0	–	0	5.8	0	0	0	0	–	–	–	0	1.3	0.12	–
1C Miniature, Not Pkd:46 g	–	–	0	0	–	0	–	–	0	0	0	0.01	0	0.01	0	–	–	0.01
Milk Chocolate, Plain 1 oz:28 g	146	2.2	9	5.1	–	–	15.9	75.6	0.02	0.1	0.08	–	–	–	0	63.8	0.31	–
Milk Chocolate, With Almonds 1 oz:28 g	149	2.6	10	4.5	–	–	14.4	64.4	0.02	0.11	0.22	–	–	–	0	64.1	0.45	–
Milk Chocolate, With Peanuts 1 oz:28 g	152	4	10.7	4.4	–	–	12.5	50.4	0.07	0.07	1.4	–	–	–	0	48.7	0.39	–
Molasses; Cane, Barbados 1 C:328 g	889	0	0	0	–	–	229.6	0	0.2	0.66	0.66	–	–	–	0	803.6	14.1	–
1 fl oz:41 g	111	0	0	0	–	–	28.7	0	0.02	0.08	0.08	–	–	–	0	100.5	1.76	–
Molasses; Cane, First Extraction or Light																		
1 C:328 g	827	0	0	0	–	–	213.2	0	0.23	0.2	0.66	–	–	–	0	541.2	14.1	–
1 fl oz:41 g	103	0	0	0	–	–	26.7	0	0.03	0.02	0.08	–	–	–	0	67.7	1.76	–
Molasses; Cane, Second Extraction or Medium 1 C:328 g	761	0	0	0	–	–	196.8	0	0.3	0.39	3.94	–	–	–	0	951.2	19.68	–
1 fl oz:41 g	95	0	0	0	–	–	24.6	0	0.04	0.05	0.49	–	–	–	0	118.9	2.46	–

Food Item & Measure	Cals	Protein (g)	All fat (g)	Sat/fat (g)	Pol/fat (g)	Mon/fat (g)	Carbo (g)	Vit A (IU)	Thiam (mg)	Ribo (mg)	Niac (mg)	Vit B6 (mg)	Fol (mcg)	Vit B12 (mcg)	Vit C (mg)	Ca (mg)	Iron (mg)	Zinc (mg)
Sweet Things cont.																		
Molasses; Cane, Third Extraction or Blackstrap 1 C:328 g	699	0	0	0	–	–	180.4	0	0.36	0.62	6.56	–	–	–	0	2243.5	52.81	–
1 fl oz:41 g	87	0	0	0	–	–	22.6	0	0.05	0.08	0.82	–	–	–	0	280.4	6.6	–
Peanut Bars 1 oz:28 g	144	4.9	9	2	–	–	13.2	0	0.12	0.02	2.63	–	–	–	0	12.3	0.5	–
Peanut Brittle 1 oz:28 g	118	1.6	2.9	0.6	–	–	22.7	0	0.04	0.01	0.95	–	–	–	0	9.8	0.64	–
Sugar-Coated Chocolate Discs 1 C:197 g	918	10.2	38.8	21.7	–	–	143.2	197	0.12	0.39	0.59	–	–	–	0	266	2.56	–
1 oz:28 g	130	1.5	5.5	3.1	–	–	20.4	28	0.02	0.06	0.08	–	–	–	0	37.8	0.36	–
Sugar-Coated Almonds 1 C:195 g	889	15.2	36.3	2.9	–	–	136.9	0	0.1	0.53	1.95	–	–	–	0	195	3.71	–
1 oz:28 g	128	2.2	5.2	0.4	–	–	19.7	0	0.01	0.08	0.28	–	–	–	0	28	0.53	–
Sugars; Beet or Cane, Brown																		
1 C Not Packed:145 g	541	0	0	0	–	–	139.8	0	0.01	0.04	0.29	–	–	–	0	123.3	4.93	–
1 C Packed:220 g	821	0	0	0	–	–	212.1	0	0.02	0.07	0.44	–	–	–	0	187	7.48	–
Sugars; Beet or Cane, Granulated 1 C:200 g	770	0	0	0	–	–	199	0	0	0	0	–	–	–	0	0	0.2	–
1 Tsp:4 g	15	0	0	0	–	–	4	0	0	0	0	–	–	–	0	0	0	–
Sugars; Beet or Cane, Icing																		
1 C Sifted, Spooned:100 g	385	0	0	0	–	–	99.5	0	0	0	0	–	–	–	0	0	0.1	–
1 C Unsifted, Spooned:120 g	462	0	0	0	–	–	119.4	0	0	0	0	–	–	–	0	0	0.12	–
Syrup; Maple 1 C:315 g	794	0	0	0	–	–	204.8	0	0.41	0.19	0.32	–	–	–	0	327.6	3.78	–
1 Tbsp:19.7 g	50	0	0	0	–	–	12.8	0	0.03	0.01	0.02	–	–	–	0	20.5	0.24	–
VEGETABLES																		
Alfalfa Seeds, Sprouted, Raw 1 C:33 g	10	1.3	0.2	0.1	0	0	1.3	51.2	0.03	0.04	0.16	0.01	11.9	0	2.7	10.6	0.32	0.3
1 Tbsp:3 g	1	0.1	0	0	0	0	0.1	4.7	0	0	0.01	0	1.1	0	0.3	1	0.03	0.03
Amaranth, Boiled, Drained, Wo/Salt																		
1 C:132 g	28	2.8	0.2	0.1	0.1	0.1	5.4	3656.4	0.03	0.18	0.74	0.23	75	0	54.3	275.9	2.98	1.16
½ C:66 g	14	1.4	0.1	0	0	0	2.7	1828.2	0.01	0.09	0.37	0.12	37.5	0	27.1	137.9	1.49	0.58
Artichokes (Globe or French), Boiled, Drained, Wo/Salt																		
1 Med Artichoke:120 g	60	4.2	0.2	0	0	0	13.4	212.4	0.08	0.08	1.2	0.13	61.2	0	12	54	1.55	0.59
½ C Hearts:84 g	42	2.9	0.1	0	0	0	9.4	148.7	0.05	0.06	0.84	0.09	42.8	0	8.4	37.8	1.08	0.41

Food Item & Measure	Cals	Protein (g)	All fat (g)	Sat/fat (g)	Pol/fat (g)	Mon/fat (g)	Carbo (g)	Vit A (IU)	Thiam (mg)	Ribo (mg)	Niac (mg)	Vit B6 (mg)	Fol (mcg)	Vit B12 (mcg)	Vit C (mg)	Ca (mg)	Iron (mg)	Zinc (mg)
Vegetables cont.																		
Asparagus; Boiled, Drained, Wo/Salt																		
½ C:90 g	23	2.3	0.3	0.1	0.1	0	4	746.1	0.09	0.11	0.95	0.13	88.3	0	24.4	21.6	0.59	0.43
4 Spears, ½-In Base:60 g	15	1.6	0.2	0	0.1	0	2.6	497.4	0.06	0.07	0.63	0.08	58.9	0	16.3	14.4	0.4	0.29
Asparagus; Cnd, Reg Pk, Sol and Liq																		
1 Can:411 g	58	7.4	0.8	0.2	0.4	0	9.3	1948.1	0.22	0.37	3.5	0.4	350.6	0	67.4	57.5	2.38	1.93
½ C:122 g	17	2.2	0.2	0.1	0.1	0	2.8	578.3	0.07	0.11	1.04	0.12	104.1	0	20	17.1	0.71	0.57
Aubergine, Boiled, Drained, Wo/Salt																		
1 C, 1-In Cubes:96 g	27	0.8	0.2	0	0.1	0	6.4	61.4	0.07	0.02	0.58	0.08	13.8	0	1.3	5.8	0.34	0.14
½ C, 1-In Cubes:48 g	13	0.4	0.1	0	0	0	3.2	30.7	0.04	0.01	0.29	0.04	6.9	0	0.6	2.9	0.17	0.07
Bamboo Shoots; Boiled, Drained, Wo/Salt																		
1 C, ½-In Slcs:120 g	14	1.8	0.3	0.1	0.1	0	2.3	0	0	0.06	0.36	0.12	2.8	0	0	14.4	0.29	0.56
1 Shoot:144 g	17	2.2	0.3	0.1	0.1	0	2.8	0	0.03	0.07	0.43	0.14	3.3	0	0	17.3	0.35	0.68
Bamboo Shoots; Raw 1 C, ½-In Slices:151 g	41	4	0.5	0.1	0.2	0	7.9	30.4	0.23	0.11	0.91	0.36	10.8	0	6.1	19.8	0.76	1.67
½ C, ½-In Slcs:76 g	21	2	0.2	0.1	0.1	0	4	15.2	0.11	0.05	0.46	0.18	5.4	0	3	9.9	0.38	0.84
Beetroots, Boiled, Drained, Wo/Salt																		
½ C Slices:85 g	26	0.9	0	0	0	0	5.7	11.1	0.03	0.01	0.23	0.03	45.2	0	4.7	9.4	0.53	0.21
2 Roots, 2-In Diam:100 g	31	1.1	0.1	0	0	0	6.7	13	0.03	0.01	0.27	0.03	53.2	0	5.5	11	0.62	0.25
Beetroots; Raw ½ C Slices:68 g	30	1	0.1	0	0	0	6.8	13.6	0.03	0.01	0.27	0.03	63	0	7.5	10.9	0.62	0.25
2 Roots, 2-In Diam:163 g	72	2.4	0.2	0	0.1	0	16.3	32.6	0.08	0.03	0.65	0.07	150.9	0	17.9	26.1	1.48	0.6
Broccoli; Ckd, Boiled, Drained, Wo/Salt																		
1 Spear:180 g	50	5.4	0.6	0.1	0.3	0	9.1	2498.4	0.1	0.2	1.03	0.26	90	0	134.3	82.8	1.51	0.68
½ C Chopped:78 g	22	2.3	0.3	0	0.1	0	4	1082.6	0.04	0.09	0.45	0.11	39	0	58.2	35.9	0.66	0.3
Broccoli; Raw 1 Spear:151 g	42	4.5	0.5	0.1	0.3	0	7.9	2328.4	0.1	0.18	0.96	0.24	107.2	0	140.7	72.5	1.33	0.6
½ C Chopped:44 g	12	1.3	0.2	0	0.1	0	2.3	678.5	0.03	0.05	0.28	0.07	31.2	0	41	21.1	0.39	0.18
Brussels Sprouts, Boiled, Drained, Wo/Salt																		
1 Sprout:21 g	8	0.5	0.1	0	0.1	0	1.8	151	0.02	0.02	0.13	0.04	12.6	0	13	7.6	0.25	0.07
½ C:78 g	30	2	0.4	0.1	0.2	0	6.8	560.8	0.08	0.06	0.47	0.14	46.8	0	48.4	28.1	0.94	0.26
Cabbage; Common (Green, White, Savoy, Chinese), Raw 1 Head:908 g	218	11	1.6	0.2	0.8	0.1	48.8	1144.1	0.45	0.27	2.72	0.06	514.8	0	429.5	426.8	5.08	1.63
½ C Shredded:35 g	8	0.4	0.1	0	0	0	1.9	44.1	0.02	0.01	0.11	0.03	19.9	0	16.6	16.5	0.2	0.06

Food item & Measure	Cals	Protein (g)	All fat (g)	Sat/fat (g)	Pol/fat (g)	Mon/fat (g)	Carbo (g)	Vit A (IU)	Thiam (mg)	Ribo (mg)	Niac (mg)	Vit B$_6$ (mg)	Fol (mcg)	Vit B$_{12}$ (mcg)	Vit C (mg)	Ca (mg)	Iron (mg)	Zinc (mg)
Vegetables cont.																		
Cabbage; Common, Boiled, Drained, Wo/Salt																		
1 Head:1262 g	265	12.1	3.2	0.4	1.5	0.2	60.2	1085.8	0.72	0.69	2.9	0.81	256.2	0	306.7	416.5	4.92	2.02
½ C Shredded:75 g	16	0.7	0.2	0	0.1	0	3.6	64.5	0.04	0.04	0.17	0.05	15.2	0	18.2	24.8	0.29	0.12
Cabbage; Red, Boiled, Drained, Wo/Salt																		
1 Leaf:22 g	5	0.2	0	0	0	0	1	5.9	0.01	0	0.04	0.03	2.8	0	7.6	8.1	0.08	0.03
½ C Shredded:75 g	16	0.8	0.2	0	0.1	0	3.5	20.3	0.03	0.02	0.15	0.11	9.5	0	25.8	27.8	0.26	0.11
Cabbage; Red, Raw 1 C Shredded:70 g	19	1	0.2	0	0.1	0	4.3	28	0.04	0.02	0.21	0.15	14.5	0	39.9	35.7	0.34	0.15
½ C Shredded:35 g	9	0.5	0.1	0	0	0	2.1	14	0.02	0.01	0.11	0.07	7.3	0	20	17.9	0.17	0.07
Carrots; Boiled, Drained, Wo/Salt																		
1 Carrot:46 g	21	0.5	0.1	0	0	0	4.8	11294.8	0.02	0.03	0.23	0.11	6.4	0	1.1	14.3	0.29	0.14
½ C Slices:78 g	35	0.9	0.1	0	0.1	0	8.2	19152.1	0.03	0.04	0.39	0.19	10.8	0	1.8	24.2	0.48	0.23
Carrots; Cnd, Reg Pk, Sol and Liq																		
1 Can:454 g	104	2.8	0.8	0.2	0.4	0	23	59782.7	0.09	0.12	1.91	0.51	36.8	0	12.7	113.5	2.77	1.32
½ C Slices:123 g	28	0.8	0.2	0	0.1	0	6.2	16196.6	0.02	0.03	0.52	0.14	10	0	3.4	30.8	0.75	0.36
Carrots; Raw 1 Carrot, 7½ In:72 g	31	0.7	0.1	0	0.1	0	7.3	20252.9	0.07	0.04	0.67	0.11	10.1	0	6.7	19.4	0.36	0.14
½ C Shredded:55 g	24	0.6	0.1	0	0	0	5.6	15471	0.05	0.03	0.51	0.08	7.7	0	5.1	14.9	0.28	0.11
Cauliflower; Boiled, Drained, Wo/Salt																		
½ C, 1-In Pieces:62 g	15	1.2	0.1	0	0.1	0	2.9	8.7	0.04	0.03	0.34	0.13	31.7	0	34.4	16.7	0.26	0.15
3 Florets:54 g	13	1	0.1	0	0.1	0	2.5	7.6	0.03	0.03	0.3	0.11	27.7	0	29.9	14.6	0.23	0.13
Cauliflower; Raw ½ C, 1-In Pieces:50 g	12	1	0.1	0	0	0	2.5	8	0.04	0.03	0.32	0.12	33.1	0	35.8	14.5	0.29	0.09
3 Florets:56 g	13	1.1	0.1	0	0.1	0	2.8	9	0.04	0.03	0.35	0.13	37	0	40	16.2	0.32	0.1
Celery; Boiled, Drained, Wo/Salt																		
1 C Dice:150 g	27	1.3	0.2	0.1	0.1	0.1	6	198	0.06	0.07	0.48	0.13	33	0	9.2	63	0.63	0.21
½ C Dice:75 g	14	0.6	0.1	0	0.1	0	3	99	0.03	0.04	0.24	0.06	16.5	0	4.6	31.5	0.32	0.11
Celery; Raw 1 Stalk:40 g	6	0.3	0.1	0	0	0	1.5	53.6	0.02	0.02	0.13	0.03	11.2	0	2.8	16	0.16	0.05
½ C Dice:60 g	10	0.5	0.1	0	0	0	2.2	80.4	0.03	0.03	0.19	0.05	16.8	0	4.2	24	0.24	0.08
Chard; Swiss, Raw 1 Leaf:48 g	9	0.9	0.1	—	—	—	1.8	1584	0.02	0.04	0.19	0.05	6.6	0	14.4	24.5	0.86	0.17
½ C Chopped:18 g	3	0.3	0	—	—	—	0.7	594	0.01	0.02	0.07	0.02	2.5	0	5.4	9.2	0.32	0.06

Vegetables cont.

Food Item & Measure	Cals	Protein (g)	All fat (g)	Sat fat (g)	Pol/fat (g)	Mon/fat (g)	Carbo (g)	Vit A (IU)	Thiam (mg)	Ribo (mg)	Niac (mg)	Vit B6 (mg)	Fol (mcg)	Vit B12 (mcg)	Vit C (mg)	Ca (mg)	Iron (mg)	Zinc (mg)
Chicory; Witloof, Raw																		
1 Head, 5-7 In Lng:53 g	8	0.5	0.1	0	0	0	1.7	0	0.04	0.07	0.27	0.02	59.4	0	5.3	9.5	0.27	0.33
½ C:45 g	7	0.5	0.1	0	0	0	1.4	0	0.03	0.06	0.23	0.02	50.4	0	4.5	8.1	0.23	0.28
Coleslaw 1 Tbsp:8 g	6	0.1	0.2	0	0.1	0.1	1	50.8	0.01	0	0.02	0.01	2.1	0	2.6	3.6	0.05	0.02
½ C:60 g	41	0.8	1.6	0.2	0.8	0.4	7.5	381	0.04	0.04	0.16	0.08	15.9	0	19.6	27	0.35	0.12
Collards; Boiled, Drained, Wo/Salt																		
1 C Chopped:128 g	35	1.7	0.2	-	-	-	7.9	3490.6	0.03	0.07	0.37	0.07	7.7	0	15.5	29.4	0.2	0.14
½ C Chopped:64 g	17	0.9	0.1	-	-	-	3.9	1745.3	0.01	0.03	0.19	0.03	3.8	0	7.7	14.7	0.1	0.07
Collards; Raw 1 C Chopped:36 g	11	0.6	0.1	-	-	-	2.6	1198.8	0.01	0.02	0.13	0.02	4.3	0	8.4	10.4	0.07	0.05
½ C Chopped:18 g	6	0.3	0	-	-	-	1.3	599.4	0.01	0.01	0.07	0.01	2.2	0	4.2	5.2	0.03	0.02
Corn On The Cob W/ Butter:146 g	155	4.5	3.4	1.6	0.6	1	31.9	391.3	0.25	0.1	2.18	0.32	43.8	0	6.9	4.4	0.88	0.91
Corn On The Cob, Boiled, Drained, Wo/Salt																		
½ C Cut:82 g	89	2.7	1.1	0.2	0.5	0.3	20.6	177.9	0.18	0.06	1.32	0.05	38.1	0	5.1	1.6	0.5	0.39
Kernels From 1 Ear:77 g	83	2.6	1	0.2	0.5	0.3	19.3	167.1	0.17	0.06	1.24	0.05	35.7	0	4.8	1.5	0.47	0.37
Corn; Cnd, Brine Pk, Reg Pk, Sol and Liq																		
1 Can:482 g	294	9.4	2.2	0.3	1	0.6	71.5	578.4	0.13	0.29	4.53	0.18	183.6	0	32.3	19.3	1.69	1.74
½ C:128 g	78	2.5	0.6	0.1	0.3	0.2	19	153.6	0.03	0.08	1.2	0.05	48.8	0	8.6	5.1	0.45	0.46
Corn; Frz, Kernels On Cob, Unprepared ½ C																		
Kernels:82 g	80	2.7	0.6	0.1	0.3	0.2	19.3	201.7	0.08	0.07	1.38	0.15	32.9	0	5.9	3.3	0.56	0.57
Kernels From 1 Ear:125 g	123	4.1	1	0.2	0.5	0.3	29.4	307.5	0.13	0.11	2.1	0.22	50.1	0	9	5	0.85	0.88
Cress; Garden, Raw 1 Sprig:1 g	0	0	0	0	0	0	0.1	93	0	0	0.01	0	0.8	0	0.7	0.8	0.01	0
½ C:25 g	8	0.7	0.2	0	0.1	0.1	1.4	2325	0.02	0.07	0.25	0.06	20.1	0	17.3	20.3	0.33	0.06
Cucumber, Not Pared, Raw																		
1 Cucumber, 8¼ In:301 g	39	1.6	0.4	0.1	0.2	0	8.8	135.5	0.09	0.06	0.9	0.16	41.8	0	14.2	42.1	0.84	0.69
½ C Slices:52 g	7	0.3	0.1	0	0	0	1.5	23.4	0.02	0.01	0.16	0.03	7.2	0	2.4	7.3	0.15	0.12
Leeks (Bulb and Lower Leaf-Portion), Boiled, Drained, Wo/Salt																		
1 Leek:124 g	38	1	0.3	0	0.1	0	9.5	57	0.03	0.02	0.25	0.14	30.1	0	5.2	37.2	1.36	0.07
½ C Chopped:26 g	8	0.2	0.1	0	0	0	2	12	0.01	0.01	0.05	0.03	6.3	0	1.1	7.8	0.29	0.02

Food item & Measure	Cals	Protein (g)	All fat (g)	Sat/fat (g)	Pol/fat (g)	Mon/fat (g)	Carbo (g)	Vit A (IU)	Thiam (mg)	Ribo (mg)	Niac (mg)	Vit B6 (mg)	Fol (mcg)	Vit B12 (mcg)	Vit C (mg)	Ca (mg)	Iron (mg)	Zinc (mg)
Vegetables cont.																		
Leeks (Bulb and Lower Leaf-Portion), Raw																		
1 Leek:124 g	76	1.9	0.4	0.1	0.2	0	17.6	117.8	0.07	0.04	0.5	0.29	79.5	0	14.9	73.2	2.6	0.15
½ C Chopped:26 g	16	0.4	0.1	0	0	0	3.7	24.7	0.02	0.01	0.1	0.06	16.7	0	3.1	15.3	0.55	0.03
Lettuce; Cos or Romaine, Raw																		
1 Inner Leaf:10 g	2	0.2	0	0	0	0	0.2	260	0.01	0.01	0.05	0	13.6	0	2.4	3.6	0.11	0.03
½ C Shredded:28 g	4	0.5	0.1	0	0	0	0.7	728	0.03	0.03	0.14	0.01	38	0	6.7	10.1	0.31	0.07
Lettuce; Iceberg, Raw																		
1 Head, 6-in Diam:539 g	70	5.4	1	0.1	0.5	0	11.3	1778.7	0.25	0.16	1.01	0.22	301.8	0	21	102.4	2.7	1.19
1 Leaf:20 g	3	0.2	0	0	0	0	0.4	66	0.01	0.01	0.04	0.01	11.2	0	0.8	3.8	0.1	0.04
Mushrooms, Boiled, Drained, Wo/Salt 1 Mushroom:12 g	3	0.3	0.1	0	0	0	0.6	0	0.01	0.04	0.54	0.01	2.2	0	0.5	0.7	0.21	0.1
½ C Pieces:78 g	21	1.7	0.4	0.1	0.1	0	4	0	0.06	0.23	3.48	0.07	14.2	0	3.1	4.7	1.36	0.68
Mushrooms; Cnd, Drained Solids																		
1 Mushroom:12 g	3	0.2	0	0	0	0	0.6	0	0.01	0	0.19	0.01	1.5	0	0	1.3	0.09	0.09
½ C Pieces:78 g	19	1.5	0.2	0	0.1	0	3.9	0	0.07	0.02	1.24	0.05	9.6	0	0	8.6	0.62	0.56
Mushrooms; Raw																		
1 Mushroom:18 g	5	0.4	0.1	0	0	0	0.8	0	0.02	0.08	0.74	0.02	3.8	0	0.6	0.9	0.22	0.13
½ C Pieces:35 g	9	0.7	0.2	0	0.1	0	1.6	0	0.04	0.16	1.44	0.03	7.4	0	1.2	1.8	0.43	0.26
Mushrooms; Shiitake, Dried																		
1 Mushroom:3.6 g	11	0.3	0	0	0	0	2.7	0	0.01	0.05	0.51	0.03	5.9	0	0.1	0.4	0.06	0.28
4 Mushrooms:15 g	44	1.4	0.2	0	0	0.1	11.3	0	0.05	0.19	2.12	0.14	24.5	0	0.5	1.7	0.26	1.15
Okra, Boiled, Drained, Wo/Salt																		
½ C Slices:80 g	26	1.5	0.1	0	0	0	5.8	460	0.11	0.04	0.7	0.15	36.6	0	13	50.4	0.36	0.44
8 Pods, 3-in Long:85 g	27	1.6	0.1	0	0	0	6.1	488.8	0.11	0.05	0.74	0.16	38.9	0	13.9	53.6	0.38	0.47
Onion Rings, Breaded and Fried																		
8-9 Rings:83 g	276	3.7	15.5	7	0.7	6.7	31.3	8.3	0.08	0.1	0.92	0.06	11.6	0.12	0.6	73	0.85	0.35
Onion Rings, Breaded, Par-Fried, Frz, Prepared, Heated in Oven																		
2 Rings:20 g	81	1.1	5.3	1.7	1	2.2	7.6	45	0.06	0.03	0.72	0.02	2.6	0	0.3	6.2	0.34	0.08
7 Rings:70 g	285	3.7	18.7	6	3.6	7.6	26.7	157.5	0.2	0.1	2.53	0.05	9.1	0	1	21.7	1.18	0.29

268 COMPOSITION OF VEGETARIAN FOODS

Vegetables cont.

Food Item & Measure	Cals	Protein (g)	All fat (g)	Sat/fat (g)	Pol/fat (g)	Mon/fat (g)	Carbo (g)	Vit A (IU)	Thiam (mg)	Ribo (mg)	Niac (mg)	Vit B6 (mg)	Fol (mcg)	Vit B12 (mcg)	Vit C (mg)	Ca (mg)	Iron (mg)	Zinc (mg)
Onions, Boiled, Drained, Wo/Salt																		
1 Tbsp Chopped:15 g	7	0.2	0	0	0	0	1.5	0	0.01	0	0.02	0.02	2.3	0	0.8	3.3	0.04	0.03
½ C Chopped:105 g	46	1.4	0.2	0	0.1	0	10.7	0	0.04	0.02	0.17	0.14	15.8	0	5.5	23.1	0.25	0.22
Onions; Raw 1 Tbsp Chopped:10 g	4	0.1	0	0	0	0	0.9	0	0	0	0.01	0.01	1.9	0	0.6	2	0.02	0.02
½ C Chopped:80 g	30	0.9	0.1	0	0.1	0	6.9	0	0.03	0.02	0.12	0.09	15.2	0	5.1	16	0.18	0.15
Onions; Spring (Includes Tops and Bulb),																		
Raw 1 Tbsp Chopped:6 g	2	0.1	0	0	0	0	0.4	23.1	0	0	0.03	–	3.8	–	1.1	4.3	0.09	0.02
½ C Chopped:50 g	16	0.9	0.1	0	0.1	0	3.7	192.5	0.03	0.04	0.26	–	32	–	9.4	36	0.74	0.2
Parsley; Raw ½ C Chopped:30 g	10	0.7	0.1	–	–	–	2.1	1560	0.02	0.03	0.21	0.05	54.9	0	27	39	1.86	0.22
10 Sprigs:10 g	3	0.2	0	–	–	–	0.7	520	0.01	0.01	0.07	0.02	18.3	0	9	13	0.62	0.07
Parsnips, Boiled, Drained, Wo/Salt																		
1 Parsnip, 9-In Long:160 g	130	2.1	0.5	0.1	0.1	0.2	31.3	0	0.13	0.08	1.16	0.15	93.1	0	20.8	59.2	0.93	0.42
½ C Slices:78 g	63	1	0.2	0	0	0.1	15.2	0	0.06	0.04	0.56	0.07	45.4	0	10.1	28.9	0.45	0.2
Parsnips; Raw 1 C Slices:133 g	100	1.6	0.4	0.1	0.1	0.2	23.9	0	0.12	0.07	0.93	0.12	88.8	0	22.6	47.9	0.78	0.78
½ C Slices:67 g	50	0.8	0.2	0	0	0.1	12.1	0	0.06	0.03	0.47	0.06	44.8	0	11.4	24.1	0.4	0.4
Peas, Boiled, Drained, Wo/Salt																		
1 C:160 g	134	8.6	0.4	0.1	0.2	0	25	955.2	0.41	0.24	3.23	0.35	101.3	0	22.7	43.2	2.46	1.9
½ C:80 g	67	4.3	0.2	0	0.1	0	12.5	477.6	0.21	0.12	1.62	0.17	50.6	0	11.4	21.6	1.23	0.95
Peas, Cnd, Reg Pk, Sol and Liq																		
1 Can (303 X 406):482 g	236	14.4	1.4	0.2	0.6	0.1	43.2	1826.8	0.54	0.35	4.06	0.31	137.4	0	52.5	66.8	5.35	3.37
½ C:124 g	61	3.7	0.4	0.1	0.2	0	11.1	470	0.14	0.09	1.04	0.08	35.3	0	13.5	22.3	1.38	0.87
Peas, Raw 1 C:145 g	117	7.9	0.6	0.1	0.3	0.1	21	928	0.39	0.19	3.03	0.25	94.3	0	58	36.3	2.13	1.8
½ C:72 g	58	3.9	0.3	0.1	0.1	0	10.4	460.8	0.19	0.1	1.5	0.12	46.8	0	28.8	18	1.06	0.89
Peppers; Hot Chilli, Green, Cnd, Pods, Excluding Seeds, Sol and Liq																		
1 Pepper:73 g	18	0.7	0.1	0	0	0	4.5	8681.2	0.01	0.04	0.58	0.11	7.3	0	49.6	5.1	0.37	0.12
½ C Chopped:68 g	17	0.6	0.1	0	0	0	4.2	8086.6	0.01	0.03	0.54	0.1	6.8	0	46.2	4.8	0.34	0.12
Peppers; Sweet, Green, Boiled, Drained, Wo/Salt 1 Pepper:73 g	20	0.7	0.2	0	0.1	0	4.9	432.2	0.04	0.02	0.35	0.17	11.7	0	54.3	6.6	0.34	0.09
½ C Chopped:68 g	19	0.6	0.1	0	0.1	0	4.6	402.6	0.04	0.02	0.32	0.16	10.9	0	50.6	6.1	0.31	0.08

Food Item & Measure	Cals	Protein (g)	All fat (g)	Sat/fat (g)	Pol/fat (g)	Mon/fat (g)	Carbo (g)	Vit A (IU)	Thiam (mg)	Ribo (mg)	Niac (mg)	Vit B6 (mg)	Fol (mcg)	Vit B12 (mcg)	Vit C (mg)	Ca (mg)	Iron (mg)	Zinc (mg)
Vegetables cont.																		
Peppers; Sweet, Green, Raw 1 Pepper:74 g	20	0.7	0.1	0	0.1	0	4.8	467.7	0.05	0.02	0.38	0.18	16.3	0	66.1	6.7	0.34	0.09
½ C Chopped:50 g	14	0.5	0.1	0	0.1	0	3.2	316	0.03	0.02	0.25	0.12	11	0	44.7	4.5	0.23	0.06
Pimiento; Cnd 1 Slice:1 g	0	0	0	0	0	0	0	3.3	0	0	0	0	0	0	0.2	0.1	0.01	0
1 Tbsp:12 g	2	0.1	0	0	0	0	0.5	39.5	0.04	0.01	0.01	0	0.1	0	0.2	1.1	0.06	0.02
Potato Crisps, With Salt Added 1 oz:28.4 g	149	1.8	10.1	2.6	5.2	1.8	14.7	0	0.04	0.01	1.19	0.14	12.8	0	11.8	6.8	0.34	0.3
10 Crisps:20 g	105	1.3	7.1	1.8	3.6	1.3	10.4	0	0.03	0	0.84	0.1	9	0	8.3	4.8	0.24	0.21
Potato Salad 1 C:250 g	358	6.7	20.5	3.6	9.3	6.2	27.9	522.5	0.19	0.15	2.23	0.35	16.8	0	25	47.5	1.63	0.78
½ C:125 g	179	3.4	10.3	1.8	4.7	3.1	14	261.3	0.1	0.08	1.11	0.18	8.4	0	12.5	23.8	0.81	0.39
Potatoes; Baked, Flesh and Skin, Wo/Salt Potato, 2-⅓ × 4-¾ in:202 g	220	4.7	0.2	0.1	0.1	0	51	0	0.22	0.07	3.32	0.7	22.2	0	26.1	20.2	2.75	0.65
Potatoes; Boiled, Ckd In Skin, Flesh, Wo/Salt ½ C:78 g	68	1.5	0.1	0	0	0	15.7	0	0.08	0.02	1.12	0.23	7.8	0	10.1	3.9	0.24	0.23
Potato, 2-½-In Diam:136 g	118	2.5	0.1	0	0.1	0	27.4	0	0.14	0.03	1.96	0.41	13.6	0	17.7	6.8	0.42	0.41
Potatoes; Boiled, Ckd In Skin, Skin, Wo/Salt Skin From 1 Potato:34 g	27	1	0	0	0	0	5.9	0	0.01	0.01	0.42	0.08	3.3	0	1.8	15.3	2.06	0.15
Potatoes; Cnd, Drained Solids 1 Potato, 1-In Diam:35 g	21	0.5	0.1	0	0	0	4.8	0	0.02	0	0.32	0.07	2.2	0	1.8	1.8	0.44	0.1
½ C:90 g	54	1.3	0.2	0.1	0.1	0	12.3	0	0.06	0.01	0.82	0.17	5.6	0	4.6	4.5	1.13	0.25
Potatoes; Frz, French-Fried, Par-Fried, Home-Prep, Heated In Oven, Wo/Salt 10 Chips:50 g	111	1.7	4.4	2.1	0.3	1.8	17	0	0.06	0.02	1.15	0.12	8.3	0	5.5	4.5	0.67	0.21
9-oz Pkg:198 g	440	6.9	17.3	8.2	1.3	7	67.1	0	0.24	0.06	4.55	0.46	32.7	0	21.6	17.8	2.63	0.81
Potatoes; Frz, French-Fried, Par-Fried, Restaurant-Prep, Fried In Veg Oil 1 C:57 g	180	2.3	9.4	3.9	0.6	4.6	22.6	0	0.1	0.02	1.85	0.13	16.5	0	5.9	10.8	0.43	0.22
10 Chips:50 g	158	2	8.3	3.4	0.5	4	19.8	0	0.09	0.01	1.63	0.12	14.5	0	5.2	9.5	0.38	0.19
Potatoes; Mashed, Dehyd, Prep From Flakes Wo/Milk, Whole Milk and Butter Added 1 C:210 g	237	4	11.8	7.2	0.5	3.3	31.5	378	0.23	0.11	1.41	0.02	15.5	0	20.4	102.9	0.46	0.38
½ C:105 g	119	2	5.9	3.6	0.3	1.7	15.8	189	0.12	0.05	0.7	0.01	7.8	0	10.2	51.5	0.23	0.19

Vegetables cont.

Food item & Measure	Cals	Protein (g)	All fat (g)	Sat/fat (g)	Pol/fat (g)	Mon/fat (g)	Carbo (g)	Vit A (IU)	Thiam (mg)	Ribo (mg)	Niac (mg)	Vit B6 (mg)	Fol (mcg)	Vit B12 (mcg)	Vit C (mg)	Ca (mg)	Iron (mg)	Zinc (mg)
Potatoes; Mashed, Home-Prepared, Whole Milk and Margarine Added 1 C:210 g	223	4	8.9	2.2	2.5	3.7	35.1	354.9	0.18	0.08	2.27	0.47	16.6	0	12.8	54.6	0.55	0.57
½ C:105 g	111	2	4.4	1.1	1.3	1.9	17.6	177.5	0.09	0.04	1.13	0.24	8.3	0	6.4	27.3	0.27	0.28
Pumpkin, Boiled, Drained, Wo/Salt 1 C Mashed:245 g	49	1.8	0.2	0.1	0.1	0	12	2650.9	0.08	0.19	1.01	0.11	20.8	0	11.5	36.8	1.4	0.56
½ C Mashed:122 g	24	0.9	0.1	0.1	0	0	6	1320	0.04	0.1	0.5	0.05	10.4	0	5.7	18.3	0.7	0.28
Radishes; Raw ½ C Slices:58 g	10	0.4	0.3	0	0.1	0	2.1	4.6	0	0.03	0.17	0.04	15.7	0	13.2	12.2	0.17	0.17
10 Radishes:45 g	8	0.3	0.2	0	0	0	1.6	3.6	0	0.02	0.14	0.03	12.2	0	10.3	9.5	0.13	0.14
Salad, Vegetable, Tossed, Without Dressing 1½ C:207 g	33	2.6	0.1	0	0.1	0	6.7	2351.5	0.06	0.1	1.14	0.17	76.6	0	48	26.9	1.3	0.43
¾ C:104 g	17	1.3	0.1	0	0	0	3.4	1181.4	0.03	0.05	0.57	0.08	38.5	0	24.1	13.5	0.66	0.22
Sauerkraut; Cnd, Sol and Liq 1 C:236 g	45	2.2	0.3	0.1	0.1	0	10.1	42.5	0.05	0.05	0.34	0.31	55.9	0	34.7	70.8	3.47	0.45
½ C:118 g	22	1.1	0.2	0	0.1	0	5.1	21.2	0.02	0.03	0.17	0.15	28	0	17.4	35.4	1.73	0.22
Spinach, Boiled, Drained, Wo/Salt 1 C:180 g	41	5.4	0.5	0.1	0.2	0	6.8	14742	0.17	0.42	0.88	0.44	262.4	0	17.6	244.8	6.43	1.37
½ C:90 g	21	2.7	0.2	0	0.1	0	3.4	7371	0.09	0.21	0.44	0.22	131.2	0	8.8	122.4	3.21	0.68
Swede, Boiled, Drained, Wo/Salt ½ C Cubes:85 g	29	0.9	0.2	0	0.1	0	6.6	0	0.06	0.03	0.54	0.08	13.2	0	18.6	35.7	0.4	0.26
½ C Mashed:120 g	41	1.3	0.2	0	0.1	0	9.3	0	0.09	0.04	0.76	0.11	18.6	0	26.3	50.4	0.56	0.36
Swede; Raw 1 C Cubes:140 g	50	1.7	0.3	0	0.1	0	11.4	0	0.13	0.06	0.98	0.14	28.7	0	35	65.8	0.73	0.48
½ C Cubes:70 g	25	0.8	0.1	0	0.1	0	5.7	0	0.06	0.03	0.49	0.07	14.4	0	17.5	32.9	0.36	0.24
Sweet Potatoes, Baked in Skin, Wo/Salt 1 Sweet Potato, 5 In:114 g	117	2	0.1	0	0.1	0	27.7	24877.1	0.08	0.14	0.69	0.27	25.8	0	28	31.9	0.51	0.33
½ C Mashed:100 g	103	1.7	0.1	0	0.1	0	24.3	21822	0.07	0.13	0.6	0.24	22.6	0	24.6	28	0.45	0.29
Sweet Potatoes, Boiled, Wo/Skin, Wo/Salt 1 C Mashed:328 g	344	5.4	1	0.2	0.4	0	79.6	55937.1	0.17	0.46	2.1	0.8	36.4	0	56.1	68.9	1.84	0.89
½ C Mashed:164 g	172	2.7	0.5	0.1	0.2	0	39.8	27968.6	0.09	0.23	1.05	0.4	18.2	0	28	34.4	0.92	0.44
Tomato Products, Cnd; Paste, Wo/Salt Added ½ C:131 g	110	5	1.2	0.2	0.5	0.2	24.7	3233.1	0.2	0.25	4.22	0.5	29.3	0	55.4	45.9	3.92	1.05
6-oz Can:170 g	143	6.4	1.5	0.2	0.6	0.2	32	4195.6	0.26	0.32	5.48	0.65	38.1	0	71.9	59.5	5.08	1.36

Food item & Measure	Cals	Protein (g)	All fat (g)	Sat/fat (g)	Pol/fat (g)	Mon/fat (g)	Carbo (g)	Vit A (IU)	Thiam (mg)	Ribo (mg)	Niac (mg)	Vit B6 (mg)	Fol (mcg)	Vit B12 (mcg)	Vit C (mg)	Ca (mg)	Iron (mg)	Zinc (mg)
Vegetables cont.																		
Tomato Products, Cnd; Purée, Wo/Salt																		
Added 1 C:250 g	103	4.2	0.3	0	0.1	0	25.1	3402.5	0.18	0.14	4.29	0.38	27.5	0	88.3	37.5	2.33	0.55
29-oz Cn:822 g	337	13.7	1	0.1	0.4	0.1	82.4	11187.4	0.58	0.44	14.1	1.25	90.4	0	290.2	123.3	7.64	1.81
Tomato Products, Cnd; Sauce 1 C:245 g	74	3.3	0.4	0	0.2	0	17.6	2396.6	0.16	0.14	2.82	0.38	23	0	32.1	34.3	1.89	0.61
½ C:122 g	37	1.6	0.2	0	0.1	0	8.8	1194.4	0.08	0.07	1.4	0.19	11.5	0	16	17.1	0.94	0.31
Tomato Products, Cnd; Spaghetti Sauce																		
1 C:249 g	271	4.5	11.9	1.7	3.3	6.1	39.7	3055.2	0.14	0.15	3.75	0.88	53.8	0	27.9	69.7	1.62	0.52
15-½-oz Jar:439 g	479	8	20.9	3	5.7	10.7	69.9	5386.5	0.24	0.26	6.61	1.55	94.8	0	49.2	122.9	2.85	0.92
Tomatoes; Red, Ripe, Boiled, Wo/Salt																		
1 C:240 g	65	2.6	1	0.1	0.4	0.2	14	1783.2	0.17	0.14	1.8	0.23	31.2	0	54.7	14.4	1.34	0.26
½ C:120 g	32	1.3	0.5	0.1	0.2	0.1	7	891.6	0.08	0.07	0.9	0.11	15.6	0	27.4	7.2	0.67	0.13
Tomatoes; Red, Ripe, Cnd, Whole, Reg Pk																		
1 C:240 g	48	2.2	0.6	0.1	0.2	0.1	10.3	1449.6	0.11	0.07	1.76	0.22	18.7	0	36.2	62.4	1.46	0.38
½ C:120 g	24	1.1	0.3	0	0.1	0	5.2	724.8	0.05	0.04	0.88	0.11	9.4	0	18.1	31.2	0.73	0.19
Tomatoes; Red, Ripe, Raw, Yr-Round																		
Average 1 C Chopped:180 g	38	1.5	0.6	0.1	0.2	0.1	8.4	1121.4	0.11	0.09	1.13	0.14	27	0	34.4	9	0.81	0.16
1 Tomato, 2-⅗ In:123 g	26	1.1	0.4	0.1	0.2	0.1	5.7	766.3	0.07	0.06	0.77	0.1	18.5	0	23.5	6.2	0.55	0.11
Turnips, Boiled, Drained, Wo/Salt																		
½ C Cubes:78 g	14	0.6	0.1	0	0	0	3.8	0	0.02	0.02	0.23	0.05	7.2	0	9.1	17.2	0.17	0.16
½ C Mashed:115 g	21	0.8	0.1	0	0.1	0	5.6	0	0.03	0.03	0.34	0.08	10.6	0	13.3	25.3	0.25	0.23
Turnips; Raw 1 C Cubes:130 g	35	1.2	0.1	0	0.1	0	8.1	0	0.05	0.04	0.52	0.12	18.9	0	27.3	39	0.39	0.35
½ C Cubes:65 g	18	0.6	0.1	0	0	0	4.1	0	0.03	0.02	0.26	0.06	9.4	0	13.7	19.5	0.2	0.18
Vegetables; Mixed, Cnd, Sol and Liq																		
1 C:245 g	88	3.5	0.6	0.1	0.3	0	17.4	12448.5	0.08	0.1	1.18	0.19	44.1	0	9.3	51.5	1.59	1.25
½ C:122 g	44	1.7	0.3	0.1	0.2	0	8.7	6198.8	0.04	0.05	0.59	0.09	22	0	4.6	25.6	0.79	0.62
Waterchestnuts, Chinese, Cnd, Sol and Liq																		
½ C Slices:70 g	35	0.6	0	–	–	–	8.7	2.8	0.01	0.02	0.25	0.11	4.1	0	0.9	2.8	0.61	0.27
4 Waterchestnuts:28 g	14	0.3	0	–	–	–	3.5	1.1	0	0.01	0.1	0.04	1.6	0	0.4	1.1	0.24	0.11
Watercress; Raw 1 Sprig:2.5 g	0	0.1	0	0	0	0	0	117.5	0	0	0.01	0	0.2	0	1.1	3	0.01	0
½ C Chopped:17 g	2	0.4	0	0	0	0	0.2	799	0.02	0.02	0.03	0.02	1.6	0	7.3	20.4	0.03	0.02

POSTSCRIPT

Since *The New Why You Don't Need Meat* was first published, many readers have kindly expressed interest in being updated about developments in this rapidly-evolving area. On the BSE front, the situation is increasingly disquieting. At the time of writing, the total number of affected cattle has now surpassed 110,000. In September 1991 the Ministry of Agriculture's chief veterinary officer asserted 'within the next year we will see the start of a rapid decline in the number of confirmed BSE cases'[1]. Like so many other predictions about this sinister plague, it was hopelessly optimistic. Meanwhile, the number of human cases of Creutzfeldt Jakob disease has risen by 50% in Britain since 1991, to a total of 48 for the year ending April 1993. Of particular concern is the recent death of two farmers from CJD, both of whom kept cows which were BSE-infected. Since farmers probably incur more intensive exposure to the 'BSE agent' than most meat-eaters, they could be regarded as the canary in the coal mine – any further incidence of CJD in this category would be terminally ominous. In a chilling scientific study just published, Dr Stephen Dealler calculates that at least one beef eater in ten is expected to develop CJD, unless we prove immune to the infection[2]. He suggests that very low doses of the infectious agent in meat, liver, and kidney can become cumulatively fatal because we eat so much beef and live so long.

The nature of the BSE agent is still unresolved. The currently prevailing theory is that it, and other spongiform diseases, are caused by prions – a term derived from 'proteinaceous infectious particles' – small protein particles which are somehow infectious and self-replicating. Recent research indicates that prions are very similar to proteins involved in regulating interactions between nerves and muscles, and it is possible that prions are therefore faulty, or mutant, forms of such proteins. But the prion theory still has many scientific critics. Another (perhaps linked) theory given much publicity is that BSE results from chronic poisoning due to the use of organophosphorous (OP) pesticides and insecticides. These unpleasant substances – chemically related to nerve gasses and used in certain sheep dips, warble fly treatments and flea collars for pets – have been accused of damaging the human immune system and of causing ME (myalgic encephalomyelitis, also called chronic fatigue syndrome), as well as causing genetic damage. Since the agro-chemicals lobby is one of the most powerful and vocal in the country, we may not know the truth behind these allegations for many years. Meanwhile, the spectre of 'vertical transmission' – in other words, cow to calf infection – continues to haunt those involved with BSE. If proven, there are two serious implications – firstly, that the disease will not be eliminated for many years, and secondly, the safety of all dairy products must be seriously questioned.

The Meat and Livestock Commission continues to conduct itself in characteristic fashion. Shortly after *The New Why You Don't Need Meat* was first published, they agreed to meet me in debate on a national radio programme. Sadly, it may not have been as decisive a victory as they were hoping for, since they declined every subsequent invitation to debate with me on television or radio. But although they refuse to appear with me they are, at least, gratifyingly interested in my broadcasts – paying to have every word I uttered painstakingly transcribed on at least one occasion. I hope they learnt something from it.

Their propaganda to schoolchildren becomes increasingly offensive. A recent issue of *Meat In The News*, a broadsheet they describe as 'an update for students', reports the anonymous views of ten boys and girls who the Meat & Livestock Commission describes as 'ordinary people'[3]. Predictably, these views are all crude attacks on vegetarianism. 'I was like getting dizzy spells', says one girl, 'I felt really weak, I lost all the colour in my skin, all my nails started going bad . . . it was making me really ill.' Another meat-eating boy merely looks at a photograph of a vegetarian boy, and reacts with loathing, 'He looks the type of bloke that will protest at anything.'

The subtext behind that statement disturbs me greatly. Apparently vegetarians look different from 'ordinary' people. 'You can spot a vegetarian – they're not like "us"' is the message, 'vegetarians will "protest" at anything'. Protest, it is implied, is wrong. Ordinary people don't protest. Only trouble-makers protest. This is the first step down an old path which leads to segregation, bullying, victimisation, and ultimately, far worse attrocities. Such tactics disseminate prejudice and hatred, and I fear that we will see much more of them in years to come.

Peter Cox
London
November 1993

NOTES

1 *The Times*, 11 March 1993
2 *Guardian*, 30 October 1993
3 *Meat In The News*, No 2, Summer 1992

INDEX

Index